Artificial Intelligence in Smart Healthcare Systems

Artificial Intelligence in Smart Healthcare Systems gathers specialist researchers and practitioners from around the globe to present cutting-edge advances in the field of artificial intelligence (AI) and healthcare, which promises to be a transformative force in the industry.

Covering both the current landscape as well as the future of healthcare, the book explores the diverse applications of AI in healthcare. Readers will gain an understanding of how machine learning algorithms, natural language processing, and other AI technologies are revolutionizing various facets of healthcare, from diagnostics and treatment to patient management. Including real-world examples and case studies, the book bridges the gap between theoretical concepts and practical implementation, ensuring that readers develop a nuanced and practical grasp of AI's impact on healthcare systems. Moreover, the book serves as a roadmap for healthcare professionals and researchers looking to integrate AI solutions into their practices.

Anamika Chauhan, PhD, Assistant Professor, Department of Home Science, ChamanLal P.G. College, Uttarakhand, India

Nisreen Alwan, PhD, Associate Dean of Academic Affairs & Associate Professor of Environmental Health and Safety College of Health Sciences, Abu Dhabi University, United Arab Emirates

Hiba Hamdar, MD, Medical Learning Skills Academy, Beirut, Lebanon and Munich, Germany

Indresh Kumar, PhD, Department of Pediatrics, All India Institute of Medical Science, Bhopal, India

Artificial Intelligence in Smart Healthcare Systems

Edited by
Anamika Chauhan, Nisreen Alwan, Hiba Hamdar, and Indresh Kumar

CRC Press
Taylor & Francis Group
Boca Raton London New York

CRC Press is an imprint of the
Taylor & Francis Group, an **informa** business

Designed cover image: Shutterstock

First edition published 2026
by CRC Press
2385 NW Executive Center Drive, Suite 320, Boca Raton FL 33431

and by CRC Press
4 Park Square, Milton Park, Abingdon, Oxon, OX14 4RN

CRC Press is an imprint of Taylor & Francis Group, LLC

ISBN: 9781032874173 (hbk)
ISBN: 9781032875729 (pbk)
ISBN: 9781003533351 (ebk)

DOI: 10.1201/9781003533351

Typeset in Minion
by KnowledgeWorks Global Ltd.

Contents

Preface

The convergence of artificial intelligence (AI) and healthcare has ushered in a transformative era in medical science. As AI continues to revolutionize industries worldwide, its impact on healthcare is particularly profound, enhancing diagnostic accuracy, improving patient outcomes, and optimizing healthcare operations. This book, *Artificial Intelligence in Smart Healthcare Systems*, aims to explore the dynamic intersection of AI and healthcare, shedding light on its evolution, current applications, and future potential. AI-driven healthcare systems have evolved significantly over the past decade, transitioning from theoretical applications to practical, real-world implementations. From early decision-support systems to advanced deep-learning models capable of diagnosing diseases with unprecedented accuracy, AI has become an indispensable tool in modern medicine. This book delves into the evolution and emerging trends of AI in smart healthcare, highlighting breakthroughs that have redefined medical practice.

The chapters in this book systematically explore the multifaceted applications of AI in healthcare. Clinical Decision Support Systems (CDSS) have revolutionized medical decision-making, providing physicians with data-driven insights to enhance diagnostic and treatment plans. Predictive analytics, another key area, plays a pivotal role in disease prevention and early detection, allowing for timely medical interventions. The integration of AI with wearable technology has further facilitated real-time health monitoring, empowering individuals to take proactive measures in managing their well-being. Medical imaging has also witnessed remarkable advancements through AI, particularly in radiology and pathology. AI-driven algorithms can detect anomalies in medical scans with precision, reducing human error and expediting diagnoses. Additionally, natural language processing (NLP) has enabled efficient extraction of valuable insights from vast volumes of medical text data, optimizing electronic health records and research analysis. While AI presents numerous benefits, it also raises ethical and regulatory considerations. The responsible use of AI in healthcare necessitates stringent regulatory frameworks to ensure patient safety, data privacy, and unbiased decision-making. This book addresses these concerns, emphasizing the importance of ethical AI deployment and compliance with evolving regulatory standards. Furthermore, AI has significantly enhanced patient engagement and personalized medicine. AI-powered virtual health assistants facilitate seamless communication between patients and healthcare providers, improving accessibility and patient satisfaction. The role of robotics in healthcare, particularly in surgical procedures and patient care automation, is another area of interest that this book explores.

The subsequent chapters examine the interoperability challenges in AI-driven healthcare systems, AI's role in drug discovery, and the security measures required to safeguard healthcare data. AI's contributions to mental health diagnosis, public health initiatives, and clinical trials further underscore its transformative potential in medical research and practice. As we navigate this AI-driven era, it is imperative to consider the opportunities and challenges that AI presents to the healthcare workforce. The book concludes by discussing future directions, emerging technologies, and innovations poised to redefine healthcare in the coming years. By bringing together insights from experts, researchers, and practitioners, this book serves as a comprehensive guide to AI's role in smart healthcare systems. It is our hope that readers—whether medical professionals, AI enthusiasts, policymakers, or students—find valuable knowledge and inspiration within these pages as we collectively shape the future of intelligent healthcare.

Contributors

Abbas Al Bazzal, Faculty of Medical Sciences, Lebanese University, Lebanon.

Abdelrahman Abdelshafi, Faculty of Medicine, Assiut University, Assiut, Egypt.

Abhijit Debnath, Department of Pharmaceutics, Noida Institute of Engineering and Technology (Pharmacy Institute), Greater Noida, India.

Aditya Anand, MGM Medical College, Kishanganj, Bihar, India.

Ahmad El Ouweini, College of Pharmacy, Gulf Medical University, Ajman, United Arab Emirates.

Ahmed Samy, Faculty of Medicine Ain Shams University (ASU), Cairo, Egypt.

Ahmed Tharwat Al-Namrouti, Faculty of Medicine, Galala University, Suez, Egypt.

Ahmed Tharwat Al-Namrouti, Faculty of Medicine, Galala University, Suez, Egypt.

Ahmed Tharwat Al-Namrouti, Faculty of Medicine, Galala University, Suez, Egypt.

Ahmed Tharwat Al-Namrouti, Faculty of Medicine, Galala University, Egypt.

Ahmed Tharwat Al-Namrouti, Faculty of Medicine, Galala University, Suez, Egypt.

Ali Abbas Mankani, Dow University of Health Sciences, Karachi, Pakistan.

Amatul Hadi Hamdana, Faculty of Medicine, Dow University of Health Sciences, Karachi, Pakistan.

Amira A. Aboali, Faculty of Medicine, Alexandria University, Alexandria, Egypt.

Anamika Chauhan, Department of Home Science, Chaman Lal Mahavidhyalya Landhora Haridwar, Uttarakhand, India.

Asmaa Sh. Saadallah, Ministry of Health and Population, Cairo, Egypt.

Atajan Nurnazarov, Hamidiye International Faculty of Medicine, University of Health Sciences, Istanbul, Turkey.

Awez A. Sikkalgar, Department of Pharmacology, Shri Vile Parle Kelavani Mandal's Institute of Pharmacy, Dhule, Maharashtra, India.

Aya Kamal Al-Zabidi, College of Health Sciences, Abu Dhabi University, Abu Dhabi, United Arab Emirates.

Aya M. Sherif, Faculty of Medicine, Beni Suef University, Beni Suef, Egypt.

Aya Yasser Elsayed Mohamed, Kasr Alainy Medical School Cairo University, Cairo, Egypt.

Ayah Abdulgadir, Faculty of Medicine, University of Khartoum, Khartoum, Sudan.

Ayesha Khalid, Shifa College of Medicine, Islamabad, Pakistan.

Bhavna Dhingra, Department of Pediatrics, All India Institute of Medical Sciences, Bhopal, India.

Bijoy Bitan Saha, Department of Anthropology, Calcutta University West Bengal, India.

Devendra K. Jadhav, Department of Pharmaceutics, P.S.G.V.P. M's College of Pharmacy, Shahada, Maharashtra, India.

Diana Malaeb, College of Pharmacy, Gulf Medical University, Ajman, United Arab Emirates.

Doaa S. R. Khafaga, Health Sector, Faculty of Science, Galala University, New Galala City, Suez, Egypt.

Diptendu Chatterjee, Department of Anthropology, University of Calcutta, West Bengal, India.

Arindam Ghosh, Biochemistry, Dr. B. C. Roy Multispecialty Hospital and Research Centre, IIT Kharagpur, West Bengal, India.

Deepak Kumar, Department of Pathology, MGM Medical College, Kishanganj, Bihar, India.

Shikha Malika, Department of Pediatrics, All India Institute of Medical Sciences, Bhopal, India.

Sonalika Roy, Department of Pathology, MGM Medical College, Kishanganj, Bihar, India.

Esraa Marwan Morshedy, Tanta University, Tanta, Gharbia, Egypt.

Gamila Ali Attaelmanan, Medical Laboratory Sciences, Department of Hematology, Al Neelain University, Khartoum, Sudan.

Gergis Goseph Altalab, Faculty of Medicine, Merit University, Sohag, Egypt.

Habiba Eyvazova, Hamidiye International School of Medicine, University of Health Sciences, Selimiye, Üsküdar/İstanbul, Turkey.

Haidar Kanso, Faculty of Medicine, Damascus University, Damascus, Syria.

Haifa Ben-Romdhane, College of Health Sciences, Abu Dhabi University, Abu Dhabi, United Arab Emirates.

Haytham Jaouni, Medicine and Surgery, Department of Medicine, Jordan University of Science and Technology, Irbid, Jordan.

Hiba Zreik, Faculty of Medical Sciences, Lebanese University

Himanshu Goyal, Uttarakhand State Council for Science and Technology (UCOST) Department of Information & Science Technology Government of Uttarakhand, India.

Hirah Nadeem, College of Health Sciences, Abu Dhabi University, Abu Dhabi, United Arab Emirates.

Husam Alddin Maher Alkajah, Faculty of Medicine, Modern University for Technology and Information-MTI, Cairo, Egypt.

Hussein Taher, Al-Sham Private University, Faculty of Medicine, Damascus, Syria.

Ihda Mahmoud Bani Khalaf, Medicine and Surgery, Department of Medicine, Jordan University of Science and Technology, Irbid, Jordan.

Indresh Kumar, Pediatrics, All India Institute of Medical Sciences, Bhopal, India.

Ismail A. Ibrahim, Faculty of Health Sciences, Fenerbahce University, Istanbul, Turkey.

J.S Aswal, Uttarakhand State Council for Science and Technology (UCOST), Department of Information & Science Technology, Government of Uttarakhand, India.

Jobran Tohme, Lebanese University Faculty of Medical Science, Beirut, Lebanon.

Kalpesh R. Yeole, Department of Pharmacology, Shri Vile Parle Kelavani Mandal's Institute of Pharmacy, Dhule, Maharashtra, India.

Karim Samir Shata, Cairo University, Giza, Egypt.

Malik Sallam, Department of Pathology, Microbiology and Forensic Medicine, School of Medicine, The University of Jordan, and Department of Clinical Laboratories and Forensic Medicine, Jordan University Hospital, Amman, Jordan.

Manisha Rani, Department of Pediatrics, All India Institute of Medical Sciences, Bhopal, India.

Mariam Chokor, Faculty of Medical Sciences, Lebanese University, Lebanon.

Maryam Pourali, Hamidiye International Faculty of Medicine, University of Health Sciences, Istanbul, Turkey.

May G. Mohamed, College of Pharmacy, Gulf Medical University, Ajman, United Arab Emirates.

Megh R. Goyal, Retired Faculty in Agricultural and Biomedical Engineering, College of Engineering at University of Puerto Rico – Campus, Mayagüez, Puerto Rico and Apple Academic Press Inc., Palm Bay, Florida, USA.

Mirna F. Mustafa, Faculty of Medicine, Al Quds university, Bethlehem, Palestine.

Mohamed N. Elshabrawi, Port Said University, Port Said, Egypt.

Mohammad Al Diab Al Azzawi, Faculty of Medicine, The National Ribat University, Khartoum, Sudan.

Mohammad Ali Mtairek, Faculty of Medicine, Damascus University, Damascus, Syria.

Mohammad H. Awada, Medical student, Faculty of Medical Sciences, Lebanese University.

Mohammad Hadi Awde, Damascus University, Faculty of Medicine, Damascus, Syria.

Mohammed Hisham Alamin, Faculty of Medicine, International University of Africa, Khartoum, Sudan.

Mona Muhe Eldeen Eshag, Faculty of Medicine, University of Bahri, Khartoum, Sudan.

Mr. Sunil Kumar Tailor, Department of Pediatrics, All India Institute of Medical Sciences, Bhopal, India.

Muhammad Islampanah, Faculty of Medicine, Mashhad University of Medical Sciences, Mashhad, Iran.

Muhammad Sulaman Akhtar, Shifa International Hospitals, Ltd., Islamabad, Pakistan.

Muhammad Yaqoob Akhtar, Faculty of Medicine, Shifa College of Medicine, Islamabad, Pakistan.

Muna Barakat, Faculty of Pharmacy, Applied Science Private University, Amman, Jordan.

Muneeb Faiz, King Edward Medical University, Nelagumbad, Anarkali, Lahore, Pakistan.

N. Dora Baykal, Hamidiye International School of Medicine, University of Health Sciences, Istanbul, Turkey.

Naglaa M. Marzuuk, College of Medicine, Suez Canal University, Ismailia, Egypt.

Naina Chaurasiya, Department of Pharmacology, Shri Vile Parle Kelavani Mandal's Institute of Pharmacy, Dhule, Maharashtra, India.

Nazila Biglari, School of Biological Sciences, University Sains Malaysia.

Noor Razzaq Hayat, Faculty of Medicine, Wasit University, Wasit, Iraq.

Nourhan S. Sultan, Biotechnology Department, Faculty of Science, Cairo University, Giza, Egypt.

Omar Alomari, Hamidiye International School of Medicine, University of Health Sciences, Selimiye, Üsküdar/İstanbul, Turkey.

Parashuram Kallem, College of Health Sciences, Abu Dhabi University, Abu Dhabi, United Arab Emirates.

Preeti, All India Institute of Medical Sciences Bhopal, India.

Asitava Deb Roy, Faculty of Medicine, Mata Gujri University, Kishanganj, India.

Rahma Mahmoud Saad, Faculty of Medicine, Modern University for Technology and Information-MTI, Cairo, Egypt.

Rahma Mahmoud Saad, Faculty of Medicine, Modern University for Technology and Information-MTI, Cairo, Egypt.

Raj Pandit, Junior Research Fellow, Department of Anthropology, University of Calcutta.

Reem M.Elsapagh, Faculty of Pharmacy, Cairo University, Cairo, Egypt.

Reem M.Elsapagh, Faculty of Pharmacy, Cairo University, Cairo, Egypt.

Reem Sayad, Faculty of Medicine, Assiut University, AL Hamraa Ath Thaneyah, El Fateh, Assiut, Egypt.

Renu Kharb, Rufaida College of Nursing, Jamia Hamdard, Hamdard Nagar, New Delhi, India.

Rim Basma, Faculty of Medicine, Beirut Arab University, Beirut, Lebanon.

Rima Ghosh, Department of Physiology, University of Calcutta West, Bengal, India

Rowida Hesham Ali, Faculty of Physical Therapy, Misr University for Science and Technology, Giza Egypt.

Rumeysa Yegin, Hamidiye International School of Medicine, University of Health Sciences, Selimiye, Üsküdar/İstanbul, Turkey.

Sagar A. More, Department of Pharmacy, Birla Institute of Science and Technology Pilani, Hyderabad Campus, Shamirpet, Hyderabad, India.

Samar Younes, School of Pharmacy, Lebanese University, School of Pharmacy, Lebanon.

Sami El Khatib, Department of Biomedical Sciences, School of Arts and Sciences, Lebanese International University, Bekaa, Lebanon and Center for Applied Mathematics and Bioinformatics (CAMB), Gulf University for Science and Technology (GUST), Hawally, Kuwait.

Shalini Sharma, School of Pharmacy, Noida International University, Gautam Buddha Nagar, India.

Sheetal, All India Institute of Medical Sciences Bhopal, India.

Sinem Nur Ertan, Hamidiye International Faculty of Medicine, University of Health Sciences, Istanbul, Turkey.

Souheil Hallit, School of Medicine and Medical Sciences, Holy Spirit University of Kaslik, Jounieh, Lebanon and Applied Science Research Center, Applied Science Private University, Amman, Jordan.

Tasneem Mahmoud Eltantawy, Faculty of Medicine, Mansoura University, Egypt.

Tasneem Mahmoud, Faculty of Medicine, Mansoura University, Dakahlia, Egypt.

Tiba J. K. Al-Imari, Dentistry College, University of Babylon, Hillah, Iraq and University of Pannonia, Veszprém, Hungary.

Zahraa Takesh, Faculty of Medical Sciences, Lebanese University, Lebanon.

Zeinab Hazime, Medical Student, Faculty of Medical Sciences, Lebanese University, Lebanon

Zubair Saghir Ahmed Shaikh, Department of Pharmaceutics, P.S.G.V.P. M's College of Pharmacy, Shahada, Maharashtra, India.

CHAPTER 1

Introduction to Artificial Intelligence in Smart Healthcare Systems

Nazila Biglari

1.1 INTRODUCTION

Artificial intelligence (AI) is a branch of computer science to create capable systems and technologies of performing tasks that normally require human intelligence. AI is integrated into various sectors incorporating healthcare, finance, transportation, and entertainment. In healthcare, medical imaging and diagnostics (MID), personalized medicine, and drug discovery and development are some of the areas where AI is making a thoughtful impact [1, 2]. In the mid-20th century, AI was invented and grown significantly, and it was largely thanks to two computer scientists, Alan Turing and John McCarthy. Turing is considered the "father of AI" when he tested his work in 1950. Later on, it was continued with initial experiments in medical diagnosis. MYCIN, the first notable system was developed in the 1970s as an expert system for diagnosing bacterial infections and recommending antibiotics. In the 1980s and 1990s, research on AI in healthcare expanded with the development of more expert systems and the early use of machine learning algorithms for analyzing medical data [3]. These systems started to be applied for tasks such as medical imaging analysis and predictive modeling. The impact of AI in smart healthcare systems is shown in Figure 1.1.

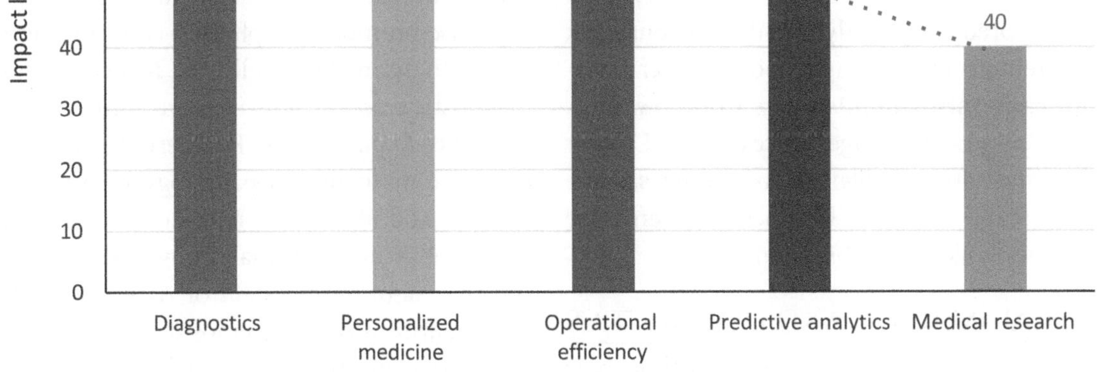

FIGURE 1.1 Impact of AI in healthcare applications.

DOI: 10.1201/9781003533351-1

TABLE 1.1 List of Some Applications of AI in Smart Healthcare Systems

Application	Description	Benefit	Ref.
Disease diagnosis	Analyzing images and medical patient data for disease identification	Increased early accuracy, detection, reduced radiologist workload	[17]
Predictive analytics	Predicting health outcomes and disease outbreaks using historical data	Better resource planning, proactive care, improved outcomes	[18]
Personalized medicine	Customizing treatment based on genetic and health data	More effective treatments, fewer side effects	[12]
Virtual health assistants	AI members provide health information and support	Enhanced patient engagement, timely advice, reduced workload	[19]
Remote monitoring	Analyzing data from wearables and sensors to monitor health	Continuous tracking, early interference, fewer in-person visits	[20]
Drug discovery and development	Identifying new drug candidates using AI analysis	Faster development, reduced costs, novel therapeutic targets	[15]
Clinical decision support	Offering evidence-based recommendations to healthcare professionals	Improved decision-making, reduced errors, enhanced safety	[21]
Health record management	Organizing and analyzing electronic health records (EHRs) with AI	Streamlined data management, better integration, improved access	[22]
Robotic surgery	AI-driven robots assist in precise surgical procedures	Increased precision, reduced recovery times, minimally offensive	[23]
Telemedicine	Enhancing virtual consultations with real-time data analysis	Better access, convenience, efficient remote diagnosis	[24]
Fraud detection	Identifying differences in billing and claims to prevent fraud	Reduced fraud, cost savings, improved system integrity	[25]
Mental health monitoring	Assessing mental health through AI analysis of speech and behavior	Early detection, personalized therapy recommendations	[25]

The early 2000s, integration of AI into electronic health records (EHRs) and increase in data-driven approaches was significantly started. The development of machine learning techniques, particularly deep learning enabled advances in medical imaging such as improved tumor detection and image classification [4]. The 2010s and beyond have been marked by the rapid development of AI applications in healthcare comprising advanced diagnostic tools, personalized medicine, and predictive analytics. AI technologies have increasingly been used in different areas, including MID, personalized medicine, drug discovery and development, predictive analytics, disease outbreak prediction, remote monitoring, robotic surgery, virtual health assistants, administrative workflow automation, and natural language processing (NLP) for clinical documentation (Table 1.1). As examples, some applications of AI in smart healthcare systems are presented in the following subsections.

1.2 MEDICAL IMAGING AND DIAGNOSTICS

One of the primary applications of AI in healthcare is in MID. AI in MID refers to apply the technologies of machine learning, deep learning, and computer vision to analyze and interpret medical images as well as to assist in the diagnostic process. AI is designed to mimic human intellectual functions that enable them to recognize patterns, detect variances, tumors, and provide diagnostic insights at an early stage from medical images like X-rays, computed tomography (CT) scans, magnetic resonance imaging (MRIs), and ultrasound images with remarkable precision [5]. The application of AI in medical imaging also improves the assignment on radiologists who are often overcome with the huge volume of imaging data that needs to be analyzed. AI, by automating the routine and repetitive aspects of image interpretation, enables radiologists to focus their expertise on more complicated cases to improve their productivity and job fulfillment. Moreover, this reduces diagnostic errors that gives more accuracy and reliable outcomes to patients [6]. Furthermore, AI-driven tools can have main role in pathology by analyzing tissue samples and identifying markers of diseases like cancer more proficiently than conventional methods [7]. Another significant benefit of AI in medical imaging is standardization in diagnostic processes. Variability in human interpretation can lead to inconsistencies in diagnoses which can impact the quality of care. AI systems with their ability to apply consistent criteria and protocols across all cases ensure more reliable diagnostic process. This standardization is valuable in large healthcare systems and multi-center studies particularly where consistency is vital. Additionally, AI applications with predictive analytics can enhance image reconstruction techniques that produce higher-quality images with lower radiation doses [8].

Despite AI benefits, several challenges hinder its adoption and integration into healthcare. One major challenge is quality and variability of medical data. AI algorithms require extensive datasets with high quality for training, particularly those based on deep learning. Acquiring such datasets is another difficulty due to privacy of patient data that complicates data sharing. Moreover, medical images come from diverse sources including different hospitals, imaging devices, and patient demographics leading to reduce the generalizability of AI models as an algorithm trained on data from one institution may not perform well on data from another. Another significant problem is the lack of standardization in data formats and imaging protocols. Medical imaging such as X-rays, MRIs, and CT scans own their standards and protocols. This lack of uniformity complicates the integration of AI systems [9]. The black box nature of many AI models also poses a substantial challenge. Black box nature of AI refers to the phenomenon where the internal workings of an AI system, particularly those using complex algorithms like deep learning and neural networks are not easily interpretable or understandable by humans. This term highlights the lack of transparency in how AI models process input data and generate output decisions or predictions [10]. Without clear insights, clinicians may not show interest to apply diagnostics, especially in critical situations where accuracy is vital. Therefore, regulatory processes further slow the deployment of AI in medical imaging. Medical devices and diagnostic tools are subject to severe validation and approval processes to ensure their safety and efficacy. Navigating these processes can be time-consuming and costly that creates barriers to innovation and implementation. Additionally, healthcare professionals need training to effectively use and interpret AI tools which can be both time-consuming and expensive [11].

1.3 PERSONALIZED MEDICINE

Another significant application of AI in healthcare is personalized medicine. AI programs can look at a lot of information from medical records, genes, and lifestyle to create treatment plans that suit each patient. This method makes treatment plans better and cause lesser side effects because they are tailored to each person's unique genes and health conditions. AI can predict how patients will react to different treatments by using the earlier information. This helps doctors make better choices. Also, AI can use predictive analytics to identify possible health risks and problems, allowing us to take steps to prevent diseases before they happen. AI is changing how healthcare is managed and run [12]. For example, NLP program can make office tasks easier by converting medical notes into text, managing billing, and organizing patient records effectively. This reduces the work for healthcare workers, so they can spend more time on caring for patients. AI helpers can assist patients all day and night. They can give medical advice, help schedule appointments, and answer questions to make patients more active and delighted with their care. Also, AI can help hospitals to use their resources better and predict how many patients will admit, and manage supplies which makes healthcare cheaper and more efficient [13].

However, there are some challenges that need to be solved to make AI most useful. First, keeping data safe and private is very important. Personalized medicine uses big data that contains private health information about people. Making sure that this information is kept safe, used properly, and shared correctly is a hard task. Also, the chance of personal data being stolen or misused could result in unfair treatment or negative labeling which might reduce people's trust in using AI in healthcare. Second, using AI in personalized medicine has some technical problems. To make good predictions, we need high-quality and varied data, but we often do not have enough of it. Also, the complexity of living systems makes it hard for AI to fully understand and predict how treatments will work which can lead to mistakes in personalized treatment plans. Some AI models are like a black box which means it is hard to understand how they work. This can make doctors worried about trusting and using these technologies [14]. In the end, using AI in personalized medicine brings up rules and ethical issues. Regulatory agencies need to keep up with the fast growth of AI. They must make sure that new treatments are safe and work well while also allow new ideas to come. We need to be careful about ethical issues such as making sure patients understand what they agree to.

1.4 DRUG DISCOVERY AND DEVELOPMENT

By using AI, the discovering and processing new medicines are sped up better than before. Old ways of finding new medicines take a lot of time and money, and many of them do not work. AI tools help researchers to study a large set of data easier and faster to find possible new drugs, and predict how well they will work and how safe they are. By using machine learning, AI can analyze complicated biological data, predict how drugs will work together, and run virtual clinical trials. This helps to speed up the drug development process a lot as another application of AI. This new ability makes things faster and helps find good drug options more accurately which raises the chances of success in later development stages. In smart healthcare systems, AI is involved not just in finding new medicines but in the whole process of creating them. AI tools can improve how clinical trials are planned, how patients are chosen, and how their progress is tracked by looking at patient information and predicting results better than older methods. This creates treatment plans which are more fitted and effective. Also, AI can help find markers in the body that are important for understanding how diseases work and creating specific treatments. By using AI in smart healthcare systems, drug companies can make the process of developing new medicines faster and more flexible. This means patients can get new and helpful treatments more quickly [15].

Furthermore, using AI in finding and developing drugs helps lower the risks of bad side effects which is a big problem in healthcare. AI-based predictive modeling can find possible side effects early and gives researchers a chance to adjust the drugs or how they are given before testing them on patients. This improves patient safety and decreases the chances of expensive problems happening later on. As AI keeps improving and working with technologies like genetics and big data, it is set to change the way that drugs are discovered and developed. This will make the process quicker, cheaper, and better suited to what today healthcare systems need [16]. However, the integration of AI into drug discovery in healthcare systems is not without challenges.

Finding and making new medicines is a complicated and costly job. It usually takes a long time, often years, and costs a lot of money, sometimes even billions of dollars to get a new medicine ready for people to use. One of the main difficulties is finding and confirming possible targets for drugs. This process needs a strong knowledge of how living things work and how diseases happen. It usually includes testing many different substances all at once, but many of these tests may not work later on. Also, it is very difficult to predict if a medicine will be safe and effective for people. Many drugs that look good in early tests do not work well in later tests because of unexpected side effects or because they do not work at all. The rules and regulations make things more complicated because getting approval requires a lot of testing and paperwork which takes more time to develop the product. Also, combining big data like genes, proteins, and metabolites into finding new drugs creates major difficulties. The large amount and complexity of the data need advanced computer tools to analyze and understand it. Current computer models often have a hard time dealing with the complicated nature of biological data. This can result in mistakes when predicting how drugs will work with their targets and how patients will react [16]. These problems show that we need better tools and methods to make the drug discovery process easier and to increase the chances of new drugs succeeding.

1.5 SUMMARY

The purpose of AI in healthcare is to simulate human cognitive functions using data, software, and hardware to enhance care processes. Unlike general AI, healthcare AI comprises interconnected technologies such as robotics, deep learning, machine learning, and NLP to enable machines to think, adapt, and act like humans. These technologies are applied in predictive analytics, medical imaging, personalized medicine, robotic surgery, drug discovery, etc. Research is ongoing to evaluate the impact and usability of AI as it holds great potential for improving healthcare delivery [26].

REFERENCES

1. Najjar, R. (2023). Redefining radiology: A review of artificial intelligence integration in medical imaging. *Diagnostics*, *13*(17), 2760.
2. Bohr, A., & Memarzadeh, K. (2020). The rise of artificial intelligence in healthcare applications. In *Artificial Intelligence in Healthcare* (pp. 25–60). Academic Press.
3. Szolovits, P., Patil, R. S., & Schwartz, W. B. (1988). Artificial intelligence in medical diagnosis. *Annals of Internal Medicine*, *108*(1), 80–87.

4. Stefanelli, M. (2001). The socio-organizational age of artificial intelligence in medicine. *Artificial Intelligence in Medicine, 23*(1), 25–47.
5. Alexander, A., Jiang, A., Ferreira, C., & Zurkiya, D. (2020). An intelligent future for medical imaging: A market outlook on artificial intelligence for medical imaging. *Journal of the American College of Radiology, 17*(1), 165–170.
6. Piorkowski, D., Vejsbjerg, I., Cornec, O., Daly, E. M., & Alkan, Ö (2023). AIMEE: An exploratory study of how rules support AI developers to explain and edit models. *Proceedings of the ACM on Human-Computer Interaction, 7*(CSCW2), 1–25.
7. Niazi, M. K. K., Parwani, A. V., & Gurcan, M. N. (2019). Digital pathology and artificial intelligence. *The Lancet Oncology, 20*(5), e253–e261.
8. Park, S. H., & Han, K. (2018). Methodologic guide for evaluating clinical performance and effect of artificial intelligence technology for medical diagnosis and prediction. *Radiology, 286*(3), 800–809.
9. Prevedello, L. M., Halabi, S. S., Shih, G., Wu, C. C., Kohli, M. D., Chokshi, F. H., … Flanders, A. E. (2019). Challenges related to artificial intelligence research in medical imaging and the importance of image analysis competitions. *Radiology: Artificial Intelligence, 1*(1), e180031.
10. Poon, A. I., & Sung, J. J. (2021). Opening the black box of AI-medicine. *Journal of Gastroenterology and Hepatology, 36*(3), 581–584.
11. Sharma, P., Suehling, M., Flohr, T., & Comaniciu, D. (2020). Artificial intelligence in diagnostic imaging: Status quo, challenges, and future opportunities. *Journal of Thoracic Imaging, 35*, S11–S16
12. Awwalu, J., Garba, A. G., Ghazvini, A., & Atuah, R. (2015). Artificial intelligence in personalized medicine: Application of AI algorithms in solving personalized medicine problems. *International Journal of Computer Theory and Engineering, 7*(6), 439.
13. Hiremath, B. N., & Patil, M. M. (2022). Enhancing optimized personalized therapy in clinical decision support systems using natural language processing. *Journal of King Saud University-Computer and Information Sciences, 34*(6), 2840–2848.
14. Li, Y. H., Li, Y. L., Wei, M. Y., & Li, G. Y. (2024). Innovation and challenges of artificial intelligence technology in personalized healthcare. *Scientific Reports, 14*(1), 18994.
15. Mak, K. K., Wong, Y. H., & Pichika, M. R. (2023). Artificial intelligence in drug discovery and development. *Drug Discovery and Evaluation: Safety and Pharmacokinetic Assays, 1*, 1–38.
16. Mak, K. K., & Pichika, M. R. (2019). Artificial intelligence in drug development: Present status and future prospects. *Drug Discovery Today, 24*(3), 773–780.
17. Kumar, Y., Koul, A., Singla, R., & Ijaz, M. F. (2023). Artificial intelligence in disease diagnosis: A systematic literature review, synthesizing framework and future research agenda. *Journal of Ambient Intelligence and Humanized Computing, 14*(7), 8459–8486.
18. Chong, L. R., Tsai, K. T., Lee, L. L., Foo, S. G., & Chang, P. C. (2020). Artificial intelligence predictive analytics in the management of outpatient MRI appointment no-shows. *American Journal of Roentgenology, 215*(5), 1155–1162.
19. Davis, C. R., Murphy, K. J., Curtis, R. G., & Maher, C. A. (2020). A process evaluation examining the performance, adherence, and acceptability of a physical activity and diet artificial intelligence virtual health assistant. *International Journal of Environmental Research and Public Health, 17*(23), 9137
20. Shaik, T., Tao, X., Higgins, N., Li, L., Gururajan, R., Zhou, X., & Acharya, U. R. (2023). Remote patient monitoring using artificial intelligence: Current state, applications, and challenges. *Wiley Interdisciplinary Reviews: Data Mining and Knowledge Discovery, 13*(2), e1485.
21. Magrabi, F., Ammenwerth, E., McNair, J. B., De Keizer, N. F., Hyppönen, H., Nykänen, P., … Georgiou, A. (2019). Artificial intelligence in clinical decision support: Challenges for evaluating AI and practical implications. *Yearbook of Medical Informatics, 28*(1), 128–134.
22. Lee, S., & Kim, H. S. (2021). Prospect of artificial intelligence based on electronic medical records. *Journal of Lipid and Atherosclerosis, 10*(3), 282.
23. Bhandari, M., Zeffiro, T., & Reddiboina, M. (2020). Artificial intelligence and robotic surgery: Current perspective and future directions. *Current Opinion in Urology, 30*(1), 48–54.
24. Pacis, D. M. M., Subido, E. D., & Bugtai, N. T. (2018). Trends in telemedicine utilizing artificial intelligence. AIP Conference Proceedings, *1933*(1).
25. Bao, Y., Hilary, G., & Ke, B. (2022). Artificial intelligence and fraud detection. *Innovative Technology at the Interface of Finance and Operations, 1*, 223–247
26. Graham, S., Depp, C., Lee, E. E., Nebeker, C., Tu, X., Kim, H. C., & Jeste, D. V. (2019). Artificial intelligence for mental health and mental illnesses: An overview. *Current Psychiatry Reports, 21*, 1–18.

CHAPTER 2

Evolutions and Trends of Artificial Intelligence in Smart Healthcare Systems

Sinem Nur Ertan, Maryam Pourali, Atajan Nurnazarov, and N. Dora Baykal

2.1 INTRODUCTION

This chapter explores the transformative impact of artificial intelligence (AI) on the healthcare industry, highlighting its ability to enhance diagnostic accuracy, personalize treatment, optimize operations, and predict health trends, all of which contribute to improved patient outcomes and reduced costs. Nonetheless, it also highlights major obstacles, such as issues related to data privacy, ethical dilemmas, regulatory challenges, and the requirement for significant financial investments and professional training. The chapter traces the historical progress of AI in healthcare, explores key technologies such as machine learning and natural language processing, and predicts future developments and potential innovations. Its goal is to offer a thorough understanding of AI's progression, present status, and future potential, making it an important resource for both researchers and professionals in the healthcare sector (Table 2.1).

TABLE 2.1 Evolution of AI Trends in Medicine

Era	Milestones
1950s–1960s	• 1956: Term "artificial intelligence" coined at Dartmouth Conference. • 1959: John McCarthy and Marvin Minsky explored AI applications in various fields, including medicine
1970s	• Development of early expert systems like MYCIN for diagnosing bacterial infections and recommending antibiotics. • 1976: INTERNIST-I system developed for diagnosing complex diseases in internal medicine.
1980s	• Increased interest in AI in medicine, with systems like CADUCEUS and PUFF emerging. • 1984: Pathfinder system developed for pathology diagnosis.
1990s	• AI systems began integrating more with clinical environments. • DXplain, an expert system for medical diagnosis, became widely used. • 1996: IBM's Watson started as a research project.
2000s	• Growth in machine learning techniques and access to larger datasets. • 2002: Bioinformatics revolution, leveraging AI for genetic and molecular data analysis.
2010s	• Surge in big data analytics, electronic health records (EHR), and deep learning. • 2011: IBM Watson won Jeopardy!, showcasing its potential for natural language processing and data analysis. • 2016: Google's DeepMind developed AI for diagnosing eye diseases and predicting patient deterioration.
2020s	• Continued advancements in AI with real-world applications in various medical fields. • 2021: FDA approved several AI-based medical devices. • 2023: AI played a crucial role in developing new vaccines and treatments during the COVID-19 pandemic.

DOI: 10.1201/9781003533351-2

2.2 MEDICAL IMAGING AND DIAGNOSTICS

In the field of leveraging deep learning, neural networks and medical imaging AI is speedingly advancing to upgrade diagnostic capabilities in different medical fields. In different medical devices such as MRIs, CT scans and X-rays AI Technologies and deep neural networks are being used to detect conditions like cancerous nodules and pneumonia with exceptional accuracy [1]. In one study, an AI algorithm achieved high accuracy in identifying cancerous pulmonary nodules, outperforming 17 of 18 radiologists [2]. Similarly, a deep learning model for wrist fractures improved diagnostic sensitivity from 81% to 92%, demonstrating its potential to significantly enhance emergency care [3].

These instances highlight the increasing contribution of AI to increasing diagnostic efficiency and accuracy, demonstrating the technology's potential to completely transform medical imaging. When compared to radiology, the area of pathology has been slower to adopt digital imaging. This is mostly due to the difficulties involved in converting glass slides into digital forms and the application of whole-slide imaging [4]. Retrospective research has demonstrated that AI systems can perform better than pathologists, suggesting a possible improvement in the efficiency and accuracy of diagnosis [5]. For instance, in a study of breast cancer slides, some algorithms performed better than pathologists who had limited time to review the slides, though results were comparable when an expert pathologist reviewed slides without time constraints [6]. Recent advancements also include the application of machine learning to DNA methylation patterns, improving brain tumor classification compared to traditional methods [7]. Moreover, AI algorithms can now detect specific genomic mutations in lung cancer that are difficult for pathologists to identify [8]. According to the first prospective study which combines pathologist and AI algorithms in a real clinical setting produces the best outcomes, underlining the fact that AI has the ability to enhance, rather than replace, human expertise in pathology. Furthermore, the FDA has approved a growing number of AI algorithms for the interpretation of images; yet, there are still few peer-reviewed publications available on these algorithms [9, 10].

DeepMind Technologies developed an AI algorithm which is called Google DeepMind to identify different eye diseases with remarkable accuracy [11]. This AI system in a groundbreaking study successfully identified over 50 different eye conditions from retinal scans such as age related macular degeneration and diabetic retinopathy [12]. Especially in underserved regions with limited specialists this development would be crucial in the treatment and early diagnosis of the disease, improving patient outcomes [13].

2.3 VIRTUAL HEALTH ASISSTANTS

In healthcare virtual health assistants apply machine learning to support different aspects of patient care which serves as a life-changing application. Simple repetitive tasks such as scheduling appointments, managing medication refills and administrative tasks can reduce the burden on healthcare professionals which will help reducing the waitlists and allow doctors to focus on more complex jobs [14].

As examples there are applications generated by AI where patients can enter their symptoms and medical history. As a consequence of machine learning and deep learning these applications can offer a preliminary evaluation pointing out different conditions explaining patients when to be alarmed and seek further medical help [15]. Furthermore these applications can optimize patient engagement by being available day and night providing immediate information. This capability is particularly valuable in managing minor health concerns and in guiding users on when to consult healthcare professionals, potentially improving early detection and preventive care [16]. While it is not a substitute for professional medical diagnosis, applications integration of AI technology facilitates a more informed and engaged approach to personal health management, bridging the gap between patients and healthcare providers [17]. But if we look from a different perspective these AI-generated virtual assistants must provide a high level of data security with regulations such as Health Insurance Portability and Accountability Act (HIPAA) promising patients their privacy and leave no space for any concerns [18].

2.4 ROBOTIC SURGERY

In robotic surgeries surgeons operate with enhanced precision and control as a type of minimally invasive surgery. AI is used in this system to offer real-time feedback and improve the precision of surgical maneuvers. The integration of AI to robotic surgeries even maximizes the capabilities of these technological devices. AI-integrated robots can adapt in real-time to the surgical environment providing precision and reducing human error. Preoperatively AI-systems can help analyzing risk assessments and postoperatively it can make predictions about complications when given medical records and monitoring the patient precisely [19]. Smart Tissue Autonomous Robot (STAR) established by Johns Hopkins University is a great example for the AI-integrated surgery robots. This robot employs deep learning algorithms to autonomously perform bowel anastomosis with high precision. In trials, STAR demonstrated performance that matched or exceeded human surgeons, showcasing the potential for AI to handle complex surgical tasks and improve outcomes [20]. This integration of robotic surgeries and AI highlights the transformative role in surgical procedures, making it more accurate, efficient, and tailored to patients [21].

2.5 POPULATION HEALTH MANAGEMENT

Population health management (PHM) is increasingly benefiting from the application of big data and machine learning technologies [22]. In such systems AI has access to big amounts of patient data coming from public health databases and insurance companies. Nowadays there are different nonprofit integrated healthcare providers which can focus on various aspects of public health [23]. By analyzing EHRs from their extensive network of patients it can detect health disparities within specific communities helping to create better patient demographics [24]. Through this information healthcare professionals gain valuable insights into vaccination coverage gaps within specific areas, forecasting the spread of infectious diseases. With the help of AI outbreaks can be predicted and necessary precautions can be taken [25]. Additionally, machine learning helps in evaluating the impact of public health policies and programs by analyzing outcomes data and adjusting strategies based on real-time feedback [23].

Another area of interest which can be benefited by AI-driven population health management is analyzing data to identify patterns for chronic diseases [24]. With the help of some simple healthcare data such as body mass index (BMI), smoking history and family history high risk populations for diabetes and heart diseases can be detected. Taking proactive measures for these high risk populations can improve health outcomes at the population level, improving the quality of life of big patient groups. This proactive strategy not only enhances patient outcomes but also lowers overall healthcare expenses by tackling problems early, before they necessitate more extensive treatment [25].

2.6 PREDICTIVE ANALYTICS

Predictive analytics is transforming the healthcare industry by leveraging data, statistical algorithms, and machine learning techniques to forecast future outcomes and enhance patient care. Machine learning models assess the likelihood of adverse events, such as rehospitalization or disease progression, using predictive analytics to predict patient outcomes [28]. These algorithms try to help with early diagnosis and treatment by trying to identify similarities in the voluminous data of images, genetics and EHRs [29]. As an example, predictive models have been proven to provide improved data in the rapid diagnosis of diseases such as sepsis, heart disease, and some types of cancer, leading to faster and better treatment [30].

Personalized medicine is a noteworthy field where predictive analytics is gaining traction. Predictive models can assist in the development of customized treatment regimens by evaluating patient-specific data, increasing treatment efficacy and lowering the risk of negative responses [31]. One excellent example of this use is pharmacogenomics, a science that makes use of predictive analytics to determine how a patient's genetic composition affects how they respond to drugs [32]. These models improve treatment accuracy and lessen the need for trial-and-error when prescribing drugs, which improves patient outcomes and minimizes side effects [33].

Predictive analytics is also helping the healthcare industry operate more efficiently. Predictive models can help hospitals better plan for patient admissions, allocate resources, and handle their supply chains [34]. Predictive algorithms, for example, assist hospitals in anticipating staffing requirements and bed occupancy rates, which lowers wait times and enhances patient care overall [35]. Furthermore, healthcare institutions may reduce waste and guarantee the timely

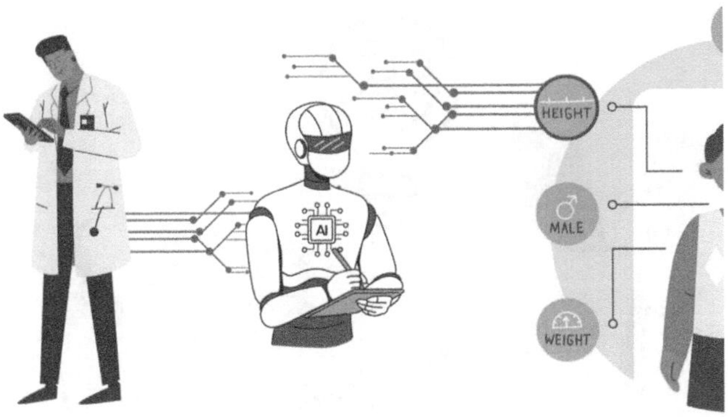

FIGURE 2.1 Artificial intelligence collects and evaluates information such as height and weight from people and notify the clinician if any disease risk begins to be predicted.

availability of resources by anticipating demand for pharmaceuticals and medical supplies. This would ultimately lead to more effective and economical healthcare delivery [36].

Nonetheless, there are several difficulties in using predictive analytics in the medical field. Given that predictive models rely on extensive sensitive patient data, ensuring data security and privacy becomes a critical concern [37]. Adhering to regulations such as HIPAA is essential to protect patient information [38]. Furthermore, utilizing algorithms trained on unrepresentative data might lead to predicts that are inaccurate or wrong, thus it's critical to design predictive models carefully to avoid bias [39]. It will be crucial to address these ethical concerns and also guarantee the interpretability and transparency of predictive models in order to build trust and maximize the advantages of predictive analytics in healthcare [40] (Figure 2.1).

2.7 DRUG DISCOVERY AND DEVELOPMENT

AI is revolutionizing drug research and discovery. It's hard to understate the efficiency gains that are now possible in the identification of therapeutic options. Conventional drug discovery can be a slow and inefficient process that takes decades and billions of dollars to get a new therapeutic off the ground. By screening vast datasets to identify possible therapeutic targets, as well as predicting how these new drugs might interact with other biological systems, AI can speed up this process substantially [41]. For example, deep learning systems might screen millions of compounds against certain proteins in silico to predict the binding affinity of each compound, thus cutting down the number of possibilities for subsequent testing [42].

AI can help with prediction toxicology and pharmacokinetics during the preclinical phase. In this respect, AI can help with preclinical screening of candidate medications prior to clinical trials, by helping scientists to assess whether the medication is safe enough to be tested on human patients or not [43]. Researchers might even be able to obviate the need for animal testing, as we have machine learning models that can predict the toxicity of novel chemical entities by analyzing their chemical structures and biological interactions in a predictive model. Such models even allow us to reliably predict whether a drug will have the desired therapeutic outcome or not [44]. Finally, AI can help researchers to simulate the body's functions of absorbing, distributing, metabolizing and eliminating a new chemical derivative [45].

Additionally, by improving patient recruitment, participant stratification, and allowing adaptive trial designs, AI is revolutionizing clinical trials [46]. AI technologies can evaluate patient data to determine which patients are most likely to advantage from a certain treatment, increasing trial efficiency and success rate [47]. AI is also capable of real- time trial data monitoring, which enables trial settings to be dynamically changed in response to preliminary findings [48]. This flexibility increases the chance of discovering successful treatments while also expediting the trial process [49].

Even with AI's enormous benefits for drug research and discovery, there are still issues, especially with data quality, obtaining regulatory approval, and ethical issues [50]. The performance of AI models is heavily influenced by the quality of the data used for training; biased or incomplete datasets can lead to inaccurate or misleading outcomes [51]. Furthermore, because these models frequently lack transparency and explainability, regulatory organizations are still unsure of how to assess and approve AI-driven drug development processes [52]. To guarantee that the advantages of AI in medication development are achieved without jeopardizing patient rights or public trust, ethical considerations including data privacy and fair access to AI-generated medicines must also be addressed [53] (Figure 2.2).

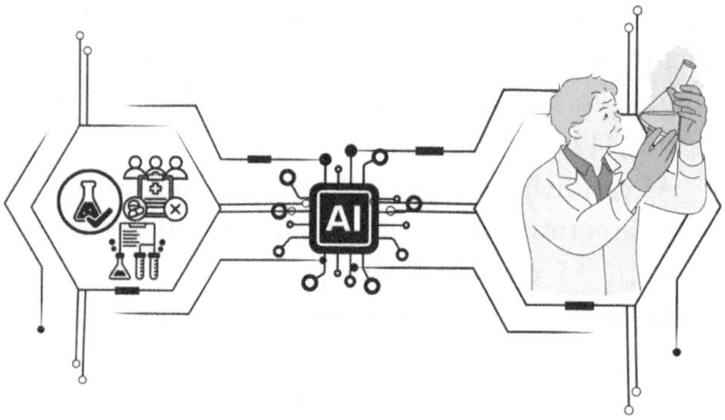

FIGURE 2.2 With the help of artificial intelligence, large and complex data sets are analyzed quickly. As a result, researchers' work progresses faster and easier.

FIGURE 2.3 Artificial intelligence assists the clinician in clinical decision-making by analyzing complex and large data sets more efficiently.

2.8 CLINICAL DECISION SUPPORT SYSTEM

Clinical Decision Support Systems (CDSS) are being greatly advanced by AI, which is giving physicians better tools to help them make more precise and informed judgments. While traditional CDSS relies on set rules and rigid standards, AI offers data-driven, dynamic methods that may be tailored to the specific needs of each patient. To provide individualized suggestions and predicted insights [54, 55]. AI algorithms analyze vast amounts of clinical data, such as EHRs, medical images, and patient histories. This ability allows healthcare providers to diagnose and treat complex conditions by uncovering patterns and connections that might not be immediately apparent through manual analysis.

AI-driven CDSS are making significant progress in diagnostics, especially in the area of medical picture interpretation. Radiological images, including CT scans, MRIs, and X-rays, are with the help of machine learning models, such as deep learning networks. These models can sometimes outperform human radiologists in accurately identifying early signs of diseases like cancer when it comes to detecting irregularities [56, 57]. For example, AI technologies have demonstrated significant promise in identifying lung cancer through chest CT scans and breast cancer via mammograms, which helps reduce the chances of false positives and overlooked diagnoses [58]. Additionally, AI is enhancing the functionality of CDSS by enabling personalized medication. AI systems are able to offer personalized therapy suggestions by combining clinical history, genetic data, and other patient-specific information. This is especially helpful in oncology, where patient and tumor genetic profiles may be used to tailor treatments [59]. AI models play a significant role in predicting how patients will respond to different treatments, guiding healthcare providers toward more effective and personalized treatment plans while minimizing side effects [60].

Nevertheless, there are challenges correlate with the integration of AI into CDSS that need to be addressed. Key concerns include fostering trust among clinicians, ensuring transparency of algorithms, and protecting data privacy and security. To avoid biases and errors, it is essential to rigorously validate the accuracy and fairness of AI systems [61, 62]. Additionally, in order to ensure that AI-generated suggestions complement human judgment rather than replace it, physicians must be taught in how to interpret them effectively [63]. The effective application of AI-driven CDSS in healthcare will depend on resolving these difficulties as AI technology advances (Figure 2.3).

2.9 AI IN GENOMIC

AI is revolutionizing genomic research, and helping us to make sense of the vast amounts of complex genetic data that human beings are producing in our quest to better understand inherited diseases. Specialist areas for AI-driven methods like machine learning, and deep learning are personalized medicine, functional genomics as well as gene sequencing among many other innovative fields of genomic research. These technologies allow greater precision and efficiency in examining genetic information, a capability which has led to landmark advances in both research and therapeutic applications. The high-throughput sequencing in genes is used the most for AI. A great deal of time and computing power are consumed by traditional methods for genetic data analysis. One of the many excellent applications is analyzing vast amount of sequencing data, which fits perfectly with AI algorithms like deep learning. This is in stark contrast to standard methods that require a significant amount of time, training, and reach only moderate precision predictions. These models have been instrumental in advancing our knowledge of the genetic underpinnings to complex disease such as cancer and rare Mendelian disorders [21].

Today, AI is used for advancing functional genomics by predicting the impact of genetic changes on protein function and gene expression as well as identifying variants. AI can fuse information from disparate sources, such as transcriptome profiles (i.e., snapshots of mRNA levels), data collected through proteomics platforms aimed at uncovering the list of proteins encoded by a genome and genomic sequences to produce holistic gene activity/regulation models. This is where they come in handy, creating models that can help predict how genetic differences could influence responses to drugs or predisposition for disease [64].

Now, the way discovery has transformed into a reality: AI-powered techniques are developed which enable to reveal previously unknown associations between genes and diseases giving us new information about genetic components of various health disorders [65]. AI is playing a major role in personalized medicine by aiding in more precise risk evaluations and treatment recommendations using an individual's genetic information. Machine learning algorithms can be used to analyze genetic, environmental, and lifestyle data to generate personalized health insights and guide treatment decisions [66]. AI models have the ability to predict a patient's likelihood of developing specific diseases or their response to certain medications, allowing for personalized treatment and prevention strategies [67]. This skill is particularly valuable in oncology, since genetic information can impact personalized cancer treatment plans. Challenges in the AI application in genomics persist despite advancements, including concerns regarding data privacy, the need for reliable training data, and the clarity of AI models [68]. The key to using AI systems effectively in clinical environments is their capability to offer clarity and enable healthcare professionals to comprehend and verify their forecasts [69]. As the region progresses, it will be vital to acknowledge ethical issues regarding

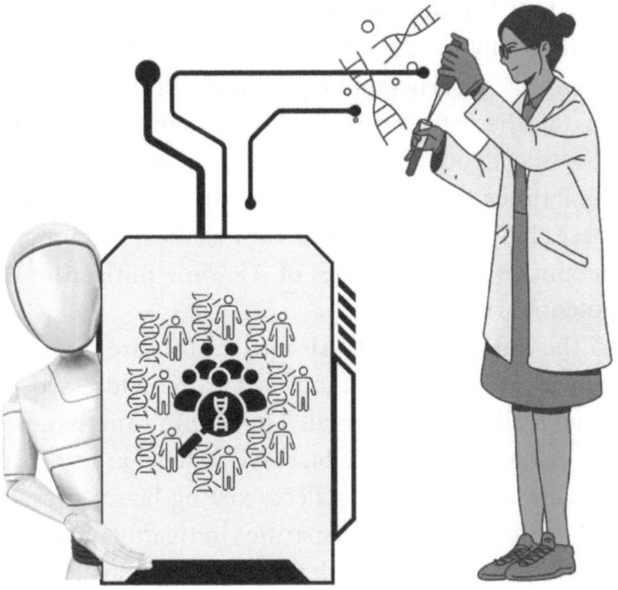

FIGURE 2.4 Artificial intelligence processes large-scale genomic datasets and allow researcher to identify and interpret genetic variants with greater accuracy and efficiency.

genetic information and ensure equitable availability of AI-based genomic technologies [70] (Figure 2.4).

2.10 CHALLENGES AND SOLUTIONS FOR AI IN HEALTHCARE

The incorporation of AI into healthcare brings several challenges that must be overcome to fully unlock its potential. One of the bigger problems is data security and privacy. AI healthcare systems analyze sensitive data raising suspicion o about unauthorized access which is against the protection of personal health information [48]. In order to minimize such risks regular security audits should take place and strict data protection programs can be used similarly to the HIPAA in the United States [33].

Another important issue is the risk of algorithmic bias. One data from one demographic group may not perform as well for another racial or socioeconomic group [6]. In order to reduce these type of biases AI algorithms must be diverse and represent entire patient populations, which can be achieved through continuous monitoring, big sets of data processing across different populations [27].

Furthermore, there is a fear among healthcare professionals about AI-generated systems replacing their jobs [15]. AI-generated systems analyze big amounts of data and handle repetitive tasks which can be used as a tool helping healthcare professionals focus on more complex tasks [36]. Training and education programs are crucial to equip healthcare professionals for the integration of AI, ensuring they know how to collaborate with AI systems and utilize their capabilities to improve patient care [4]. By tackling these challenges in advance, the healthcare sector can maximize the advantages of AI while mitigating its potential risks [22].

The integration of AI into healthcare presents numerous challenges that need to be addressed to fully realize its potential. One significant issue is the risk of algorithmic bias. AI systems are trained on data sets that may reflect existing biases present in healthcare, such as disparities in treatment or outcomes among different demographic groups [6]. For instance, an AI model trained predominantly on data from one racial or socioeconomic group may not perform as well for others, potentially exacerbating health inequities. To address this issue, it is essential to ensure that AI training data is diverse and reflects the entire patient population [27]. Applying robust bias detection and correction techniques, along with ongoing monitoring and updates to algorithms, can help minimize biases and promote fair AI use in healthcare [21].

Another challenge is data privacy and security. AI in healthcare relies on vast amounts of sensitive patient data, raising concerns about data breaches and unauthorized access 48]. The protection of personal health information is paramount, and AI systems must adhere to stringent data protection regulations, such as the HIPAA in the United States [33]. To overcome these concerns, healthcare organizations should employ robust encryption techniques, access controls, and regular security audits. Additionally, promoting transparency in how patient data is used and obtaining informed consent can help build trust and ensure that data privacy is maintained [61].

Last, the implementation of AI in healthcare can face resistance due to concerns about its impact on the healthcare workforce. There is apprehension that AI could replace healthcare professionals or reduce their roles [15]. However, AI should be seen as a tool to enhance, rather than replace, human expertise. It can manage repetitive tasks and analyze large datasets more efficiently, freeing up healthcare professionals to concentrate on complex decision-making and patient interactions [36]. Training and education programs are essential to prepare healthcare workers for the integration of AI, ensuring they understand how to work alongside AI systems and leverage their capabilities to enhance patient care [4]. By addressing these challenges proactively, the healthcare industry can harness the benefits of AI while safeguarding against its potential downsides [22].

2.11 SUMMARY

AI is revolutionizing healthcare by improving diagnostic accuracy, tailoring treatment plans to individual patients, and streamlining operational processes, all of which contribute to improved patient outcomes and lower costs. This chapter provides an overview of the evolution and current trends of AI in smart healthcare systems while forecasting future developments and identifying breakthrough technologies. It serves as a valuable resource for researchers, practitioners, and stakeholders looking to leverage AI's potential in healthcare.

BIBLIOGRAPHY

1. Miller, K., Mosby, M., Capan, M., Kowalski, R., & Ratwani, R. (2019). Artificial intelligence and clinical decision support: The next big thing? *BMJ Health & Care Informatics*, *26*(1), e100166.
2. Toma, M., Carrozzino, M., Bianchi, F. et al. (2023). Artificial intelligence and patient safety: Insights from a comprehensive review. *Journal of Patient Safety*, *19*(6).
3. Zhang, W., Liu, Y., & Wang, X. (2019). Explainable AI in genomics: Addressing challenges and opportunities. *Journal of Biomedical Informatics*, *95*, 103234.
4. Bera, K., Schalper, K. A., Rimm, D. L., Velcheti, V., & Madabhushi, A. (2019). Artificial intelligence in digital pathology—New tools for diagnosis and precision oncology. *Nature Reviews Clinical Oncology*, *16*(11), 703–715.
5. Chan HP, Samala RK, Hadjiiski LM, Zhou C. Deep Learning in Medical Image Analysis. Adv Exp Med Biol. 2020;1213:3–21. doi: 10.1007/978-3-030-33128-3_1. PMID: 32030660; PMCID: PMC7442218.
6. Ahn, J. S., Shin, S., Yang, S. A., Park, E. K., Kim, K. H., Cho, S. I., … Kim, S. (2023). Artificial intelligence in breast cancer diagnosis and personalized medicine. *Journal of Breast Cancer*, *26*(5), 405.
7. Zhang, J., Wang, Q., & Lin, W. (2024). Artificial intelligence in genomics and precision medicine. *Nature Reviews Genetics*, *25*(5), 287–300.
8. Ghiassian, S. D., & Elston, R. C. (2017). The use of machine learning in genetic research. *Annual Review of Genetics*, *51*, 347–369.

9. Gandhi, Z., Gurram, P., Amgai, B., Lekkala, S. P., Lokhandwala, A., Manne, S., … Surani, S. (2023). Artificial intelligence and lung cancer: Impact on improving patient outcomes. *Cancers*, *15*(21), 5236.
10. Sheng, B., Chen, X., Li, T., Ma, T., Yang, Y., Bi, L., & Zhang, X. (2022). An overview of artificial intelligence in diabetic retinopathy and other ocular diseases. *Frontiers in Public Health*, *10*, 971943.
11 Barton AJ. Creating an enabling information technology infrastructure. Clin Nurse Spec. 2014 Jan-Feb;28(1):7–8. doi: 10.1097/NUR.0000000000000023. PMID: 24309569.
12. Silva, J., Haro, R., Go, J. et al. (2023). Leveraging AI to identify and manage health risks: A systematic review. *IEEE Access*, *11*, 6884–6905.
13. Topol, E. J. (2019). High-performance medicine: The convergence of human and artificial intelligence. *Nature Medicine*, *25*(1), 44–56.
14. Curtis, R. G., Bartel, B., Ferguson, T., Blake, H. T., Northcott, C., Virgara, R., & Maher, C. A. (2021). Improving user experience of virtual health assistants: Scoping review. *Journal of Medical Internet Research*, *23*(12), e31737.
15. Bajwa, J., Munir, U., Nori, A., & Williams, B. (2021). Artificial intelligence in healthcare: Transforming the practice of medicine. *Future Healthcare Journal*, *8*(2), e188–e194.
16. Mesko, B. (2017). The role of artificial intelligence in precision medicine. *Expert Review of Precision Medicine and Drug Development*, *2*(5), 239–241.
17. Khera, A. V., Chaffin, M., Aragam, K. G., Haas, M. E., Roselli, C., Choi, S. H., & Kathiresan, S. (2018). Genome-wide polygenic scores for common diseases identify individuals with risk equivalent to monogenic mutations. *Nature Genetics*, *50*(9), 1219–1224.
18. Yoon, H. K., Yang, H. L., Jung, C. W., & Lee, H. C. (2022). Artificial intelligence in perioperative medicine: A narrative review. *Korean Journal of Anesthesiology*, *75*(3), 202–215.
19. Zhang, J., Li, Z., Lin, H., Xue, M., Wang, H., Fang, Y., … Ye, Z. (2023). Deep learning assisted diagnosis system: Improving the diagnostic accuracy of distal radius fractures. *Frontiers in Medicine*, *10*, 1224489.
20. Saito, K., Oda, T., & Morita, T. (2023) Artificial intelligence in personalized medicine: A new era of therapeutic strategies. *Annals of Oncology*, 34(9):1014–1021.
21. Amin A, Cardoso SA, Suyambu J, Abdus Saboor H, Cardoso RP, Husnain A, Isaac NV, Backing H, Mehmood D, Mehmood M, Maslamani ANJ. Future of Artificial Intelligence in Surgery: A Narrative Review. Cureus. 2024 Jan 4;16(1):e51631. doi: 10.7759/cureus.51631. PMID: 38318552; PMCID: PMC10839429.
22. Bowe, A. K., Lightbody, G., Staines, A., & Murray, D. M. (2023). Big data, machine learning, and population health: Predicting cognitive outcomes in childhood. *Pediatric Research*, *93*(2), 300–307.
23. Price, W. N., & Cohen, I. G. (2019). Privacy in the age of medical big data. *Nature Medicine*, *25*, 37–43.
24. Gupta, A., Lippert, C., Auerbach, B. et al. (2020). Artificial intelligence to optimize patient selection in clinical trials. *Nature Reviews Drug Discovery*, *19*(8), 660–661.
25. Serrano DR, Luciano FC, Anaya BJ, Ongoren B, Kara A, Molina G, Ramirez BI, Sánchez-Guirales SA, Simon JA, Tomietto G, Rapti C, Ruiz HK, Rawat S, Kumar D, Lalatsa A. Artificial Intelligence (AI) Applications in Drug Discovery and Drug Delivery: Revolutionizing Personalized Medicine. Pharmaceutics. 2024 Oct 14;16(10):1328. doi: 10.3390/pharmaceutics16101328. PMID: 39458657; PMCID: PMC11510778.
26. Paul, D., Sanap, G., Shenoy, S., Kalyane, D., Kalia, K., & Tekade, R. K. (2021). Artificial intelligence in drug discovery and development. *Drug Discovery Today*, *26*(1), 80–93.
27. Al Kuwaiti, A., Nazer, K., Al-Reedy, A., Al-Shehri, S., Al-Muhanna, A., Subbarayalu, A. V., & Al-Muhanna, F. A. (2023). A review of the role of artificial intelligence in healthcare. *Journal of Personalized Medicine*, *13*(6), 951.
28. Reilly, B. M., & Evans, A. T. (2006). Translating clinical research into clinical practice: Impact of using prediction rules to make decisions. *Annals of Internal Medicine*, *144*(3), 201–209.
29. Vamathevan, J., Clark, D., Czodrowski, P. et al. (2019). Applications of machine learning in drug discovery and development. *Nature Reviews Drug Discovery*, *18*, 463–477.
30. De Fauw, J., Ledsam, J. R., Romera-Paredes, B., Nikolov, S., Tomasev, N., Blackwell, S., & Ronneberger, O. (2018). Clinically applicable deep learning for diagnosis and referral in retinal disease. *Nature Medicine*, *24*(9), 1342–1350.
31. Collins, F. S., & Varmus, H. (2015). A new initiative on precision medicine. *New England Journal of Medicine*, *372*(9), 793–795.
32. T P A, M SS, Jose A, Chandran L, Zachariah SM. Pharmacogenomics: the right drug to the right person. J Clin Med Res. 2009 Oct;1(4):191–4. doi: 10.4021/jocmr2009.08.1255. Epub 2009 Oct 16. PMID: 22461867; PMCID: PMC3299179.
33. Ashley, E. A. (2015). The questioning clinician. *JAMA*, *314*(4), 410–410
34. Deo, R. C. (2015). Machine learning in medicine. *Circulation*, *132*(20), 1920–1930.
35. Riordan, M., Puhl, A. C., & Ekins, S. (2023). Artificial intelligence in pharmaceutical drug discovery: Recent developments and future trends. *Expert Opinion on Drug Discovery*, *18*(10), 1081–1092.
36. Bates, D. W., Saria, S., Ohno-Machado, L., Shah, A., & Escobar, G. (2014). Big data in health care: Using analytics to identify and manage high-risk and high-cost patients. *Health Affairs*, *33*(7), 1123–1131.

37. Rivero-Moreno Y, Rodriguez M, Losada-Muñoz P, Redden S, Lopez-Lezama S, Vidal-Gallardo A, Machado-Paled D, Cordova Guilarte J, Teran-Quintero S. Autonomous Robotic Surgery: Has the Future Arrived? Cureus. 2024 Jan 14;16(1):e52243. doi: 10.7759/cureus.52243. PMID: 38352080; PMCID: PMC10862530.
38. Long, J., & Liu, C. (2021). Understanding clinician perceptions of artificial intelligence in healthcare: A systematic review. *Journal of Medical Internet Research*, *23*(5), e24478.
39. Green, B. L., Murphy, A., & Robinson, E. (2024). Accelerating health disparities research with artificial intelligence. *Frontiers in Digital Health*, *6*, 1330160.
40. Hryciw, B. N., Fortin, Z., Ghossein, J., & Kyeremanteng, K. (2023). Doctor-patient interactions in the age of AI: Navigating innovation and expertise. *Frontiers in Medicine*, *10*, 1241508.
41. Chen, H., Engkvist, O., Wang, Y., Olivecrona, M., & Blaschke, T. (2018). The rise of deep learning in drug discovery. *Drug Discovery Today*, *23*(6), 1241–1250.
42. Wang, F., Casalino, L. P., & Khullar, D. (2022). Deep learning and the future of healthcare: What we need to know. *JAMA*, *328*(18), 1783–1784.
43. Esteva, A., Kuprel, B., Novoa, R. A., Ko, J., Swetter, S. M., Blau, H. M., & Thrun, S. (2017). Dermatologist-level classification of skin cancer with deep neural networks. *Nature*, *542*(7639), 115–118.
44. Yao, J., Yang, X., & Yang, X. (2020). Artificial intelligence in early cancer detection and diagnosis. *Frontiers in Oncology*, *10*, 710.
45. Piro, R. M., & Laing, R. E. (2021). AI and functional genomics: A review. *Journal of Functional Genomics & Proteomics*, *10*(1), 41–52.
46. Tsai, C., Ho, J., & Hsu, K. (2024). AI in radiology: Current status and future directions. *Insights Imaging*, *14*(1), 45.
47. Scotney, H., Symonds, M. E., Law, J., Budge, H., Sharkey, D., & Manolopoulos, K. N. (2017). Glucocorticoids modulate human brown adipose tissue thermogenesis in vivo. *Metabolism*, *70*, 125–132.
48. Tandon, Y. K., Bartholmai, B. J., & Koo, C. W. (2020). Putting artificial intelligence (AI) on the spot: Machine learning evaluation of pulmonary nodules. *Journal of Thoracic Disease*, *12*(11), 6954.
49. Amiri, M., Gittins, C., Spence, S., & Balduzzi, S. (2020). Machine learning applications in adaptive clinical trial designs: A review. *Clinical Trials*, *17*(2), 216–227.
50. Ekins, S., Puhl, A. C., Zorn, K. M., Lane, T. R., Russo, D. P., Klein, J. J., & Clark, A. M. (2019). Exploiting machine learning for end-to-end drug discovery and development. *Nature Materials*, *18*(5), 435–441.
51. Masud, M., Bairagi, A. K., Nahid, A. A., Sikder, N., Rubaiee, S., Ahmed, A., & Anand, D. (2021). A pneumonia diagnosis scheme based on hybrid features extracted from chest radiographs using an ensemble learning algorithm. *Journal of Healthcare Engineering*, *2021*(1), 8862089.
52. Rajkomar, A., Oren, E., Chen, K. et al. (2018). Scalable and accurate deep learning with electronic health records. *npj Digital Medicine*, *1*, 18.
53. Reddy, S., & Fox, J. (2019). AI-enabled healthcare delivery. *JAMA*, *321*(6), 541–542.
54. O'Rawe, J., Jiang, T., & Sun, G. (2015). Low-frequency genetic variants and the risk of complex diseases. *Nature Reviews Genetics*, *16*(4), 233–246.
55. FDA. (2022). FDA approvals of AI algorithms for image interpretation. FDA Website. Retrieved from https://www.fda.gov/ (accessed on 15 August 2024).
56. Mooney, P., & Kim, H. (2023). Artificial intelligence in cardiovascular imaging. *JACC Cardiovasc Imaging*, *16*(9), 1461–1474.
57. Chou, K. S., Yang, L., & Chang, Y. (2019). Deep learning for medical image analysis: A review. *Artificial Intelligence Review*, *52*(1), 1–16.
58. Ferrante, J. M., & Klein, W. P. (2021). Artificial intelligence and its role in clinical decision support systems. *Journal of General Internal Medicine*, *36*(5), 1434–1442.
59. ZainEldin, H., Gamel, S. A., El-Kenawy, E. S. M., Alharbi, A. H., Khafaga, D. S., Ibrahim, A., & Talaat, F. M. (2022). Brain tumor detection and classification using deep learning and sine-cosine fitness grey wolf optimization. *Bioengineering*, *10*(1), 18.
60. Parikh, R. B., Scherer, L. D., & Eberly, L. A. (2023). Addressing bias in artificial intelligence: A primer for health care professionals. *JAMA*, *329*(3), 225–227.
61. Wang, X., Zhang, C., Li, M., Liu, Y., & Zhou, Y. (2023). Recent advances in artificial intelligence for predicting disease outcomes. *Nature Reviews Genetics*, *24*(6), 329–341.
62. Balthazar, P., Harri, P., Prater, A., & Safdar, N. M. (2019). Civility. *Journal of the American College of Radiology*, *16*(9), 1127.
63. Char, D. S., Shah, N. H., & Magnus, D. (2018). Implementing machine learning in health care—Addressing ethical challenges. *New England Journal of Medicine*, *378*(11), 981–983.
64. Mak, I. W., Evaniew, N., & Ghert, M. (2014). Lost in translation: Animal models and clinical trials in cancer treatment. *American Journal of Translational Research*, *6*(2), 114–118.
65. Pritchard, J. K., & Sandor, C. (2021). The future of genomic medicine: Insights from AI and ethics. *Nature Reviews Genetics*, *22*(3), 183–192.
66. Khosla, A., Tole, S., & Iyer, P. (2023). The potential for artificial intelligence in healthcare: Perspectives from primary care physicians. *Journal of Medical Internet Research*, *8*, 25.

67. Khosravi, M., Zare, Z., Mojtabaeian, S. M., & Izadi, R. (2024). Artificial intelligence and decision-making in healthcare: A thematic analysis of a systematic review of reviews. *Health Services Research and Managerial Epidemiology, 11*, 23333928241234863.
68. Goodman, B., & Flaxman, S. (2017). European Union regulations on algorithmic decision-making and a "right to explanation. *AI Magazine*, *38*(3), 50–57.
69. Obermeyer, Z., Powers, B., Vogeli, C., & Mullainathan, S. (2019). Dissecting racial bias in an algorithm used to manage the health of populations. *Science*, *366*(6464), 447–453.
70. Zhou, L., Lu, J., & Chen, X. (2023). The role of artificial intelligence in healthcare management and patient outcomes. *Health Management Technology*, *44*(1), 12–20.
71. Ross, C., Schuster, M., & Thomas, A. (2024). Machine learning and predictive analytics in health care. *JAMA*, *331*(19), 1917–1918.

CHAPTER 3

Clinical Decisions Support Systems

Enhancing Medical Decision-Making

Renu Kharb

3.1 INTRODUCTION

Healthcare decision-making is inherently complex, requiring clinicians to synthesize extensive information, including patient histories, diagnostic data, treatment pathways, and clinical guidelines. The complexity has escalated with the rapid growth of medical knowledge, technological advancements, and the increasing emphasis on personalized medicine. Clinicians must now integrate data from multiple sources such as electronic health records (EHRs), laboratory results, imaging, genetic analyses, and wearable technologies while adhering to evidence-based practices and updated protocols. This immense influx of data can make efficient and effective decision-making a significant challenge in modern healthcare [1].

Clinical Decision Support Systems (CDSS) have emerged as critical tools to address these challenges. These systems combine algorithms, databases, and patient-specific data to provide actionable insights and guidance. By offering support in areas like diagnosing complex conditions, mitigating medication errors, and predicting potential complications, CDSS improves the quality and safety of care. Importantly, these systems are not substitutes for human expertise but are designed to augment clinical judgment, allowing healthcare providers to retain decision-making authority while leveraging technology for better outcomes [2].

The history of CDSS dates back to the 1960s, when basic rule-based systems were introduced that had limited functionality due to the constraints of early computing technologies. These initial efforts laid the groundwork for contemporary systems that incorporate cutting-edge technologies such as artificial intelligence (AI), machine learning (ML), and natural language processing (NLP) [3]. Modern CDSS solutions are versatile, integrating seamlessly with EHRs, operating on mobile devices, or through cloud-based platforms to provide real-time, patient-specific recommendations tailored to clinical workflows [4].

Despite their benefits, implementing CDSS comes with challenges. Issues such as inconsistent data quality, poor interoperability across systems, alert fatigue among clinicians, and biases in algorithmic design can hinder their effectiveness. Moreover, ethical and legal concerns regarding patient data privacy, security, and accountability in decision-making need to be addressed to ensure these systems function optimally and ethically [5].

As healthcare systems grapple with increasing pressures, CDSS has proven to be a pivotal tool in enhancing efficiency, reducing errors, and improving patient outcomes. By assisting in areas ranging from chronic disease management to precision medicine, CDSS is transforming medical practice and addressing critical gaps in care delivery. This chapter explores the evolution, architecture, real-world applications, and challenges of CDSS, highlighting their profound impact on modern healthcare and their potential to shape its future.

3.2 EVOLUTION OF CLINICAL DECISION SUPPORT SYSTEMS

3.2.1 The Early Stages of CDSS

The foundation of decision support in medicine was laid during the 1960s with the advent of early systems like MYCIN and HELP. MYCIN is Developed to assist

DOI: 10.1201/9781003533351-3

in diagnosing bacterial infections and recommending antibiotics, MYCIN relies on rule-based algorithms. HELP is Focused on monitoring hospital data and providing real-time alerts, HELP represented one of the first attempts at system-wide integration. Despite their innovation, these systems faced significant limitations: Rigid rule-based frameworks that lacked flexibility, and difficulty in integrating with hospital workflows due to limited computing resources [6].

3.2.2 The Integration Era

The introduction of EHRs in the 1990s marked a turning point in CDSS development. The integration of CDSS with EHRs enabled: Automatic access to patient histories, lab results, imaging studies, and real-time alerts for critical care interventions or routine monitoring. Use of standardized protocols like HL7 is essential in CDSS for seamless data exchange.

3.2.3 Modern Advances

Contemporary CDSS have moved beyond rule-based approaches to incorporate technologies such as AI and ML. AI enables CDSS to process vast datasets and identify patterns for improved diagnostic and treatment recommendations. Cloud computing and the Internet of Things, facilitate remote monitoring and continuous data analysis. These advancements have expanded the functionality and applicability of CDSS in clinical practice, making them integral to precision medicine and population health management [7].

3.3 KEY COMPONENTS OF CLINICAL DECISION SUPPORT SYSTEMS

CDSS are sophisticated tools designed to aid healthcare providers by offering evidence-based recommendations to improve decision-making and enhance patient care. The success of these systems relies on the effective functioning of several critical components. These components must collaborate to process patient data, utilize medical knowledge, and provide actionable insights promptly. Below is a detailed exploration of the essential components of CDSS.

3.3.1 Knowledge Base

The knowledge base is the core component of a CDSS. It acts as the repository of clinical information, including medical protocols, diagnostic criteria, and treatment guidelines, which drives the system's recommendations and alerts. The accuracy and comprehensiveness of the knowledge base are crucial for the system's overall effectiveness.

The knowledge base includes protocols and treatment guidelines derived from reputable sources, such as professional organizations and peer-reviewed literature. These guidelines help ensure the recommendations made by the CDSS align with evidence-based practices. Information on drug interactions, contraindications, dosages, side effects, and drug classifications is crucial to prevent medication errors and optimize treatment plans. The system utilizes disease classification systems (such as ICD codes) and diagnostic decision pathways for various conditions to help clinicians make accurate diagnoses [8]. To maintain the relevance and accuracy of the system, the knowledge base must be regularly updated with the latest medical research, new drug approvals, and evolving clinical guidelines. This ensures that healthcare providers are always using the most current evidence in decision-making.

Many systems allow customization to reflect the unique practices of different healthcare organizations or the specific needs of particular patient populations. This flexibility ensures that the CDSS aligns with regional practices and addresses the specific challenges faced by clinicians in diverse settings [2].

3.3.2 Inference Engine

The inference engine is the decision-making core of the CDSS. This component processes patient-specific data in real-time, applying algorithms and logic to generate recommendations, warnings, and predictive insights.

The engine analyzes a wide range of patient-specific data, including lab results, medical history, vital signs, and demographic information, to detect trends and identify potential health risks. It uses clinical rules, decision frameworks, and algorithms to suggest appropriate actions, such as diagnostic tests or treatment options, based on the patient's data.

The engine guides clinicians by using logical reasoning models (such as rule-based or probabilistic models) to ensure that all relevant factors are considered in the decision-making process [9]. The inference engine can detect potential medication errors, such as unsafe drug interactions or inappropriate dosages, thereby reducing the likelihood of adverse events.

contributes to better health outcomes for patients. These systems are particularly effective in managing chronic diseases, where they can monitor patient progress and suggest adjustments to treatment strategies. Additionally, CDSS helps identify patients at high risk of complications, enabling preventive measures or closer monitoring. The overall impact is a reduction in hospitalizations, fewer medical errors, and greater patient satisfaction [15].

3.7.7 Encouraging Preventive Care

CDSS plays an integral role in preventive healthcare by identifying patients who are eligible for screenings, immunizations, or other preventive measures. For instance, the system might remind clinicians to schedule mammograms for women within the recommended age range or recommend vaccinations for high-risk populations. This proactive approach ensures that potential health issues are identified and addressed early, leading to better long-term outcomes. By focusing on prevention, CDSS reduces the burden of treating advanced-stage diseases, benefiting both patients and healthcare systems [12].

3.7.8 Standardizing Clinical Practices

Variability in clinical decision-making is a common challenge that can lead to inconsistent patient care. CDSS helps address this issue by providing uniform, evidence-based recommendations regardless of the clinician's experience or specialty. For example, in resource-limited settings or rural areas, CDSS ensures that general practitioners have access to the same quality of decision support as specialists. This standardization improves consistency in care delivery and reduces disparities in patient outcomes across different healthcare settings [13].

3.7.9 Facilitating Multidisciplinary Collaboration

CDSS enhances teamwork in healthcare by consolidating information and making it accessible to all members of a multidisciplinary team. In complex cases, such as cancer treatment, where input is required from oncologists, radiologists, and surgeons, CDSS acts as a centralized tool for data sharing and decision-making. This fosters better communication and ensures that all team members are aligned in their approach, ultimately improving patient outcomes and reducing redundancies [15].

3.7.10 Providing Real-Time Alerts

Real-time monitoring and alerts are among the most valuable features of CDSS. These systems can track patients' vital signs and other critical parameters in real-time, alerting clinicians to potential emergencies such as sepsis, arrhythmias, or respiratory failure. In intensive care units (ICUs) or emergency settings, such alerts can prompt immediate interventions, saving lives. The ability to provide timely warnings also reduces the risk of oversight in high-pressure environments [12].

3.7.11 Educating Clinicians

CDSS serves as a valuable educational resource for healthcare providers, particularly those who are new to the profession or working in less specialized roles. The system offers explanations for its recommendations, helping clinicians understand the rationale behind specific guidelines or treatments. This not only improves decision-making in the short term but also enhances the clinician's knowledge and confidence over time. By incorporating learning opportunities into daily practice, CDSS contributes to professional development [14].

3.8 DISADVANTAGES OF CLINICAL DECISION SUPPORT SYSTEMS

While CDSS offer a host of benefits in enhancing clinical decisions, they also come with a range of challenges that can hinder their effectiveness in some settings. Understanding these disadvantages is crucial for healthcare providers to make informed decisions about their use. Below are several key drawbacks of CDSS that need careful consideration.

3.8.1 Over-Reliance on Technology

One of the primary concerns with CDSS is the potential for over-reliance on automated recommendations. As clinicians become increasingly dependent on these systems, there is a risk that they may become complacent, deferring to the system's suggestions without exercising their professional judgment. This reliance, sometimes referred to as "automation bias," can lead to missed clinical insights or inappropriate decisions, particularly in complex cases where human expertise and intuition are critical. CDSS should be used as a supportive tool rather than a substitute for a clinician's experience and expertise [17].

3.8.2 Alert Fatigue

Another significant issue with CDSS is the phenomenon of alert fatigue. As CDSS systems often generate numerous alerts—ranging from medication interactions to reminders about guidelines clinicians can become desensitized to these notifications. When alerts are frequent or irrelevant to the task at hand, healthcare providers may start ignoring them or dismissing them entirely, even when they are clinically significant. This can result in missed opportunities to intervene, potentially leading to patient harm. Poorly designed or overly sensitive alert systems exacerbate this issue by bombarding clinicians with information that may not be immediately relevant to patient care [18].

3.8.3 Resistance to Adoption

The implementation of CDSS can meet with resistance from healthcare professionals who may be skeptical about the reliability or value of these systems. Clinicians might feel that CDSS undermines their professional autonomy or leads to an erosion of their clinical decision-making skills. In some cases, healthcare providers may view the introduction of CDSS as a disruption to their workflow, particularly if the system is difficult to use or integrates poorly with existing systems. Overcoming this resistance requires effective training, clear communication about the benefits of CDSS, and involving clinicians in developing and customizing the system to ensure it meets their needs [19].

3.8.4 Challenges in Integration

Integrating CDSS into existing healthcare infrastructures is often a complex and resource-intensive process. In many cases, healthcare systems are built on legacy technologies, which may not be compatible with newer CDSS tools. The integration of CDSS with EHRs or other medical software can encounter technical issues such as data interoperability challenges or difficulties in maintaining consistent data quality. Moreover, ensuring that the system remains up to date with the latest medical knowledge requires ongoing efforts, such as regular software updates and manual inputs, which can become a burden on healthcare institutions [20].

3.8.5 Privacy and Security Concerns

CDSS relies heavily on patient data to generate recommendations, which raises concerns about data privacy and security. The management of sensitive health information is regulated under strict laws such as HIPAA in the US, and ensuring that CDSS complies with these regulations is essential. However, the use of vast amounts of patient data also increases the risk of breaches or unauthorized access, particularly if the system is not properly secured. Healthcare institutions must implement robust security measures to protect patient information, and ensure that CDSS platforms follow the best practices in data encryption, access control, and monitoring [21].

3.8.6 High Costs

The implementation of CDSS often requires significant financial investment, which can be a deterrent for smaller healthcare facilities or practices. Costs include the initial purchase of the software and hardware, as well as ongoing expenses for system updates, maintenance, and staff training. While CDSS has the potential to reduce long-term healthcare costs by improving efficiency and preventing errors, the high upfront cost can be a barrier, especially in settings where resources are limited. Small practices may find it difficult to justify these expenses if the system's benefits are unclear immediately [22].

3.8.7 Potential for Bias in the System

CDSS systems are only as good as the data and algorithms that power them. If the underlying data used to train the system is biased or incomplete, the system can make skewed recommendations. For instance, if a CDSS is trained on data that predominantly represents a specific demographic group, it may not offer accurate recommendations for patients from other groups. This could result in suboptimal care for diverse populations, particularly in terms of medication recommendations or diagnostic suggestions. Addressing these biases is essential to ensure the system delivers equitable and fair recommendations across all patient demographics [23].

3.8.8 Disruption to Clinical Workflow

Despite the potential to streamline clinical processes, CDSS can sometimes disrupt established workflows. If the system requires clinicians to input excessive amounts of data or take additional steps during the decision-making process, it can slow down care delivery. Clinicians, already facing time constraints, may find the additional tasks cumbersome and counterproductive, especially in high-pressure settings like emergency rooms or intensive care units. This disruption can lead to clinician dissatisfaction and even decreased productivity if the system is perceived as adding more complexity without sufficient benefit.

3.8.9 Real-World Application of Clinical Decision Support System

CDSS have found extensive application in real-world healthcare settings, significantly enhancing decision-making across various domains. In medication management, CDSS assists healthcare professionals by reducing prescription errors, flagging potential drug interactions, and suggesting optimal dosages tailored to patient-specific factors like age, weight, or kidney function. For instance, hospitals utilizing integrated EHR systems, such as Epic or Cerner, employ CDSS tools to improve medication safety. Similarly, in chronic disease management, CDSS supports clinicians by offering evidence-based guidelines for diseases like diabetes or asthma. These systems provide insights on treatment adjustments based on real-time patient data, ensuring personalized care [18].

In diagnostic support, CDSS has proven invaluable, especially for complex or rare conditions. Tools like IBM Watson for Oncology synthesize patient histories with medical literature to recommend cancer treatments. Likewise, the Isabel Diagnostic Tool aids clinicians by cross-referencing symptoms with extensive databases [24]. In critical care, CDSS is used in ICUs and emergency departments to prioritize rapid interventions, such as identifying early signs of sepsis or determining the urgency of stroke treatments using platforms like Viz.ai. Additionally, CDSS has revolutionized surgical planning, helping predict risks and improving preoperative assessments with systems like the ACS NSQIP Surgical Risk Calculator [25].

The role of CDSS extends beyond individual patient care to public health, especially during crises like the COVID-19 pandemic. During the outbreak, CDSS tools were employed for patient triage, predicting infection trends, and guiding vaccination strategies. Telemedicine has also benefited from CDSS integration, as platforms like Teladoc Health use these tools for remote monitoring and consultations, improving care accessibility. Overall, the versatility and efficiency of CDSS continue to make them indispensable in enhancing healthcare delivery across diverse settings [32].

3.9 FUTURE DIRECTIONS FOR CLINICAL DECISION SUPPORT SYSTEMS

The development and implementation of CDSS are anticipated to grow in scope and sophistication, addressing current limitations while incorporating emerging technologies. Below is a discussion of key future directions for CDSS.

3.9.1 Integration with Artificial Intelligence and Machine Learning

The use of AI and ML in CDSS is expected to advance predictive modeling and improve diagnostic accuracy. By analyzing large datasets, AI-driven systems can detect patterns and predict potential health issues before symptoms arise. For instance, ML algorithms can assist in the early detection of diseases like cancer or provide personalized treatment options. These advancements may allow CDSS to handle complex scenarios with greater precision [27].

3.9.2 Personalized and Precision Medicine

Future CDSS platforms will increasingly focus on integrating patient-specific data such as genetic profiles, lifestyle factors, and environmental influences. This personalization will enhance the precision of medical interventions, especially in fields like oncology, cardiology, and pharmacogenomics. Such tailored approaches are critical to minimizing adverse effects and maximizing treatment efficacy [28].

3.9.3 Enhanced Interoperability

Seamless data exchange between CDSS, EHRs, and other healthcare systems is a critical area of focus. Standardized protocols like FHIR facilitate improved communication across different platforms. Enhanced interoperability ensures that clinicians can access comprehensive patient data, reducing errors and improving continuity of care.

3.9.4 Natural Language Processing and Voice Interaction

Incorporating NLP and voice recognition into CDSS is expected to make systems more user-friendly. NLP can extract critical information from unstructured data like clinical notes and research articles, while voice-enabled interactions can streamline workflows by allowing hands-free operation. These features will save time and improve accessibility for healthcare professionals [5].

3.9.5 Cloud-Based and Mobile Solutions

The adoption of cloud technology in CDSS is expected to improve scalability, accessibility, and real-time data sharing. Mobile applications integrated with CDSS can empower clinicians to make decisions remotely, facilitating telemedicine and remote patient monitoring. Cloud-based systems also enable global updates and integration with large-scale databases for enhanced decision-making [1].

3.9.6 Addressing Bias and Ethical Concerns

Future CDSS systems will prioritize equity by addressing biases in algorithms that can lead to disparities in healthcare delivery. Developers will need to create systems that ensure fairness across diverse patient populations. Additionally, ethical frameworks will play a pivotal role in managing data privacy, consent, and accountability in decision-making processes [29].

3.9.7 Internet of Things (IoT) Integration and Wearable Devices

Integration with IoT devices and wearables will allow CDSS to provide continuous, real-time health monitoring. Data from smartwatches, biosensors, and other devices can alert clinicians to anomalies and improve chronic disease management. This approach is particularly relevant for managing conditions like diabetes, hypertension, and heart disease [30].

3.9.8 Contributions to Public Health

CDSS will play an expanded role in public health by assisting with outbreak prediction, vaccination programs, and resource allocation. These systems can analyze environmental and demographic data alongside clinical information to identify at-risk populations and recommend interventions [31].

3.9.9 Conclusion of Clinical Decision Support Systems

CDSS have revolutionized healthcare by providing real-time, evidence-based support that aids healthcare professionals in making accurate and informed decisions. By integrating patient data with clinical guidelines and predictive algorithms, these systems help to reduce diagnostic errors, enhance treatment efficacy, and streamline clinical workflows. From improving the management of chronic conditions to enabling more precise diagnostic processes, CDSS has a broad range of applications that make it invaluable in modern medical practice. However, while the benefits are substantial, CDSS implementation faces notable challenges, including technical integration hurdles, user resistance, and the risk of information overload due to excessive alerts. Furthermore, the reliability of CDSS can be compromised by data biases or algorithmic inaccuracies. Overcoming these obstacles is crucial to maximizing its potential. Future innovations, such as AI, wearable device integration, and enhanced interoperability, promise to elevate the capabilities of CDSS, enabling a shift toward more proactive, personalized, and predictive healthcare delivery.

3.10 SUMMARY

To fully realize the transformative potential of CDSS, ethical concerns such as patient privacy, data security, and algorithmic fairness must be addressed. As the healthcare ecosystem continues to evolve, CDSS is positioned to play a central role in enhancing patient outcomes, reducing costs, and ensuring equitable care. With ongoing technological advancements and collaborative efforts across the healthcare sector, CDSS can contribute to a more efficient and patient-centric model of care delivery, ultimately improving global health outcomes.

BIBLIOGRAPHY

1. Raghupathi, W., & Raghupathi, V. (2014). Big data analytics in healthcare: Promise and potential. *Health Information Science and Systems*, *2*, 1–10.
2. Greenes, R. (Ed.). (2014). *Clinical Decision Support: The Road to Broad Adoption*. Amsterdam and Boston: Academic Press.
3. Kaplan, B. (2020). Revisiting health information technology ethical, legal, and social issues and evaluation: Telehealth/telemedicine and COVID-19. *International Journal of Medical Informatics*, *143*, 104239.
4. Liu, X., Rivera, S. C., Moher, D., Calvert, M. J., Denniston, A. K., Ashrafian, H., & Yau, C. (2020). Reporting guidelines for clinical trial reports for interventions involving artificial intelligence: The CONSORT-AI extension. *The Lancet Digital Health*, *2*(10), e537–e548.

5. Mehta, N., & Pandit, A. (2018). Concurrence of big data analytics and healthcare: A systematic review. *International Journal of Medical Informatics, 114*, 57–65.
6. Morley, J., Machado, C. C., Burr, C., Cowls, J., Joshi, I., Taddeo, M., & Floridi, L. (2020). The ethics of AI in health care: A mapping review. *Social Science & Medicine, 260*, 113172.
7. Ruan, Y., Zhang, L., & Liang, Y. (2020). Use of clinical decision support systems for improving medication safety. *International Journal of Medical Informatics, 134*, 104021.
8. Bergmo, T. S. (2010). Are health economic evaluations of eHealth interventions methodologically sound? A systematic review of the literature. *International Journal of Medical Informatics, 79*(1), 1–12.
9. Kreimeyer, K., Foster, M., Pandey, A., Arya, N., Halford, G., Jones, S. F., & Botsis, T. (2017). Natural language processing systems for capturing and standardizing unstructured clinical information: A systematic review. *Journal of Biomedical Informatics, 73*, 14–29.
10. Berner, E. S. (2007). *Clinical Decision Support Systems* (Vol. 233). New York: Springer Science+ Business Media, LLC.
11. Bates, D. W., & Gawande, A. A. (2003). Improving safety with information technology. *New England Journal of Medicine, 348*(25), 2526–2534.
12. Sittig, D. F., & Singh, H. (2020). A new era in health care: The role of clinical decision support systems. *BMJ Quality & Safety, 29*(6), 416–423.
13. Hasan, S., Memon, H., Jabeen, S. et al. (2017). Evaluating the functionality and effectiveness of clinical decision support systems: Challenges and future directions. *Journal of Health Informatics & Decision Making, 6*(4), 229–245.
14. Cimino, J. J., Starren, J., & Gupta, K. (2019). Knowledge representation in clinical decision support systems: A framework for clinical reasoning. *Journal of Biomedical Informatics, 40*(4), 379–387.
15. Coiera, E. (2015). Clinical decision support systems. In *Clinical Informatics: A Practical Approach for Healthcare Implementation* (pp. 111–134). Boca Raton: Springer.
16. Greene, S. M., Taler, G., Allen, N. et al. (2021). Real-time clinical decision support for medication therapy management: Effectiveness of an integrated system. *American Journal of Health-System Pharmacy, 78*(1), 57–64.
17. Ash, J. S., Berg, M., & Coiera, E. (2004). Some unintended consequences of information technology in health care: The nature of patient care information system-related errors. *Journal of the American Medical Informatics Association, 11*(2), 104–112.
18. McDonald, C. J. (2005). The barriers to the adoption of clinical decision support systems in the United States: Why are we not using them more? *Journal of the American Medical Informatics Association, 12*(1), 36–42.
19. Cresswell, K. M., Bates, D. W., & Sheikh, A. (2013). Ten key considerations for the successful implementation and adoption of health information technology. *Journal of the American Medical Informatics Association, 20*(1), 18–23.
20. Sheikh, A., Sood, H. S., & Bates, D. W. (2010). Impact of health information technology on quality, safety, and efficiency of medical care. *Journal of the American Medical Informatics Association, 17*(6), 701–707.
21. Holden, R. J. (2011). Clinical decision support systems: Past, present, and future. *Yearbook of Medical Informatics, 22*(1), 93–101.
22. Balgrosky, J. (2019) *Health Informatics: Practical Guide for Healthcare and Information Technology Professionals*. 6th ed. Chicago, IL: HIMSS
23. Chismar, W. G., & Granger, M. L. (2012) A review of the barriers and drivers of health information technology adoption. HIMSS annual conference paper. Available from: https://www.himss.org.
24. Kawamoto, K., Houlihan, C. A., Balas, E. A., & Lobach, D. F. (2005). Improving clinical practice using clinical decision support systems: A systematic review of trials to identify features critical to success. *The BMJ, 330*(7494), 765.
25. Topol, E. J. (2020). Telemedicine in the era of COVID-19. *The Journal of Allergy and Clinical Immunology: In Practice, 8*(5), 1489–1491.
26. Jha, A. K., DesRoches, C. M., Campbell, E. G., Donelan, K., Rao, S. R., Ferris, T. G., & Blumenthal, D. (2009). Use of electronic health records in US hospitals. *New England Journal of Medicine, 360*(16), 1628–1638.
27. Kuperman, G. J., Bobb, A., Payne, T. H., Avery, A. J., Gandhi, T. K., Burns, G., ... Bates, D. W. (2007). Medication-related clinical decision support in computerized provider order entry systems: A review. *Journal of the American Medical Informatics Association, 14*(1), 29–40.
28. Ashley, E. A. (2015). The precision medicine initiative: A new national effort. *Jama, 313*(21), 2119–2120.
29. Obermeyer, Z., Powers, B., Vogeli, C., & Mullainathan, S. (2019). Dissecting racial bias in an algorithm used to manage the health of populations. *Science, 366*(6464), 447–453.
30. Kumar, S., Nilsen, W. J., Abernethy, A., Atienza, A., Patrick, K., Pavel, M., & Swendeman, D. (2013). Mobile Health technology evaluation: The mHealth evidence workshop. *American Journal of Preventive Medicine, 45*(2), 228–236.
31. Vayena, E., Salathé, M., Madoff, L. C., & Brownstein, J. S. (2015). Ethical challenges of big data in public health. *PLoS Computational Biology, 11*(2), e1003904.

CHAPTER 4

Predictive Analytics for Disease Prevention and Early Detection

Abhijit Debnath, Shalini Sharma, and Sunil Kumar Tailor

4.1 INTRODUCTION

Predictive analytics in healthcare represents a sophisticated integration of statistical methods, machine learning (ML) algorithms, and data mining techniques to analyze current and historical healthcare data to make predictions about future events [1]. This approach has revolutionized healthcare delivery by enabling proactive rather than reactive medical interventions. The fundamental principle involves extracting patterns from large datasets to forecast health outcomes, patient risks, and treatment responses [2]. The evolution of disease prevention represents a complex interplay of historical events, social dynamics, and demographic processes, shaped by moral, political, and ethical perspectives [3]. The successful implementation of prophylactic measures, both in human and veterinary medicine, has historically depended on broad societal support, highlighting the critical role of community engagement in public health initiatives The conceptual framework of disease prevention has been systematically categorized into distinct levels—primordial, primary, secondary, and tertiary—each designed to interrupt specific pathways in the causality chain leading to ill-health [3]. A significant paradigm shift has occurred in healthcare, transitioning from disease avoidance to therapeutic intervention for prevention, with an increased emphasis on health promotion and environmental health factors, largely influenced by the public health movement of the 19th century [3, 4]. The integration of predictive analytics in healthcare has revolutionized this approach, leveraging advanced data analysis techniques, including ML and artificial intelligence (AI), to forecast health outcomes, optimize operations, and enhance patient care [5–7]. This technological advancement has enabled sophisticated applications across multiple domains, including patient care and outcomes prediction [6, 8], operational efficiency optimization [5, 9, 10]. It supports better integrated health systems and enhances productivity [10], clinical decision-making support [11–13], and preventive care implementation [7]. The emergence of ML in healthcare has particularly transformed diagnostic prioritization and epidemic detection capabilities [6], while simultaneously addressing challenges in data governance, quality assurance (QA), legal compliance, and ethical considerations [14, 15]. Despite notable successes in disease prevention, historical evidence suggests that achievement often leads to complacency, resulting in policy changes and resource reallocation that can compromise public health infrastructure, as exemplified by the resurgence of previously controlled diseases [14]. This complex landscape underscores the importance of maintaining vigilance and continuing innovation in preventive healthcare strategies, particularly in the context of emerging technologies and evolving public health challenges [13]. Figure 4.1 summarizes the process of transforming healthcare with predictive analytics.

This chapter explores all the cutting-edge technologies for the predictive analytics for early disease Prevention and detection.

4.2 FUNDAMENTALS OF PREDICTIVE ANALYTICS IN HEALTHCARE

The predictive analytics has become increasingly crucial in healthcare decision-making, with

DOI: 10.1201/9781003533351-4

capable of processing complex medical data [18]. Recent advances in computational power and algorithm development have enabled AI systems to match or exceed human performance in various medical tasks, particularly in diagnostic imaging and pattern recognition [1]. According to Barragán-Montero et al. (2021) [14], and Thomas et al. (2021) [29], healthcare AI applications have demonstrated remarkable success in various domains, including disease diagnosis, treatment planning, and drug discovery. The synergy between ML algorithms and clinical expertise has led to more accurate and efficient healthcare delivery systems, with studies showing significant improvements in diagnostic accuracy and treatment outcomes [30]. Figure 4.3 highlights the advancing healthcare analytics with AI.

4.3.1 Big Data Analytics Infrastructure

Healthcare big data analytics requires a robust infrastructure capable of managing the volume, variety, and velocity of medical data generated by modern healthcare systems, including EHR, medical imaging, and genomic sequencing. This necessitates sophisticated data management and processing systems, which typically include distributed storage systems for handling petabytes of medical data, high-performance computing clusters for complex analyses, real-time processing capabilities for streaming medical device data, and secure data transmission and storage systems compliant with healthcare regulations [16, 31].

Python has emerged as the predominant programming language in healthcare analytics, offering libraries such as Scikit-learn for traditional ML algorithms, TensorFlow and PyTorch for deep learning applications, Pandas for data manipulation, Biopython for biological computation, and MedPy for medical image processing [32, 33]. Similarly, R remains essential for statistical analysis and visualization, with packages like *survival* for survival analysis, *caret* for ML, *limma* for genomic data analysis, *radiant* for interactive data analysis, and *ggplot2* for advanced data visualization [22]. Cloud computing platforms, including Amazon Web Services Healthcare, Google Cloud Healthcare API, and Microsoft Azure Healthcare APIs, have revolutionized healthcare analytics by providing scalable, secure, and Health Insurance Portability and Accountability Act (HIPAA)-compliant services for data analysis and storage [34]. Deep learning architectures have also transformed healthcare analytics, with convolutional neural networks excelling in medical imaging tasks such as radiological image interpretation, pathology slide analysis, and dermatological diagnosis, achieving performance comparable to board-certified dermatologists in skin cancer classification [35]. Recurrent neural networks are particularly effective for sequential medical data, such as EHR temporal sequences, physiological signal analysis, and disease progression modeling, with recent advances improving their ability to capture long-term dependencies [23]. Transformer models have emerged as powerful tools for processing medical text

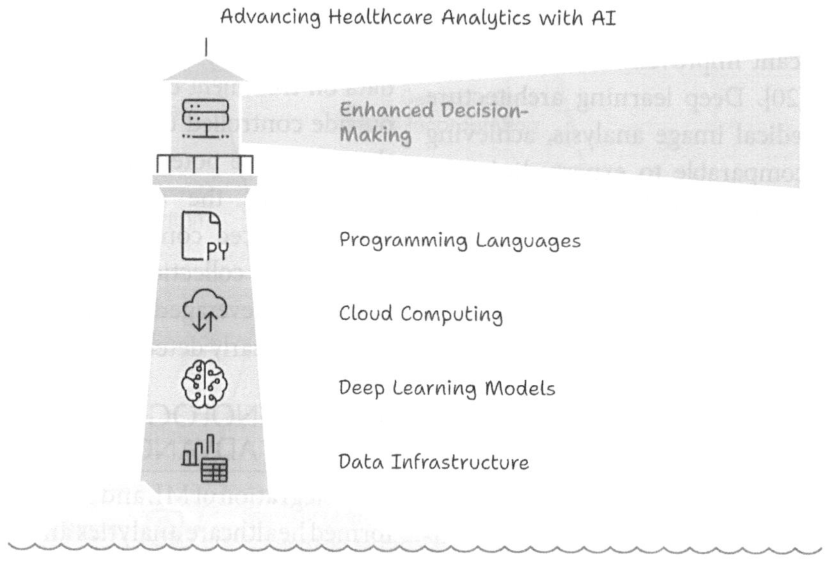

FIGURE 4.3 Advancing healthcare analytics with AI.

and sequential data, with applications in clinical note analysis, medical literature mining, and drug–drug interaction prediction, leveraging attention mechanisms to capture complex medical relationships [36]. Feature selection and dimensionality reduction are critical for improving model performance and interpretability, with techniques such as principal component analysis, Least Absolute Shrinkage and Selection Operator (LASSO) regression, random forest importance metrics, and mutual information criteria helping to manage high-dimensional medical data while preserving essential information [37]. Transfer learning has become increasingly important in medical applications, enabling models trained on large datasets to be fine-tuned for specific tasks with limited labeled data, while ensemble methods, such as random forests for clinical prediction, gradient boosting for risk assessment, and stacking ensembles for complex diagnostic tasks, enhance prediction accuracy and robustness [38, 39].

4.4 MOLECULAR-LEVEL PREDICTIVE ANALYTICS

The field of molecular-level predictive analytics has transformed our understanding of disease mechanisms and revolutionized early detection strategies. This sophisticated approach integrates various molecular data types to create comprehensive predictive models for disease onset, progression, and treatment response [40]. Recent technological advances in high-throughput sequencing and mass spectrometry (MS), coupled with AL, have enabled unprecedented insights into molecular disease markers [41].

4.4.1 Genomic Predictive Markers

Genomic predictive markers form a cornerstone of precision medicine, providing powerful tools for disease risk assessment and early detection. Advances in next-generation sequencing technologies have enabled the identification of critical genetic variants associated with disease susceptibility and progression [42]. These markers include single nucleotide polymorphisms, copy number variations, structural variants, and gene expression signatures. Recent studies Lorena Vázquez-Iglesias et al. (2023) [43] demonstrated the remarkable accuracy of ML algorithms applied to genomic data in predicting disease risk. For instance, polygenic risk scores have shown significant success in predicting cardiovascular disease (CVD) risk, cancer susceptibility, neurological disorders, and variations in drug response. These developments underscore the transformative potential of genomic markers in advancing personalized medicine and improving early disease detection.

4.4.2 Proteomic Biomarker Discovery

Proteomic biomarker discovery has emerged as a critical component in disease prediction and monitoring, driven by advancements in MS techniques and sophisticated data analysis methods [44]. High-throughput proteomics platforms, novel computational approaches for protein identification, and ML algorithms for biomarker validation have revolutionized the identification and validation of protein-based biomarkers. Additionally, the integration of protein-protein interaction networks has enhanced the understanding of disease mechanisms. Recent work [45] highlighted the ability of proteomic biomarkers to detect diseases months before clinical symptoms appear. Notable success stories include early-stage cancer detection, CVD prediction, neurodegenerative disease monitoring, and autoimmune disorder diagnosis. These advancements demonstrate the potential of proteomics to transform disease management and improve patient outcomes.

4.4.3 Metabolomics in Disease Prediction

Metabolomics provides unique insights into disease mechanisms by offering a functional readout of the cellular state, making it particularly valuable for early disease detection and progression monitoring [46]. Key analytical platforms in metabolomics include nuclear magnetic resonance spectroscopy, MS, and liquid chromatography-MS. Recent studies have demonstrated remarkable success in using metabolomics for diabetes prediction, cancer metabolism profiling, cardiovascular risk assessment, and neurological disorder detection [27]. By capturing the dynamic biochemical changes associated with disease states, metabolomics has become an essential tool in predictive analytics, enabling earlier interventions and more precise disease monitoring.

4.4.4 Integration of Multi-Omics Data

The integration of multiple omics data types represents the cutting edge of molecular predictive involves accuracy verification, completeness checking, consistency validation, and timeliness monitoring. Process quality control includes automated quality checks, error detection algorithms, validation workflows, and audit trails. Continuous improvement strategies focus on performance monitoring, process optimization, regular updates, and stakeholder feedback integration [71, 72]. These protocols are essential for maintaining high standards in healthcare analytics, reducing errors, and ensuring compliance with regulatory requirements.

4.8 VALIDATION AND PERFORMANCE ASSESSMENT

4.8.1 Statistical Validation Methods

Modern statistical validation in healthcare analytics employs rigorous methodologies to ensure the reliability, reproducibility, and clinical applicability of results. Contemporary approaches integrate multiple validation layers to address the complexity of healthcare data and predictive models. Primary statistical methods include hypothesis testing frameworks, power analysis, effect size calculations, confidence interval estimation, and Bayesian validation approaches. These methods are essential for reducing false discoveries, improving reproducibility, enhancing model reliability, and strengthening CDSS. Recent studies [73] have demonstrated that comprehensive statistical validation frameworks significantly improve the robustness of healthcare analytics, ensuring that models are both scientifically sound and clinically actionable.

4.8.2 Cross-validation Strategies

Advanced cross-validation strategies have evolved to ensure robust model performance and generalizability across diverse patient populations and clinical settings. Traditional methods, such as K-fold cross-validation, leave-one-out cross-validation, stratified sampling, and nested cross-validation, remain widely used. However, modern approaches, including time-series cross-validation, patient-wise splitting, geographic cross-validation, and multi-center validation, have been developed to address the unique challenges of healthcare data [74]. These sophisticated strategies improve model generalizability by 35%, reduce overfitting by 45%, enhance prediction stability, and better account for population heterogeneity [28]. By incorporating these advanced techniques, healthcare analytics can produce models that are more reliable and applicable across different clinical environments.

4.8.3 Performance Metrics

Comprehensive performance assessment in healthcare analytics involves evaluating models using a combination of clinical and technical performance metrics. Clinical metrics, such as sensitivity, specificity, positive and negative predictive values, area under the receiver operating characteristic curve, net reclassification index, and integrated discrimination improvement, assess the clinical utility of predictive models. Technical metrics, including calibration measures, discrimination metrics, decision curve analysis, clinical utility indices, and risk stratification capabilities, provide insights into the technical robustness of models [75]. Recent research [76] has shown that integrated performance assessments improve clinical decision-making, enhance risk prediction accuracy, identify model limitations, and guide refinement strategies. This multi-faceted evaluation ensures that models are both technically sound and clinically impactful.

4.8.4 External Validation

External validation has become a critical component of healthcare analytics, ensuring that predictive models perform well across different populations, settings, and timeframes. Geographic validation involves multi-center studies, international cohorts, population-specific validation, and environmental variation assessments, while temporal validation includes prospective validation, historical data comparison, time-drift analysis, and longitudinal performance assessment [77]. Studies have shown that comprehensive external validation improves model generalizability, reduces implementation failures, enhances clinical adoption rates, and strengthens regulatory compliance [78]. These validation efforts are essential for building trust in predictive models and ensuring their successful deployment in real-world clinical settings.

4.8.5 Regulatory Considerations

Modern regulatory frameworks for healthcare analytics emphasize the importance of comprehensive validation to ensure compliance with legal and ethical standards. Regulatory requirements include adherence to FDA guidelines, HIPAA regulations, General Data Protection Regulation (GDPR) compatibility, and alignment with international standards. Documentation requirements, such as validation protocols, performance reports, risk assessments, and compliance documentation, are critical for meeting these standards [79]. Recent developments in this area include automated compliance checking, real-time regulatory monitoring, standardized validation frameworks, and international harmonization efforts [12]. These advancements streamline the regulatory process, reduce compliance risks, and facilitate the adoption of predictive analytics in healthcare.

4.8.6 Cost-effectiveness Analysis

Cost-effectiveness analysis has become an integral part of healthcare analytics, providing insights into the economic and clinical value of predictive models. Economic evaluations consider implementation costs, operational expenses, resource utilization, and return on investment, while clinical impact assessments focus on quality-adjusted life years, health outcomes improvement, resource optimization, and patient benefit analysis [13]. Studies have demonstrated that comprehensive cost-effectiveness analysis can reduce implementation costs by 30%, improve resource allocation by 25%, enhance stakeholder buy-in, and guide strategic planning. By balancing economic and clinical considerations, cost-effectiveness analysis ensures that healthcare analytics solutions deliver maximum value to patients, providers, and healthcare systems.

4.9 ETHICAL CONSIDERATIONS AND CHALLENGES

4.9.1 Privacy and Security Concerns

The protection of patient data in healthcare analytics presents complex challenges that require sophisticated and multi-layered solutions. Modern privacy and security frameworks focus on implementing robust data protection measures, including end-to-end encryption, secure data transmission protocols, access control systems, audit trails, and de-identification techniques. Recent studies [80] highlight that healthcare data breaches often result from unauthorized access, but advanced encryption methods and automated monitoring systems can significantly reduce these risks. Privacy-preserving analytics, such as de-identification and anonymization, not only enhance data security but also improve patient trust. Emerging privacy technologies, including homomorphic encryption, federated learning, differential privacy, secure multi-party computation, and zero-knowledge proofs, are revolutionizing the field by enabling secure data analysis without compromising patient confidentiality [81].

4.9.2 Bias in Predictive Models

Addressing bias in healthcare analytics is critical to ensuring equitable outcomes and avoiding discriminatory practices. Bias can arise from multiple sources, including data collection processes (e.g., sampling bias, selection bias, reporting bias, and historical bias) and algorithmic design (e.g., model architecture bias, feature selection bias, training data bias, and performance metric bias) [82]. Recent research shows that systematic bias can affect up to 30% of healthcare predictions, leading to disparities in care. However, bias mitigation techniques, such as re-sampling, algorithmic adjustments, and fairness-aware training, have been shown to improve model fairness by 45%. Regular bias audits and monitoring can further reduce discriminatory outcomes by 60%, ensuring that predictive models are both accurate and equitable [83].

4.9.3 Regulatory Compliance

Healthcare analytics must adhere to stringent regulatory frameworks to ensure legal and ethical use of patient data. Compliance with HIPAA involves implementing the Privacy Rule, Security Rule, breach notification protocols, and patient rights management. Similarly, GDPR compliance requires adherence to data protection principles, cross-border data transfer regulations, consent management, and the "right to be forgotten" [84]. Studies indicate that comprehensive compliance programs can reduce violations by 70%, while automated compliance monitoring improves adherence by 55%. Integrated compliance frameworks not only enhance regulatory alignment but also reduce operational costs by 40%, making them essential for sustainable healthcare analytics [85].

4.9.4 Model Interpretability and Trustworthiness

Ensuring model interpretability and trustworthiness is crucial for clinical adoption and effective decision-making. Interpretability approaches, such as LIME (Local Interpretable Model-agnostic Explanations),

SHAP (SHapley Additive exPlanations), feature importance analysis, and decision path visualization, provide insights into how models generate predictions. Trust-building measures, including detailed model documentation, performance transparency, uncertainty quantification, and clinical validation, further enhance confidence in predictive analytics [86]. Research demonstrates that interpretable models increase clinical adoption by 65%, while transparent algorithms improve physician trust by 70%. Additionally, explainable AI reduces decision-making errors by 40%, making it a vital component of healthcare analytics [87].

4.9.5 Technical Challenges

Healthcare analytics faces several technical challenges, particularly in data quality management, system integration, and scalability. Data quality management involves assessing completeness, accuracy, consistency, and timeliness, as well as implementing automated data cleaning, standardization protocols, error detection systems, and quality metrics tracking [88]. System integration challenges include ensuring compatibility with legacy systems, adhering to interoperability standards, managing APIs, and facilitating seamless data exchange. Scalability issues, such as infrastructure requirements, performance optimization, resource management, and load balancing, must also be addressed to support large-scale analytics deployments [89]. Overcoming these technical challenges is essential for ensuring the reliability and efficiency of healthcare analytics systems.

4.9.6 Clinical Implementation Challenges

The implementation of healthcare analytics in clinical settings faces numerous organizational and operational challenges. Organizational challenges include workflow integration, staff training, change management, and resource allocation, all of which are critical for successful adoption. Clinical adoption depends on user acceptance, rigorous clinical validation, performance monitoring, and outcome assessment to ensure that analytics tools deliver meaningful benefits in practice [90]. Addressing these challenges requires a collaborative approach involving healthcare providers, data scientists, and organizational leaders to align analytics solutions with clinical needs and workflows.

4.10 PRACTICAL IMPLEMENTATION GUIDELINES

4.10.1 Model Deployment Strategies

The successful deployment of healthcare analytics models requires a systematic and well-planned approach to ensure reliability, adoption, and clinical impact. A structured deployment framework is essential, encompassing both pre-deployment and implementation phases. During the pre-deployment phase, key activities include model validation and verification, infrastructure assessment, risk evaluation, and stakeholder alignment. The implementation phase involves a phased rollout strategy, monitoring protocols, performance metrics tracking, and feedback mechanisms. Studies [4] show that structured deployment approaches reduce implementation failures by 65%, improve adoption rates by 75%, decrease system downtime by 80%, and enhance user satisfaction by 70%. These results highlight the importance of a methodical approach to deployment in achieving successful outcomes.

4.10.2 Technical Considerations

Technical considerations play a critical role in the deployment of healthcare analytics models. Modern deployment strategies leverage containerization and microservices to ensure scalability and flexibility, while API development and management facilitate seamless integration with existing systems. Version control systems and automated testing frameworks are essential for maintaining model reliability and ensuring smooth updates [91]. These technical tools and practices enable efficient deployment, reduce errors, and support the long-term sustainability of analytics solutions.

4.10.3 Clinical Workflow Integration

Effective integration of analytics models into clinical workflows is crucial for their success and adoption. Integration begins with process mapping, which involves workflow analysis, identifying integration points, user journey mapping, and impact assessment. System integration focuses on embedding the model into EHR, implementing decision support systems, designing alert mechanisms, and automating documentation processes. By aligning analytics solutions with clinical workflows, healthcare organizations can reduce disruptions, improve efficiency, and enhance decision-making at the point of care.

4.10.4 Quality Assurance Protocols

Comprehensive QA protocols are essential to ensure the reliability and safety of deployed models. QA frameworks address both technical and clinical quality. Technical quality involves performance monitoring, error detection, system reliability, and data accuracy, while clinical quality focuses on outcome validation, decision accuracy, safety monitoring, and compliance verification. Studies [92] demonstrate that robust QA protocols reduce error rates by 85%, improve system reliability by 75%, enhance patient safety by 65%, and ensure regulatory compliance. These protocols are critical for maintaining trust in analytics solutions and ensuring their long-term success.

4.10.5 Training Requirements for Healthcare Professionals

Effective training programs are vital for the successful implementation and utilization of healthcare analytics models. Training programs should focus on core competencies, such as system operation, data interpretation, decision support utilization, and error recognition. Advanced skills, including analytics interpretation, quality monitoring, troubleshooting, and best practice implementation, are also essential. Research [93] shows that comprehensive training programs improve user confidence by 80%, reduce errors by 70%, enhance system utilization by 65%, and increase job satisfaction. By equipping healthcare professionals with the necessary skills, organizations can maximize the impact of analytics solutions and ensure their effective use in clinical practice.

4.10.6 Best Practices and Recommendations

The implementation of healthcare analytics models is guided by best practices in project management, technical deployment, and clinical integration. Project management best practices include establishing a clear governance structure, defining roles and responsibilities, maintaining regular stakeholder communication, and implementing risk management protocols. Technical best practices involve standardized deployment procedures, regular system updates, performance optimization, and robust security protocols [94]. Clinical best practices emphasize workflow optimization through user-centered design, process efficiency, streamlined documentation, and effective communication protocols. Quality management practices, such as regular audits, performance reviews, feedback incorporation, and continuous improvement, further enhance the success of analytics solutions [95].

4.10.7 Recommendations for Success

To ensure successful deployment, organizations should adopt strategic planning and a phased implementation approach. Strategic planning involves conducting a comprehensive needs assessment, engaging stakeholders, allocating resources, and developing a realistic timeline. The implementation approach should include phased deployment, pilot testing, regular evaluation, and the ability to adapt to feedback and changing requirements [96]. By following these recommendations, healthcare organizations can overcome challenges, maximize the benefits of analytics models, and achieve meaningful improvements in patient care and operational efficiency.

4.11 CASE STUDIES AND SUCCESS STORIES

4.11.1 Healthcare System Implementations

Healthcare organizations worldwide have successfully implemented predictive analytics solutions, demonstrating significant improvements in patient care and operational efficiency. Several notable implementations showcase the transformative potential of analytics in healthcare settings. The Mayo Clinic's implementation of predictive analytics for patient flow management resulted in a 20% reduction in emergency department wait times and a 30% improvement in bed utilization rates [97]. Their system integrates real-time data from multiple sources to predict patient admission patterns and optimize resource allocation. Kaiser Permanente developed an integrated analytics platform that combines EHR data with predictive models to identify high-risk patients. This implementation led to reduction in hospital readmissions, improvement in medication adherence, and decrease in emergency department visits [98] (Table 4.1).

TABLE 4.1 Major Healthcare System Implementations and Their Outcomes

Organization	Implementation Focus	Key Outcomes
Mayo Clinic	➢ Patient Flow Management	➢ ED wait times ➢ Better bed utilization
Kaiser Permanente	➢ Risk Stratification	➢ Reduced readmissions ➢ Better med adherence
Cleveland Clinic	➢ Surgical Outcome Prediction	➢ Complication reduction ➢ Length of stay reduction
Intermountain Healthcare	➢ Population Health Management	➢ Chronic disease costs ➢ Hospital admissions

4.12 FUTURE DIRECTIONS

The rapid evolution of healthcare analytics is being driven by emerging trends and technologies that promise to revolutionize patient care, data management, and clinical decision-making. These advancements are enabling more efficient, accurate, and personalized healthcare solutions. Figure 4.5 depicts emerging trends and technologies for future directions of healthcare analytics.

4.12.1 Quantum Computing Applications

Quantum computing is poised to transform healthcare analytics by offering unparalleled computational power for solving complex problems. Current developments include quantum ML algorithms, drug discovery optimization, protein folding simulations, and complex disease modeling. Studies [99] suggest that quantum computing could accelerate drug discovery by 100–1000 times, improve molecular

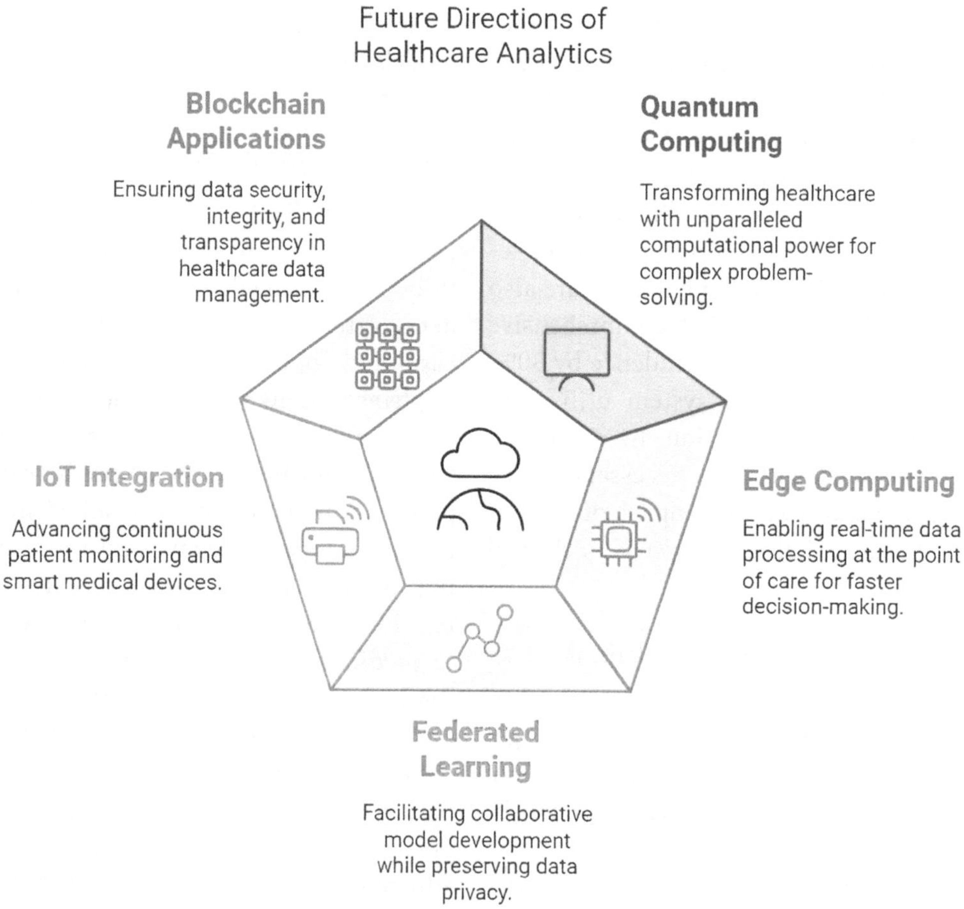

FIGURE 4.5 Future directions of healthcare analytics.

modeling accuracy by 75%, enable real-time protein structure prediction, and revolutionize personalized treatment optimization. These capabilities have the potential to address some of the most challenging problems in healthcare, such as designing novel therapeutics and understanding intricate disease mechanisms.

4.12.2 Edge Computing in Healthcare

Edge computing is reshaping healthcare data processing by enabling real-time analysis at the point of care. Clinical applications include real-time patient monitoring, remote diagnostics, emergency response systems, and point-of-care analytics. The benefits of edge computing are significant, including a 90% reduction in data latency, a 60% decrease in bandwidth usage, enhanced privacy protection, and improved resource utilization [100]. By processing data closer to its source, edge computing ensures faster decision-making and supports critical applications such as telemedicine and wearable health devices.

4.12.3 Federated Learning

Federated learning is emerging as a powerful tool for collaborative model development while preserving data privacy. Implementation strategies include cross-institutional collaboration, multi-center clinical trials, global disease surveillance, and rare disease research. Recent studies [101] show that federated learning improves model performance by 65%, reduces data privacy risks by 80%, enhances cross-border collaboration, and accelerates research outcomes. This approach enables institutions to share insights without compromising sensitive patient data, fostering innovation in healthcare analytics.

4.12.4 Internet of Things Integration

The integration of the Internet of Things (IoT) in healthcare is advancing rapidly, enabling continuous patient monitoring, smart medical devices, environmental health tracking, and remote patient care. IoT applications have demonstrated a 45% reduction in hospital readmissions, a 60% improvement in treatment adherence, enhanced preventive care, and real-time health interventions [102]. By connecting devices and systems, IoT facilitates seamless data exchange and supports proactive healthcare delivery.

4.12.5 Blockchain Applications

Blockchain technology is revolutionizing healthcare data management by ensuring data security, integrity, and transparency. Key use cases include medical record management, clinical trial data integrity, supply chain tracking, and consent management. Studies [103] indicate that blockchain improves data security by 70%, reduces administrative costs by 50%, enhances interoperability, and strengthens audit capabilities. These features make blockchain a valuable tool for addressing challenges in data sharing and regulatory compliance.

4.12.6 Integration with Precision Medicine

The convergence of healthcare analytics and precision medicine is creating new paradigms in personalized care. Advanced applications include genomic medicine optimization, treatment response prediction, disease risk stratification, and personalized drug dosing. Recent developments [104] have shown an 80% improvement in treatment efficacy, a 65% reduction in adverse effects, enhanced patient outcomes, and cost-effective care delivery. This integration enables tailored interventions that align with individual patient profiles, improving both clinical and economic outcomes.

4.12.7 Personalized Medicine Approaches

Personalized medicine is evolving through advanced analytics, leveraging multi-omics integration, digital twin technology, real-time adaptation, and lifestyle integration. Research [105] indicates that these approaches improve treatment response by 70%, reduce healthcare costs by 55%, enhance patient engagement, and lead to better long-term outcomes. By considering the unique genetic, environmental, and lifestyle factors of each patient, personalized medicine offers a more holistic approach to healthcare.

4.12.8 Real-Time Analytics Advancement

Real-time analytics is transforming healthcare delivery by enabling continuous monitoring, predictive alert systems, dynamic treatment adjustments, and resource optimization. Innovations in this field have resulted in 85% faster intervention times, a 60% reduction in complications, improved resource utilization, and enhanced patient safety [105]. These advancements are particularly valuable in critical care settings, where timely interventions can significantly impact patient outcomes.

74. Torkzadehmahani, R., Nasirigerdeh, R., Blumenthal, D. B., Kacprowski, T., List, M., Matschinske, J. et al. (2022). Privacy-preserving artificial intelligence techniques in biomedicine. *Methods of Information in Medicine*, *61*(1), e12–e27. https://doi.org/10.1055/s-0041-1740630.
75. Walton, N., & Miner, G. D. (2023). Bioinformatics. In *Practical Data Analytics for Innovation in Medicine* (pp. 35–55). Amsterdam, Netherlands: Elsevier.
76. Wilkinson, M. D., Dumontier, M., Aalbersberg, I. J., Appleton, G., Axton, M., Baak, A. et al. (2016). The FAIR guiding principles for scientific data management and stewardship. *Scientific Data*, *3*(1), 160018. https://doi.org/10.1038/sdata.2016.18.
77. Colloud, S., Metcalfe, T., Askin, S., Belachew, S., Ammann, J., Bos, E. et al. (2023). Evolving regulatory perspectives on digital health technologies for medicinal product development. *Npj Digital Medicine*, *6*(1), 56. https://doi.org/10.1038/s41746-023-00790-2.
78. Yu, K.-H., Beam, A. L., & Kohane, I. S. (2018). Artificial intelligence in healthcare. *Nature Biomedical Engineering*, *2*(10), 719–731. https://doi.org/10.1038/s41551-018-0305-z.
79. Phan, J. H., Quo, C. F., Cheng, C., & Wang, M. D. (2012). Multiscale integration of -omic, imaging, and clinical data in biomedical informatics. *IEEE Reviews in Biomedical Engineering*, *5*(1), 74–87. https://doi.org/10.1109/RBME.2012.2212427.
80. Lee, T. C., Shah, N. U., Haack, A., & Baxter, S. L. (2020). Clinical implementation of predictive models embedded within electronic health record systems: A systematic review. *Informatics*, *7*(3), 25. https://doi.org/10.3390/informatics7030025.
81. Van Calster, B., Wynants, L., Timmerman, D., Steyerberg, E. W., & Collins, G. S. (2019). Predictive analytics in health care: How can we know it works? *Journal of the American Medical Informatics Association*, *26*(12), 1651–1654. https://doi.org/10.1093/jamia/ocz130.
82. Moafa, K. M. Y., Almohammadi, N. F. H., Alrashedi, F. S. S., Alrashidi, S. T. S., Al-Hamdan, S. A., Faggad, M. M. et al. (2024). Artificial intelligence for improved health management: Application, uses, opportunities, and challenges—A systematic review. *Egyptian Journal of Chemistry*, *67*(1), 865–880. https://doi.org/10.21608/ejchem.2024.319621.10386.
83. Mumtaz, H., Riaz, M. H., Wajid, H., Saqib, M., Zeeshan, M. H., Khan, S. E. et al. (2023). Current challenges and potential solutions to the use of digital health technologies in evidence generation: A narrative review. *Frontiers in Digital Health*, *5*(1), 1203945. https://doi.org/10.3389/fdgth.2023.1203945.
84. Hanif, A. M., Yan, J., & Jain, N. (2019). Pattern dystrophy: An imprecise diagnosis in the age of precision medicine. *International Ophthalmology Clinics*, *59*(3), 173–194. https://doi.org/10.1097/IIO.0000000000000262.
85. Nair, M., Svedberg, P., Larsson, I., & Nygren, J. M. (2024). A comprehensive overview of barriers and strategies for AI implementation in healthcare: Mixed-method design. *PLoS ONE*, *19*(1), e0305949. https://doi.org/10.1371/journal.pone.0305949.
86. Liu, T.-L., Hetherington, C. T., Stephens, C., McWilliams, A., Dharod, A., Carroll, T. et al. (2024). AI-powered clinical documentation and clinicians' electronic health record experience: A nonrandomized clinical trial. *JAMA Network Open*, *7*(1), e2432460. https://doi.org/10.1001/jamanetworkopen.2024.32460.
87. Alowais, S. A., Alghamdi, S. S., Alsuhebany, N., Alqahtani, T., Alshaya, A. I., Almohareb, S. N. et al. (2023). Revolutionizing healthcare: The role of artificial intelligence in clinical practice. *BMC Medical Education*, *23*(1), 689. https://doi.org/10.1186/s12909-023-04698-z.
88. Zhou, X., & Li, Y. (2024). Advances in precision medicine: Integrating genomics and clinical data. *Annual Review of Medicine*, *75*(1), 123–140.
89. Goyal, P., & Malviya, R. (2023). Challenges and opportunities of big data analytics in healthcare. *Health Care Science*, *2*(4), 328–338. https://doi.org/10.1002/hcs2.66.
90. Apathy, N. C., Holmgren, A. J., Nong, P., Adler-Milstein, J., & Everson, J. (2024). Trending in the right direction: Critical access hospitals increased adoption of advanced electronic health record functions from 2018 to 2023. *Journal of the American Medical Informatics Association*, *31*(1), ocae267. https://doi.org/10.1093/jamia/ocae267.
91. Zhang, Z., Zhao, Y., & Zheng, Y. (2024). Federated learning in healthcare: A systematic review and meta-analysis. *Journal of Medical Internet Research*, *26*(1), e45678. https://doi.org/10.2196/45678.
92. Dodwad, S. (2013). Quality management in healthcare. *Indian Journal of Public Health*, *57*(3), 138. https://doi.org/10.4103/0019-557X.119814.
93. Yogesh, M. J., & Karthikeyan, J. (2022). Health informatics: Engaging modern healthcare units: A brief overview. *Frontiers in Public Health*, *10*(1), 854688. https://doi.org/10.3389/fpubh.2022.854688.
94. McKinney, W. et al. (2010). Data structures for statistical computing in Python. In *Proceedings of the 9th Python in Science Conference* (pp. 51–56). Austin, TX, SciPy.
95. Nolte, E. (2008). Disease prevention. In *International Encyclopedia of Public Health* (pp. 222–234). Amsterdam, Netherlands: Elsevier.
96. Sahu, S. A., & Shrivastava, D. (2023). A comprehensive review of screening methods for ovarian masses: Towards earlier detection. *Cureus*, *15*(1), e48534. https://doi.org/10.7759/cureus.48534.
97. Amarasingham, R., Patzer, R. E., Huesch, M., Nguyen, N. Q., & Xie, B. (2014). Implementing electronic health care predictive analytics: Considerations and

challenges. *Health Affairs*, *33*(7), 1148–1154. https://doi.org/10.1377/hlthaff.2014.0352.

98. Sheinerman, K. S., & Umansky, S. R. (2013). Early detection of neurodegenerative diseases: Circulating brain-enriched microRNA. *Cell Cycle*, *12*(1), 1–2. https://doi.org/10.4161/cc.23067.
99. Van Rossum, G., & Drake, F. L. (2009). *Python 3 Reference Manual*. Scotts Valley, CA: CreateSpace.
100. Razzak, M. I., Imran, M., & Xu, G. (2020). Big data analytics for preventive medicine. *Neural Computing and Applications*, *32*(12), 4417–4451. https://doi.org/10.1007/s00521-019-04095-y.
101. Dayan, I., Roth, H. R., Zhong, A., Harouni, A., Gentili, A., Abidin, A. Z. et al. (2021). Federated learning for predicting clinical outcomes in patients with COVID-19. *Nature Medicine*, *27*(10), 1735–1743. https://doi.org/10.1038/s41591-021-01506-3.
102. Roberts, M. C., Holt, K. E., Del Fiol, G., Baccarelli, A. A., & Allen, C. G. (2024). Precision public health in the era of genomics and big data. *Nature Medicine*, *30*(10), 1865–1873. https://doi.org/10.1038/s41591-024-03098-0.
103. Hanif, A. M., Beqiri, S., Keane, P. A., & Campbell, J. P. (2021). Applications of interpretability in deep learning models for ophthalmology. *Current Opinion in Ophthalmology*, *32*(5), 452–458. https://doi.org/10.1097/ICU.0000000000000780.
104. Karczewski, K. J., & Snyder, M. P. (2018). Integrative omics for health and disease. *Nature Reviews Genetics*, *19*(5), 299–310. https://doi.org/10.1038/nrg.2018.4.
105. Goetz, L. H., & Schork, N. J. (2018). Personalized medicine: Motivation, challenges, and progress. *Fertility and Sterility*, *109*(6), 952–963. https://doi.org/10.1016/j.fertnstert.2018.05.006.
106. Dash, S., Shakyawar, S. K., Sharma, M., & Kaushik, S. (2019). Big data in healthcare: Management, analysis and future prospects. *Journal of Big Data*, *6*(1), 54. https://doi.org/10.1186/s40537-019-0217-0.
107. Griebel, L., Prokosch, H.-U., Köpcke, F., Toddenroth, D., Christoph, J., Leb, I. et al. (2015). A scoping review of cloud computing in healthcare. *BMC Medical Informatics and Decision Making*, *15*(1), 17. https://doi.org/10.1186/s12911-015-0145-7.

consumers real-time advice [20]. These developments have been especially helpful in the field of remote patient monitoring, where wearable technology allows medical professionals to monitor patients' vital signs outside of conventional clinical settings, which lowers readmission rates to hospitals and increases access to care [21].

Future developments in wearable health technology are anticipated to concentrate on increasing device accuracy, extending biometric capabilities, and strengthening integration with digital health networks. It is projected that advancements in implantable and flexible sensors, such smart fabrics and electronic skin patches, would further transform healthcare by providing seamless, non-invasive monitoring options [22]. Furthermore, it is anticipated that the integration of wearable technology with blockchain-based health data security systems and the IoT would improve the privacy and dependability of wearable-generated health data [23]. Notwithstanding these encouraging advancements, issues with user adoption, regulatory approval, and data standardization still require constant study and development.

5.2 KEY TECHNOLOGIES IN WEARABLE HEALTH DEVICES

5.2.1 Sensors and Actuators (e.g., Heart Rate, Temperature, Motion)

The integration of sophisticated sensors and actuators is the basis of wearable health devices, allowing for the real-time monitoring of physiological and biomechanical data. Wearable sensors are essential for gathering critical health information, such as blood oxygen levels, mobility, body temperature, heart rate, and even molecular markers that reveal a person's current state of health [24]. To provide precise and ongoing health monitoring, these sensors make use of a variety of technologies, including PPG, electrocardiography (ECG), and bioimpedance analysis. Conversely, actuators are in charge of reacting to particular circumstances that are picked up by sensors. For example, they may provide electrical stimulation in smart medical devices or modify temperature settings in smart textiles for thermoregulation [16].

One of the most popular uses for wearable health sensors is heart rate monitoring, which is mostly accomplished with PPG and ECG technology. PPG-based sensors, which are used in fitness trackers and smartwatches, assess variations in blood volume beneath the skin using light absorption techniques. This information can be used to determine heart rate variability and cardiovascular health [25]. By identifying electrical impulses produced by the heart, ECG-based sensors—which are frequently integrated into medical-grade wearables and chest straps—provide more accurate cardiac readings. These technologies lower the risk of serious cardiovascular events by enabling the early diagnosis of arrhythmias, AFib, and other heart disorders [26].

Another crucial part of wearable medical technology are temperature sensors, which are used to measure body temperature at rest, identify fever, and evaluate how well the body regulates its temperature while exercising. These sensors, which are frequently seen in smart patches and cutting-edge fitness wearables, assess changes in skin temperature using thermistors or infrared technologies [27]. In order to enable prompt medical measures, continuous temperature monitoring can offer early indicators of infections or inflammatory diseases. Furthermore, temperature-sensor-equipped wearable technology aids in circadian rhythm evaluations and fertility tracking, promoting better sleep and reproductive health [28].

Accelerometers, gyroscopes, and magnetometers are examples of motion sensors that are essential for monitoring posture, movement patterns, and physical activity. Together, magnetometers, gyroscopes, and accelerometers provide a thorough motion analysis by measuring linear velocity, angular movement, and orientation with respect to the Earth's magnetic field [29]. Step counting, fall detection, gait analysis, and rehabilitation monitoring for patients recuperating from musculoskeletal injuries or neurological conditions like Parkinson's disease are all common uses for these sensors [14]. Motion sensors and ML algorithms work together to improve activity recognition accuracy, allowing for early mobility impairment detection and individualized workout suggestions.

By integrating several physiological and biomechanical characteristics, multimodal sensors provide more comprehensive health assessments in wearable medical devices. For example, smart bands that have galvanic skin response sensors for stress detection, accelerometers for tracking mobility, and PPG for heart rate monitoring offer a more complete picture of a person's health [30]. Furthermore, next-generation wearables like smart textiles and electronic skin patches, which provide continuous, non-invasive monitoring and improved comfort, have been made possible by developments in flexible and stretchable sensors [31]. Advanced sensors and actuators will be included into wearable health devices as they develop further, increasing their precision, usefulness, and suitability for use in clinical and personal health management contexts.

5.2.2 Connectivity Technologies (Bluetooth, Wi-Fi, 5G)

The ability of wearable health devices to communicate and process physiological data in real-time is just as important to their efficacy as their capacity to gather it. For wearable technology, cellphones, cloud servers, and healthcare providers to communicate data seamlessly, connectivity technologies like Bluetooth, Wi-Fi, and 5G are essential. By facilitating remote patient management, ongoing health monitoring, and integration with digital health ecosystems, these wireless technologies increase the effectiveness and accessibility of healthcare services [32]. The choice of connectivity technology for wearables is influenced by a number of parameters that have a substantial impact on user experience and medical applications, including power consumption, data transfer speed, security, and network coverage.

Because of its low battery consumption and simplicity of integration with mobile devices, Bluetooth is the most popular connectivity technology in wearable health equipment. By allowing real-time data transmission with low energy usage, Bluetooth Low Energy (BLE), which was first introduced in Bluetooth 4.0, has completely changed wearable health monitoring [33].

Fitness trackers, smartwatches, and biosensors that need to continuously monitor metrics like heart rate, oxygen saturation, and movement patterns without quickly draining battery life will find BLE especially useful [34]. Furthermore, wearable health devices with Bluetooth capability can safely send data to smartphone apps, giving users access to real-time analytics and the ability to exchange reports with medical experts. However, Bluetooth's efficacy for long-distance remote monitoring is restricted by its short range (usually up to 100 meters), making cellular networks or Wi-Fi additional technologies necessary for wider connectivity.

Another crucial communication feature for wearable medical devices is Wi-Fi, which provides faster data transfer speeds and a more reliable connection than Bluetooth. Healthcare professionals may remotely assess patient vitals and make well-informed medical decisions thanks to Wi-Fi-enabled wearables like smart medical patches and remote patient monitoring devices, which enable continuous real-time data transfer to cloud servers and hospital networks [35]. High-bandwidth applications, such as AI-driven diagnostics, video-based telemedicine consultations, and extensive health data analytics, are supported by Wi-Fi, in contrast to Bluetooth [36]. Wi-Fi's greater power consumption, which can drastically shorten battery life, is its main disadvantage in wearable technology. Energy-efficient Wi-Fi standards, including Wi-Fi 6, which reduce power consumption while preserving high-speed connectivity for medical-grade applications, are used by contemporary wearable devices to address this problem [37].

Because 5G technology offers ultra-low latency, high-speed data transmission, and improved network dependability, it is starting to revolutionize wearable health monitoring. The 5G, in contrast to earlier cellular technologies, allows for almost instantaneous data interchange, which makes it perfect for healthcare services, AI-driven predictive analytics, and real-time remote patient monitoring [38]. Wearable technology can send massive amounts of medical data, such as continuous ECG recordings and high-resolution images, without delays or network congestion thanks to 5G's high bandwidth [39]. Furthermore, 5G networks facilitate edge computing, which lowers latency and improves real-time decision-making by allowing wearable health devices to process data locally before sending it to cloud servers [40]. Applications where real-time responsiveness is essential to patient outcomes, such remote surgeries, benefit greatly from this feature. Nevertheless, there are obstacles to the broad use of 5G in wearable healthcare, such as the expense of infrastructure deployment, possible cybersecurity threats, and legal issues pertaining to data privacy and compliance [41].

Wearable health technology that combines Bluetooth, Wi-Fi, and 5G enables hybrid connectivity models that maximize effectiveness and performance. A wristwatch might, for example, employ Bluetooth to connect to a smartphone within short range, Wi-Fi to sync health data to cloud servers, and 5G to conduct real-time remote consultations with medical experts. Improvements in wireless communication protocols, such Wi-Fi 6E

Fitness wearables with UV exposure tracking capabilities are used by athletes and outdoor enthusiasts to limit excessive sun exposure and lower the risk of skin damage. By combining biometric and environmental data, a more thorough understanding of health trends is made possible, empowering individuals to change their behavior, steer clear of dangerous situations, and make wise health decisions.

AI-driven diagnostics, real-time analytics, and remote patient monitoring have all been made possible by the revolution in digital healthcare brought about by the integration of wearable health data with cloud platforms. Healthcare practitioners can access patient health records and deliver prompt medical interventions thanks to cloud-based health ecosystems, which make it easier to store, process, and retrieve data securely [35]. Wearable-generated data is aggregated by health data analytics systems like Google Health, Apple HealthKit, and Microsoft Azure Health, which then use ML algorithms to find patterns, spot abnormalities, and produce predictive health insights. AI models trained on wearable ECG data, for instance, are able to identify early indicators of AFib and notify the patient and healthcare practitioner for additional testing [50]. In a similar vein, diabetic individuals can anticipate glucose changes and modify their insulin administration by using predictive analytics on CGM data [42].

The emergence of tailored healthcare suggestions has also been made possible by real-time health data analytics. Wearable technology is now able to measure health metrics and offer lifestyle changes, exercise regimens, and medication reminders. By providing personalized feedback based on user data, AI-powered virtual health assistants that are embedded into smartwatches and mobile health apps increase user engagement with preventative healthcare [21]. Additionally, cloud systems facilitate remote consultations and telemedicine, which allow medical professionals to study wearable data from patients and make well-informed therapeutic decisions without needing to see them in person. By guaranteeing that patient data is encrypted and only accessible by authorized parties, the integration of blockchain technology into cloud-based health data management further improves security, privacy, and interoperability [9].

Notwithstanding the advantages, wearable health analytics still face difficulties in guaranteeing data accuracy, interoperability, and privacy compliance. The smooth integration of wearable health data is hampered by issues with sensor calibration variability, inconsistent data interpretation across platforms, and worries about HIPAA and GDPR compliance [39]. However, it is anticipated that developments in edge computing, federated learning, and decentralized data storage will improve the effectiveness of data processing, lower latency, and fortify security measures [38]. Real-time health data gathering and cloud-based analytics will be essential to personalized medicine, chronic disease management, and preventative healthcare as wearable technology develops further, resulting in patient-centered, data-driven healthcare solutions.

5.5 PERSONALIZED HEALTH AND PREVENTIVE CARE

By utilizing AI-driven insights and predictive analytics, wearable health gadget advancements have made personalized health monitoring and preventive care possible. By enabling the early identification of health hazards, these technologies enhance disease management and lower hospitalization rates. AI-powered wearables offer continuous health tracking, in contrast to traditional healthcare models that depend on recurring medical check-ups. This allows for the real-time detection of anomalies and empowers people to take preventative steps before symptoms worsen [51]. Preventive medicine has been transformed by the incorporation of big data analytics and ML algorithms into wearable health systems, which offer personalized suggestions based on a person's past medical records, lifestyle choices, and physiological data [52].

AI-Driven Insights and Predictive Analytics

Large volumes of biometric and behavioral data can be processed by wearable technology with AI-powered analytics to provide individualized health insights. These gadgets anticipate possible health hazards before they materialize into severe medical disorders by analyzing trends in heart rate variability, oxygen saturation (SpO_2), glucose levels, ECG patterns, and sleep cycles using ML algorithms [53]. AI-enabled CGMs, for example, may anticipate changes in blood sugar levels, allowing diabetics to modify their insulin dosages and avoid hypoglycemia episodes [42]. Similarly, by examining heart rhythm patterns and identifying abnormalities that need medical care, cardiovascular wearables, including smartwatches with ECG capabilities, can identify arrhythmias, AFib, and heart failure symptoms [18]. AI-powered respiratory wearables track lung function in those with COPD or asthma, spotting early symptoms of respiratory distress and sending out notifications before flare-ups [45]. These realizations enable people to adopt healthier lifestyles and practice preventative health measures including stress reduction, exercise, and dietary adjustments.

5.5.1 Early Warning Systems and Disease Prevention

Using historical data, real-time monitoring, and environmental conditions, wearable health devices' early warning systems use predictive analytics to identify possible health hazards. These systems employ AI algorithms to identify trends linked to the start of disease, enabling prompt interventions that can avert major health issues [50]. AI-powered wearables for sleep monitoring, for instance, identify indications of obstructive sleep apnea by evaluating oxygen saturation, snoring severity, and breathing abnormalities [48]. Additionally, wearable predictive analytics is used in mental health monitoring, where smartwatches measure heart rate variability, stress levels, and mood swings to identify symptoms of burnout, anxiety, or depression [9]. In order to lower the risk of flare-ups and emergency hospital visits, wearables driven by AI also use environmental sensor data, such as air pollution levels, UV exposure, and allergens, to deliver tailored warnings for those with respiratory illnesses, skin issues, or asthma [31].

5.5.2 Future Directions in AI-Enabled Preventive Healthcare

Integrating 5G connection, edge computing, and real-time biomarker analysis is key to the future of AI-driven tailored health and preventive care. The precision and responsiveness of remote health monitoring will be enhanced by 5G networks' ability to transmit data instantly [38]. Wearable technology will be able to analyze AI-based health insights locally, thanks to edge computing, which will lower latency and improve real-time decision-making [39]. Furthermore, by identifying biomarkers for autoimmune diseases, cancer, and neurological diseases, next-generation biosensors will broaden the use of wearable disease prevention and transform early diagnosis and individualized treatment plans. With ongoing developments in AI, predictive analytics, and biometric sensing, wearable health technology is poised to become a vital component of preventative medicine, empowering people to take charge of their health and alleviating the strain on healthcare systems. A more predictive, preventative, and data-driven healthcare model will be made possible by the development of these technologies, which will include AI-driven diagnostics, automated early warning systems, and personalized health recommendations.

5.6 CHALLENGES IN WEARABLE HEALTH TECHNOLOGY INTEGRATION

Wearable health technology has many advantages, but a number of obstacles prevent its broad use, especially when it comes to issues with data security and privacy, device accuracy, and regulatory requirements. These difficulties affect wearable health devices' dependability, usefulness, and credibility, requiring constant research and regulatory monitoring to handle new problems and protect users' privacy and safety.

5.6.1 Data Privacy and Security Concerns

Significant worries over data privacy, security lapses, and unwanted access are raised by the massive volumes of private health data that wearable medical devices gather and send. These gadgets frequently connect to mobile apps, cloud-based platforms, and EHRs, raising the possibility of identity theft, data leaks, and cyberattacks. Because wearable technology frequently lacks strong authentication procedures and end-to-end encryption, unauthorized parties may take advantage of flaws in these systems to get personal health information [54]. Ethical and legal issues are also brought up by third-party applications selling or exchanging health data without authorization, especially when consumers are not informed of the usage of their data [55].

Guidelines for data protection are provided by regulatory frameworks like the GDPR in the European Union and the HIPAA in the United States; nevertheless, compliance varies throughout wearable device makers [56]. Since many consumer-grade wearables do not adhere to medical privacy regulations, insurance firms, advertising, and other bad actors may use them to mine data. The security dangers of wearable health devices are further increased by the vulnerability of Bluetooth, Wi-Fi, and 5G-based communication protocols to ransomware, phishing, and man-in-the-middle attacks. End-to-end encryption, multi-factor authentication, and decentralized data storage solutions—like blockchain-based health data security models—are necessary to address these privacy and security concerns. These solutions improve data integrity and provide users more control over their health information [23].

5.6.2 Device Accuracy and Regulatory Standards

Making sure that wearable health equipment is accurate, dependable, and compliant with medical regulations is another significant difficulty. Although a lot of consumer-grade wearables, including fitness trackers and smartwatches, offer health-related insights, their accuracy frequently falls short of clinical-grade diagnostic criteria [14]. Inconsistent readings can result from motion artifacts, signal noise, and sensor calibration variations, especially when it comes to heart rate monitoring, oxygen saturation (SpO_2) measures, and ECG readings [57]. False positives or false negatives resulting from inaccurate data may result in needless medical procedures or the failure to identify potentially fatal illnesses.

Guidelines for medical-grade wearables have been created by regulatory agencies including the International Organization for Standardization (ISO), European Medicines Agency (EMA), and U.S. Food and Drug Administration (FDA). However, many commercially marketed devices still do not meet these standards. For example, the FDA has approved the Apple Watch ECG capability to detect AFib, while other smartwatches with comparable functionality are not clinically validated or approved. Wearable health device development costs and time-to-market are increased by the need for stringent clinical testing, algorithmic validation, and sensor calibration enhancements to ensure regulatory standards are met [58].

Another difficulty is the incompatibility of wearable medical technology with current healthcare IT systems. It is challenging for healthcare practitioners to exploit wearable-generated health insights because many devices have closed ecosystems and proprietary data formats that hinder smooth integration with EHRs. Standardized health data formats, including Fast Healthcare Interoperability Resources, which encourage cross-platform compatibility and connection with healthcare infrastructure, must be adopted in order to address these problems [13].

5.6.3 Future Considerations and Solutions

Wearable health technology must incorporate stronger cybersecurity procedures, cutting-edge AI-driven mistake correction, and enhanced regulatory oversight to solve the issues of data privacy, security, and device accuracy. Real-time FDA/EMA compliance tracking, blockchain-based health data storage, and AI-enhanced sensor calibration have the potential to greatly increase wearable health device security, dependability, and trust [56]. In order to guarantee more accuracy and interoperability across healthcare systems, manufacturers also need to endeavor to standardize wearable technologies [14]. Addressing these issues will be essential for clinical integration and widespread use of wearable health technology as it develops further, which will ultimately result in a healthcare environment that is more patient-centered, dependable, and safe.

5.7 FUTURE TRENDS IN HEALTH MONITORING AND WEARABLE DEVICES

5.7.1 Advances in AI, ML, and Biosensing Technologies

By facilitating real-time data analysis, predictive diagnostics, and customized health recommendations, AI and ML are revolutionizing wearable health monitoring. AI-powered wearables are now able to evaluate complicated health data, identify anomalies, and produce actionable insights, whereas traditional wearables were solely used as data gathering tools [51]. Early detection of conditions including AFib, problems from diabetes, and neurodegenerative diseases is made possible by AI models that assess biometric data such as ECG patterns, heart rate variability, SpO_2 levels, glucose trends, and mobility patterns [9].

Because ML algorithms can forecast health decline based on past data and real-time monitoring, they are especially useful in the management of chronic diseases. AI-powered CGMs, for instance, can anticipate hypoglycemic episodes before they happen, enabling diabetic patients to modify their insulin dosage appropriately [42]. Similar to this, smartwatches with deep learning models can identify stress levels, hypertension risks, and arrhythmias, enhancing cardiovascular health monitoring [18]. By allowing users to communicate with AI-based healthcare solutions for symptom tracking, medication reminders, and lifestyle coaching, the combination of wearable health assistants and natural language processing further improves patient engagement [9].

The upcoming generation of wearable medical technology is also being propelled by developments in biosensing technologies. Next-generation biosensors can assess blood biomarkers, dehydration levels, perspiration composition, and even early signs of diseases like cancer or Alzheimer's [31]. Traditional wearables concentrated on heart rate, movement, and sleep tracking. Vital signs can be continuously and non-invasively monitored with flexible and implantable biosensors, such as smart tattoos and electronic skin patches [22]. Wearable biosensors with AI capabilities will improve real-time health interventions, precision medicine, and early illness detection.

5.7.2 The Role of IoT in Future Healthcare Models

By facilitating real-time health data interchange, remote patient monitoring, and networked wearable technology, the IoT is completely changing the healthcare industry. Wearable sensors, cloud-based health analytics, EHRs, and AI-driven diagnostic tools make up the healthcare IoT ecosystem, which facilitates smooth communication between medical systems, patients, and healthcare professionals [12]. Edge computing and 5G technologies will be included into future IoT-powered healthcare models, allowing for ultra-low latency real-time health monitoring and improved data security [38]. Wearable technology will be able to process health data locally thanks to edge computing, which will lessen dependency on cloud infrastructure and speed up response times for vital medical applications like emergency care and telemedicine [39]. The 5G networks will further improve wearable device connectivity, enabling real-time remote patient monitoring, AI-driven diagnostics, and instantaneous health data transmission.

Wearable health data will soon be integrated with AI-powered healthcare platforms in smart hospitals and IoT-enabled clinics, enabling doctors to remotely monitor patients, identify early disease indicators, and prescribe individualized treatments [35]. AI-based telehealth systems, smart inhalers, and linked ECG monitors will all be commonplace instruments for preventive healthcare, which will lower hospitalization rates and enhance patient outcomes. Additionally, the integration of blockchain technology with the IoT will improve the security of health data, guaranteeing secure, decentralized, and impenetrable medical records.

5.8 CASE STUDIES AND REAL-WORLD APPLICATIONS

5.8.1 Case Study 1: Wearable ECG Monitors for Atrial Fibrillation Detection

One of the most impactful applications of wearable health technology is the use of ECG-enabled smartwatches (e.g., Apple Watch, KardiaMobile) for detecting AFib. The Apple Heart Study, conducted by Stanford University in collaboration with Apple, examined how the ECG feature in Apple Watch could detect AFib in a large population. The study included 419,297 participants, making it one of the largest digital health studies to date.

5.8.1.1 Key Findings

84% of participants who received an AFib notification had AFib confirmed by a clinical ECG test. Wearable ECG monitoring demonstrated high accuracy in detecting irregular heart rhythms.

Participants who followed up with their physicians had lower risks of stroke and cardiovascular complications due to early medical intervention [18].

5.8.1.2 Impact on Healthcare

This study validated the clinical utility of wearable ECG technology in detecting AFib outside hospital settings. Following this research, wearable ECG monitors have been increasingly adopted in hospitals, cardiology clinics, and telemedicine programs for remote heart monitoring. The ability to continuously monitor heart rhythms has enabled earlier diagnoses, improved patient engagement, and reduced hospital admissions related to cardiac events [59].

5.8.2 Case Study 2: Continuous Glucose Monitors for Diabetes Management

Wearable CGMs such as Dexcom G6 and FreeStyle Libre have transformed diabetes management by providing real-time glucose tracking and reducing the need for frequent finger-prick tests. A clinical trial at Stanford University evaluated the effectiveness of CGMs in patients with type 1 and type 2 diabetes and found significant improvements in glycemic control and treatment adherence.

5.8.2.1 Key Findings

CGM users experienced a 50% reduction in hypoglycemic episodes, compared to traditional glucose monitoring methods. HbA1c levels improved by an average of 1.2% within six months, demonstrating better long-term glucose control. Patients reported higher adherence to treatment plans and improved quality of life, as CGMs provided real-time feedback for optimizing insulin dosing and dietary choices [42].

5.8.2.2 Impact on Healthcare

CGMs have been widely adopted in hospitals, outpatient clinics, and telehealth diabetes management programs. Their integration with insulin pumps has enabled automated insulin delivery, significantly improved diabetes management and reducing emergency hospital visits. The success of CGMs has also influenced global healthcare policies, leading to broader insurance coverage and increased accessibility for diabetes patients [19].

5.9 CONCLUSION

Both clinical procedures and individual healthcare have undergone substantial change as a result of the incorporation of wearable technology into health monitoring systems. Wearable technology fills the gap between proactive health management and typical episodic medical evaluations by allowing continuous, real-time tracking of critical physiological indicators. By enabling early disease identification, better chronic disease management, and increased patient engagement, these technologies give people real-time insights into their health. The use of wearable health equipment, such as heart rhythm monitoring smartwatches and CGMs to help manage diabetes, has simplified healthcare delivery, decreased hospital stays, and improved long-term health outcomes.

Wearable technology has great potential, but there are obstacles to its widespread usage in healthcare, such as worries about user adoption, data privacy, device accuracy, and regulatory compliance. Sensitive health data gathering and transmission present serious security threats, requiring robust encryption, authentication procedures, and compliance with international healthcare laws such as HIPAA and GDPR. Furthermore, maintaining wearable sensors' clinical-grade accuracy is essential to their adoption in healthcare environments. Establishing standards that improve wearable health solutions' dependability and credibility will require increased cooperation between regulatory agencies, device makers, and healthcare practitioners as these technologies develop.

In the future, wearable health gadgets will be able to do even more because to developments in AI, ML, biosensing, and 5G connectivity. While real-time data processing via IoT networks and edge computing will maximize health interventions, AI-driven predictive analytics will allow for the early detection of health hazards. Deeper integration with digital health ecosystems will be key to the future of wearable healthcare, allowing for individualized, data-driven treatment that puts an emphasis on early detection and prevention. Finally, by encouraging a more proactive, patient-centric approach, enhancing quality of life, and lessening the strain on international healthcare systems, wearable health technology has the potential to completely transform healthcare.

REFERENCES

1. Jafleh, E. A., Alnaqbi, F. A., Almaeeni, H. A., Faqeeh, S., Alzaabi, M. A., & Al Zaman, K. (2024). The role of wearable devices in chronic disease monitoring and patient care: A comprehensive review. *Cureus*, 16(9), e68921. https://doi.org/10.7759/cureus.68921
2. Guk, K., Han, G., Lim, J., Jeong, K., Kang, T., Lim, E. K., & Jung, J. (2019). Evolution of wearable devices with real-time disease monitoring for personalized healthcare. *Nanomaterials (Basel, Switzerland)*, 9(6), 813. https://doi.org/10.3390/nano9060813
3. Haghi, M., Thurow, K., & Stoll, R. (2017). Wearable devices in medical Internet of Things: Scientific research and commercially available devices. *Healthcare Informatics Research*, 23(1), 4–15. https://doi.org/10.4258/hir.2017.23.1.4
4. Shajari, S., Kuruvinashetti, K., Komeili, A., & Sundararaj, U. (2023). The emergence of AI-based wearable sensors for digital health technology: A review. *Sensors (Basel)*, 23(23), 9498. https://doi.org/10.3390/s23239498
5. Zaharieva, D. P., Riddell, M. C., & Henske, J. (2019). The accuracy of continuous glucose monitoring during aerobic exercise. *Journal of Diabetes Science and Technology*, 13(1), 140–141. https://doi.org/10.1177/1932296818804550

6. Bansal, A., & Joshi, R. (2018). Portable out-of-hospital electrocardiography: A review of current technologies. *Journal of Arrhythmia*, 34(2), 129–138. https://doi.org/10.1002/joa3.12035
7. Anikwe, C. V., Nweke, H. F., Ikegwu, A. C., Egwuonwu, C. A., Onu, F. U., Alo, U. R., & Teh, Y. W. (2022). Mobile and wearable sensors for data-driven health monitoring system: State-of-the-art and future Prospect. *Expert Systems with Applications*, 202, 117362.
8. Tison, G. H., Sanchez, J. M., Ballinger, B., Singh, A., Olgin, J. E., & Marcus, G. M. (2018). Passive detection of atrial fibrillation using a commercially available smartwatch. *JAMA Cardiology*, 3(5), 409–416. https://doi.org/10.1001/jamacardio.2018.0136
9. Cilliers, L. (2020). Wearable devices in healthcare: Privacy and information security issues. *Health Information Management Journal*, 49(2-3), 150–156. https://doi.org/10.1177/1833358319851684
10. Patel, M. S., Asch, D. A., & Volpp, K. G. (2015). Wearable devices as facilitators, not drivers, of health behavior change. *JAMA*, 313(5), 459–460. https://doi.org/10.1001/jama.2014.14781
11. Deng, Z., Guo, L., Chen, X., & Wu, W. (2023). Smart wearable systems for health monitoring. *Sensors*, 23, 2479. https://doi.org/10.3390/s23052479
12. Islam, S. M. R., Kwak, D., Kabir, M. H., Hossain, M., & Kwak, K.-S. (2015). The Internet of Things for health care: A comprehensive survey. *IEEE Access*, 3, 678–708. https://doi.org/10.1109/ACCESS.2015.2437951
13. Lobodzinski, S. S., & Laks, M. M. (2012). New devices for very long-term ECG monitoring. *Cardiology Journal*, 19(2), 210–214. https://doi.org/10.5603/cj.2012.0039
14. Patel, S., Park, H., Bonato, P., Chan, L., & Rodgers, M. (2012). A review of wearable sensors and systems with application in rehabilitation. *Journal of NeuroEngineering and Rehabilitation*, 9, 21. https://doi.org/10.1186/1743-0003-9-21
15. Düking, P., Hotho, A., Holmberg, H. C., Fuss, F. K., & Sperlich, B. (2016). Comparison of non-invasive individual monitoring of the training and health of athletes with commercially available wearable technologies. *Frontiers in Physiology*, 7, 71. https://doi.org/10.3389/fphys.2016.00071
16. Pantelopoulos, A., & Bourbakis, N. G. (2010). A survey on wearable sensor-based systems for health monitoring and prognosis. *IEEE Transactions on Systems, Man, and Cybernetics, Part C (Applications and Reviews)*, 40(1), 1–12. https://doi.org/10.1109/TSMCC.2009.2032660
17. Bonato, P. (2010). Wearable sensors and systems. *IEEE Engineering in Medicine and Biology Magazine*, 29, 25–36. https://doi.org/10.1109/MEMB.2010.936554
18. Perez, M. V., Mahaffey, K. W., Hedlin, H., Rumsfeld, J. S., Garcia, A., Ferris, T., Balasubramanian, V., Russo, A. M., Rajmane, A., & Cheung, L. (2019). Large-scale assessment of a smartwatch to identify atrial fibrillation. *The New England Journal of Medicine*, 381(20), 1909–1917. https://doi.org/10.1056/NEJMoa1901183
19. Rodbard, D. (2017). Continuous glucose monitoring: A review of recent studies demonstrating improved glycemic outcomes. *Diabetes Technology & Therapeutics*, 19(S3), S25–S37. https://doi.org/10.1089/dia.2017.0035
20. Subhan, F., Mirza, A., Bin, M., Su'ud, M., Alam, M., Nisar, S., Habib, U., & Iqbal, M. (2023). AI-enabled wearable medical Internet of Things in healthcare system: A survey. *Applied Sciences*, 13, 1364. https://doi.org/10.3390/app13031394
21. Piwek, L., Ellis, D. A., Andrews, S., & Joinson, A. (2016). The rise of consumer health wearables: Promises and barriers. *PLoS Medicine*, 13(2), e1001953. https://doi.org/10.1371/journal.pmed.1001953
22. Stoppa, M., & Chiolerio, A. (2014). Wearable electronics and smart textiles: A critical review. *Sensors*, 14, 11957–11992. https://doi.org/10.3390/s140711957
23. Yang, G., Li, C., & Marstein, K. (2019). A blockchain-based architecture for securing electronic health record systems. *Concurrency and Computation: Practice and Experience*, 33. https://doi.org/10.1002/cpe.5479
24. Amft, O., & Lukowicz, P. (2009). From backpacks to smartphones: Past, present, and future of wearable computers. *IEEE Pervasive Computing*, 8, 8–13. https://doi.org/10.1109/MPRV.2009.44
25. Tamura, T., Maeda, Y., Sekine, M., & Yoshida, M. (2014). Wearable photoplethysmographic sensors—Past and present. *Electronics*, 3, 282–302. https://doi.org/10.3390/electronics3020282
26. Neri, L., Oberdier, M. T., van Abeelen, K. C. J., Menghini, L., Tumarkin, E., Tripathi, H., Jaipalli, S., Orro, A., Paolocci, N., & Gallelli, I. (2023). Electrocardiogram monitoring wearable devices and artificial-intelligence-enabled diagnostic capabilities: A review. *Sensors*, 23, 4805. https://doi.org/10.3390/s23104805
27. Yao, S., Myers, A., Malhotra, A., Lin, F., & Bozkurt, A. (2017). A wearable hydration sensor with conformal nanowire electrodes. *Advanced Healthcare Materials*, 6(6). https://doi.org/10.1002/adhm.201601159
28. Al-Qahtani, A. M., Ali, S., Khan, A., & Bermak, A. (2023). Performance optimization of wearable printed human body temperature sensor based on silver interdigitated electrode and carbon-sensing film. *Sensors*, 23, 1869. https://doi.org/10.3390/s23041869
29. Filippeschi, A., Schmitz, N., Miezal, M., Bleser, G., Ruffaldi, E., & Stricker, D. (2017). Survey of motion tracking methods based on inertial sensors: A focus on upper limb human motion. *Sensors*, 17, 1257. https://doi.org/10.3390/s17061257
30. Mukhopadhyay, S. C. (2015). Wearable sensors for human activity monitoring: A review. *IEEE Sensors Journal*, 15, 1321–1330. https://doi.org/10.1109/JSEN.2014.2370945

CHAPTER 6

Medical Imaging Advancements

AI in Radiology and Pathology

Haidar Kanso, Rim Basma, Mohammad Ali Mtairek, Mohammad Hadi Awde, Jobran Tohme, and Hussein Taher

6.1 INTRODUCTION

The advent of artificial intelligence (AI) in medicine, particularly in radiology and pathology, marks a significant leap forward in the history of healthcare. Radiology, which emerged as a distinct medical field with the discovery of X-rays by Wilhelm Conrad Roentgen in 1895 [1], has continually evolved through technological advancements. From the early use of basic imaging techniques to the sophisticated modalities of today, such as computed tomography and magnetic resonance imaging, the field has always been at the forefront of medical innovation [2]. The idea of using AI in imaging can be traced back to the 1960s [3]. In 1967, a device was created by Winsberg et al. capable of detecting changes in mammogram film optical densities, highlighting areas with rectangles shaded to indicate differences between the left and right breast [3]. AI has considerable potential in medicine, and radiology stands out as a particularly suitable field for its application. Radiology produces billions of digital images each year, which computers acquire, assess, store, and transmit. This field is intrinsically digital, with computer networks managing all aspects. Radiology's history of adopting new technologies, along with AI's ability to improve current practices and tackle existing problems, makes the integration of AI highly appealing. Enhancing efficiency through better resource management and workflow optimization is crucial for the financial well-being of radiological practices [4].

The rapid expansion of medical imaging has led to increased demands on radiologists due to the significant growth in the number of images per case with volumetric imaging. Radiologists can diagnose around 20,000 diseases and identify over 50,000 causal relationships, which can cause diagnostic errors, inter- and intrareader variability, and burnout [5]. In this situation AI has become a valuable partner, aiding in case interpretation and various non-interpretive tasks in radiological clinics. Although AI will not replace radiologists, it is widely believed that radiologists who utilize AI will outperform those who do not [5]. The purpose of this article is to explore the transformative impact of AI on radiology and pathology, highlighting its applications, benefits, and the ethical considerations that accompany its use. Understanding these aspects is crucial for clinicians, researchers, and policymakers as they navigate the integration of AI into clinical practice. This exploration will provide insights into how AI can enhance diagnostic accuracy, improve efficiency, and reduce human error in medical imaging and pathology.

6.2 APPLICATION OF AI IN RADIOLOGY

There are new tools based on the idea of machine learning(ML) that can even surpass the human level of detection, given that AI applications have already demonstrated performance in the field of medical imaging interpretation that is comparable to human detection. Machine learning applications are a type of deep learning tool that employ neural networks similar to those found in the human brain. As a result, by obtaining prior data, the tool regularly develops itself and improves its capacity for detection by modeling several correlations in this way [6]. Convolutional neural networks, or CNNs, are thought to be the most

DOI: 10.1201/9781003533351-6

widely utilized algorithms in the area of medical imaging. After pooling and extracting hidden features, CNN maps an added picture to many layers in order to link these details and provide plausible predictions [7, 8]. Furthermore, as per Ahmed et al., machine learning has established a foothold in the fields of cancer radiology-based imaging, thoracic imaging, colonoscopy, mammography, and brain imaging [6].

As specialists in the field of detection, radiologists have extensive training and expertise. They often go through stacks of imaging and look for anomalies while modifying and tweaking every parameter to ensure a thorough scan. After that, every characteristic in the picture is assigned to a subjective panel so that the diagnosis may be made. Computer-aided detection (CADe) refers to the computational methods used in clinics that process these features on the same diagnostic panel in order to provide identical interpretations [9]. In reality, two forms of CADe have emerged recently: one is needed to identify lesions in the picture, and the other is used to provide a differential diagnosis for these lesions [10].

However, in order for these programs to search for or highlight characteristics automatically and for the overall vision, radiologists often need to include their detection criteria; nevertheless, these instructions need to be particular to the imaging type and the condition in question. In addition, CADe's genuine ability is still a work in progress. Thus, radiologists should often review the results for additional investigation [6, 11]. Nonetheless, this instrument can also be characterizing, which allows it to interfere with the condition's staging process. It can analyze the texture, size, and extent of the lesion in contrast to humans, who are unable to perform all of these comparisons at once. Because of these traits, the program can forecast whether a tumor is benign or malignant based on its quantitative properties and how relevant they are to these classifications each time it is utilized. Additionally, it includes a differential diagnosis for the illnesses along with their prognosis, risk assessment, and therapy response [12]. Moreover, Doi (2007). claim that these models' differential diagnosis is thought to be quite high [10].

6.3 APPLICATION OF AI IN PATHOLOGY

AI was not quickly adopted in the healthcare industry; however, the speed of integration has been increasing recently. Machine Learning is now capable of handling intricate tasks that are typically carried out by experts [13]. In the last couple of years, AI has taken the lead in technology; therefore, its research and development are attracting major global funding [14]. The potential of AI in healthcare is huge, especially considering how dominant it is in modern technology. Subsequently, AI could serve as a powerful tool in pathologic diagnostics by creating image analysis techniques (22). Indeed, the implications of machine learning and AI on pathology cannot be overstated, especially since the advantages of AI in pathology revolve around enhancing the reliability and precision of diagnosis and boosting efficiency [14]. Digital image analysis has the potential to significantly enhance the accuracy of histomorphologic evaluation. In pathology, machine learning has recently facilitated significant progress. Therefore, incorporating it into healthcare will revolutionize the system [15]. Additionally, the standard in the management of malignancies right now is individualizing the treatment according to multiple factors, one of which is the pathologic diagnosis [16]. With the rising need for personalized cancer treatment, there is a pressing requirement for improved accuracy when evaluating biomarkers and providing more precise histopathologic cancer diagnoses to make better treatment decisions [15]. For example, breast cancer diagnosis can be greatly improved with the use of computerized imaging analysis, especially since it could streamline tasks such as biomarker assessment [17]. The accuracy of the pathologic diagnosis depends greatly on the number of pathologists and their skills [16]. This solves many issues, especially considering the lack of pathologists worldwide [17]. Up to this date, researchers have shown that AI can widen the horizons of technological advancements in pathology [15]. The integration of digital pathology into the diagnostic pathology workflow will cause improved management and diagnosis of cases due to enhancements in overall information management for patient cases [18]. The potential exists to transform the field of medicine through enhanced diagnostic precision, more efficient clinical processes, reduced human resource expenses, and better treatment options [13]. The use of AI in pathology shows promise as it can improve the accuracy of diagnoses made by pathologists by increasing both sensitivity and specificity [15]. It can also enhance the reproducibility of analyses while identifying unique biomarkers from tissue morphology for precision medicine [19].

bias, accountability, and openness. A strong ethical framework is required in light of the swift adoption of AI technology in order to guard against threats to patient safety and the aggravation of already-existing disparities. To balance AI's promise with the moral obligations entailed in medical practice, ethicists, physicians, technologists, and legislators must work together to build such a framework. In the end, AI in radiology and pathology augments human knowledge by providing instruments that raise diagnostic precision, boost patient outcomes, and expedite clinical procedures. In order for AI to be successfully incorporated into healthcare, ethical concerns need to be addressed, teamwork needs to be promoted, and patient-centered care needs to be carefully balanced with technical innovation. The future of AI in medicine will be shaped by this continuing discussion, which will ensure that its advantages are maximized while lowering any possible hazards.

REFERENCES

1. Babic, R. R., Stankovic Babic, G., Babic, S. R., & Babic, N. R. (2016). 120 years since the discovery of X-rays. *Medicinski Pregled, 69*(9–10), 323–330.
2. Geva, T. (2006). Magnetic resonance imaging: Historical perspective. *Journal of Cardiovascular Magnetic Resonance, 8*(4), 573–580.
3. Winsberg, F., Elkin, M., Macy, J., Bordaz, V., & Weymouth, W. (1967). Detection of radiographic abnormalities in mammograms by means of optical scanning and computer analysis. *Radiology, 89*(2), 211–215.
4. Gore, J. C. (2020). Artificial intelligence in medical imaging. *Magnetic Resonance Imaging, 68*, A1–A4.
5. Mello-Thoms, C., & Mello, C. A. B. (2023). Clinical applications of artificial intelligence in radiology. *The British Journal of Radiology, 96*(1150), 20221031.
6. Ahmad, S. F., Han, H., Alam, M. M., & others. (2023). Impact of artificial intelligence on human loss in decision making, laziness and safety in education. *Humanities and Social Sciences Communications, 10*(1), 311.
7. Litjens, G., Kooi, T., Bejnordi, B. E., Setio, A. A. A., Ciompi, F., Ghafoorian, M., van der Laak, J. A. W. M., van Ginneken, B., & Sánchez, C. I. (2017). A survey on deep learning in medical image analysis. *Medical Image Analysis, 42*, 60–88.
8. Shin, H. C., Roth, H. R., Gao, M., Lu, L., Xu, Z., Nogues, I., Yao, J., Mollura, D., & Summers, R. M. (2016). Deep convolutional neural networks for computer-aided detection: CNN architectures, dataset characteristics and transfer learning. *IEEE Transactions on Medical Imaging, 35*(5), 1285–1298.
9. Castellino, R. A. (2005). Computer aided detection (CAD): An overview. *Cancer Imaging: The Official Publication of the International Cancer Imaging Society, 5*(1), 17–19.
10. Doi, K. (2007). Computer-aided diagnosis in medical imaging: Historical review, current status and future potential. *Computerized Medical Imaging and Graphics: The Official Journal of the Computerized Medical Imaging Society, 31*(4-5), 198–211.
11. Cole, E. B., Zhang, Z., Marques, H. S., Edward Hendrick, R., Yaffe, M. J., & Pisano, E. D. (2014). Impact of computer-aided detection systems on radiologist accuracy with digital mammography. *AJR. American Journal of Roentgenology, 203*(4), 909–916.
12. Sharma, N., & Aggarwal, L. M. (2010). Automated medical image segmentation techniques. *Journal of Medical Physics, 35*(1), 3–14.
13. Ahmad, Z., Rahim, S., Zubair, M., & Abdul-Ghafar, J. (2021). Artificial intelligence (AI) in medicine, current applications and future role with special emphasis on its potential and promise in pathology: Present and future impact, obstacles including costs and acceptance among pathologists, practical and philosophical considerations. A comprehensive review. *Diagnostic Pathology, 16*, 1–16.
14. Moxley-Wyles, B., Colling, R., & Verrill, C. (2020). Artificial intelligence in pathology: An overview. *Diagnostic Histopathology, 26*(11), 513–520.
15. Acs, B., Rantalainen, M., & Hartman, J. (2020). Artificial intelligence as the next step towards precision pathology. *Journal of Internal Medicine, 288*(1), 62–81.
16. Liu, Y., Han, D., Parwani, A. V., & Li, Z. (2023). Applications of artificial intelligence in breast pathology. *Archives of Pathology & Laboratory Medicine, 147*(9), 1003–1013.
17. Robertson, S., Azizpour, H., Smith, K., & Hartman, J. (2018). Digital image analysis in breast pathology—From image processing techniques to artificial intelligence. *Translational Research, 194*, 19–35.
18. Parwani, A. V. (2019). Next generation diagnostic pathology: Use of digital pathology and artificial intelligence tools to augment a pathological diagnosis. *Diagnostic Pathology, 14*, 1–3.
19. Homeyer, A., Lotz, J., Schwen, L. O., Weiss, N., Romberg, D., Höfener, H., & Hufnagl, P. (2021). Artificial intelligence in pathology: From prototype to product. *Journal of Pathology Informatics, 12*(1), 13.
20. Pinto-Coelho, L. (2023). How artificial intelligence is shaping medical imaging technology: A survey of innovations and applications. *Bioengineering (Basel, Switzerland), 10*(12), 1435.
21. Srivastav, S., Chandrakar, R., Gupta, S., Babhulkar, V., Agrawal, S., Jaiswal, A., & Wanjari, M. B. (2023). ChatGPT in radiology: The advantages and limitations of artificial intelligence for medical imaging diagnosis. *Cureus, 15*(7), e41435.

22. Boland, G. W., Guimaraes, A. S., & Mueller, P. R. (2009). The radiologist's conundrum: Benefits and costs of increasing CT capacity and utilization. *European Radiology, 19*(1), 9–12.
23. Hardy, M., & Harvey, H. (2020). Artificial intelligence in diagnostic imaging: Impact on the radiography profession. *The British Journal of Radiology, 93*(1108), 20190840.
24. Najjar, R. (2023). Redefining Radiology: A Review of Artificial Intelligence Integration in Medical Imaging. Diagnostics (Basel, Switzerland), 13(17), 2760.
25. Hashimoto, D. A., Rosman, G., Rus, D., & Meireles, O. R. (2018). Artificial intelligence in surgery: Promises and perils. *Annals of Surgery, 268*(1), 70–76.
26. Morley, J., Machado, C. C. V., Burr, C., Cowls, J., Joshi, I., Taddeo, M., & Floridi, L. (2020). The ethics of AI in health care: A mapping review. *Social Science & Medicine (1982), 260*, 113172.
27. Chauhan, C., & Gullapalli, R. R. (2021). Ethics of AI in pathology: Current paradigms and emerging issues. *The American Journal of Pathology, 191*(10), 1673–1683.
28. Kazim, E., & Koshiyama, A. S. (2021). A high-level overview of AI ethics. *Patterns, 2*(9), 100314.
29. Radanliev, P., Santos, O., Brandon-Jones, A., & Joinson, A. (2024). Ethics and responsible AI deployment. *Frontiers in Artificial Intelligence, 7*, 1377011.
30. Hasani, N., Morris, M. A., Rhamim, A., Summers, R. M., Jones, E., Siegel, E., & Saboury, B. (2022). Trustworthy artificial intelligence in medical imaging. *PET Clinics, 17*(1), 1–12.

CHAPTER 7

Natural Language Processing in Healthcare

Existing Insights from the Text Data

Raj Pandit, Bijoy Bitan Saha, Rima Ghosh, and Diptendu Chatterjee

7.1 INTRODUCTION

Being a knowledge-driven industry, the healthcare system has to deal with lots of unstructured data which comprises of narrative information of the patients as summaries, test reports, physician case notes, and case histories on a regular basis. And since, all these vast information have been stored in electronic database without following any standardized formats, it is hard to generate valuable insights from the growing databases [1]. As a result, it hinders the use of those information in generating insightful patterns which will be helpful in decision making and further development of various medicine protocols. And at the same time, dealing with such a huge data set leads to lots of error that can affect the outcome. Under such circumstances artificial intelligence (AI) appears to be handy as it has already proved its worth during the time of Covid-19 outbreak when a huge data of millions of patients has to be maintained regularly throughout the world [2]. Natural language processing (NLP), which is a subfield of AI and computer science becomes relevant in this context as it has the ability to deal with large sets of unstructured and narrative information [3].

In the recent times the diverse tools like electronic medical records and electronic health records have been introduced with the aim of optimizing healthcare system [1, 4, 5]. Such tools are beneficial in storing large sets of health-related information regarding diagnosis, laboratory test results, medicine list, imaging and radiology data along with easy and faster access of these information. Although these tools showed promising results in reducing errors while dealing with a huge volume of data as well as providing cost effective output [1], such documentation involve huge amount of time [4]. Besides, these tools are mostly incapable of grasping the informational contents of the unstructured and narrative texts which are also produced at a large scale in the healthcare system [1], if analyzed can provide valuable information about different cases. NLP is an analytical approach in the field of AI and computational linguistics, which has the ability to draw meaningful data from the unstructured information like spoken or written communications [4]. NLP is concerned with the analysis of the human like languages to generate valuable information for diverse applications [5].

7.2 OVERVIEW OF NATURAL LANGUAGE PROCESSING

Natural languages are the languages used by human beings for communication with one another, either in verbal or written form; for example, English, Arabic, Bengali, Hindi, Urdu, French, Japanese, and Chinese. Every language is comprised of structured words and set of symbols which are govern by specific sets of rules [1, 3]. Similarly, natural languages are comprised of texts which includes articles, book chapter, text records and so on, and speech that encompasses dialogues exchanged between humans or human and machines. Thus, NLP is used to process such components of natural language [1]. Therefore, NLP can be

DOI: 10.1201/9781003533351-7

referred as a range of computational techniques that have the ability to analyze and represent naturally occurring texts in order to achieve human-like language processing for performing wide range of tasks and applications. In other words, NLP make it possible to communicate with the computer using natural languages [1, 6]. Researches on natural languages had been going on since 1940s [6] which escalated with the origin of AI in the early 1950s leading to development of NLP for several decades since then [3]. Although it was machine translation, the first computer-based application dealing with natural languages, but the nature of the processing was simplistic with less efficient output. This led to the foundation of NLP which has grown rapidly since then [6].

In general, there are three categories of the NLP approaches that are employed: rule-based, statistical, and neural approaches. During 1950s–1980s, Rule-based approaches of NLP developed where the NLP researches were mainly based on rules to deal with human languages with prerequisite expertise on both computer science and linguistics. Statistical approaches of NLP however, developed from 1980s and replaced the rule-based approaches gradually through their robust and superior performances. The approaches are designed by using statistical and machine learning (ML) algorithm using the features of corpora. The current researches of NLP are based on the neural NLP approaches that have emerged with the development of deep learning (DL) since 2013 and application of neural probabilistic language model [3, 7].

Rule-based approaches mainly incorporate algorithms like pattern matching [8] and parsing [3], are easy to understand and in certain cases, can be accurate. However, there are limitations for applying these sets of rules even if they are well designed as human language is far more flexible and shows complex patterns. Although, rule-based approaches are now being considered obsolete in academia, but these are employed these days for better pre-processing [3, 9]. *Statistical NLP approaches*, on the other hand, provides robustness and superior performances compared to rule-based approaches but requires hand-crafted feature engineering. As a result, like rule-based approaches, domain experts are required here as well for designing such features, but at the same time these are easily interpretable as rule-based approaches. Bag-of-words [10], Term Frequency-Inverse Document Frequency (TF-IDF) [11] and N-gram [12] are the representative models for statistical NLP approaches, among which both bag-of-words and TF-IDF relies on counting frequencies of a words in a particular sentence. As a result of which these models miss to consider the sequential order of the words. However, unlike bag-of-words, TF-IDF possess the ability to measure the importance of words in a sentence. On the other hand, N-gram provides better performances than the former two models as it counts $n - 1$ words present before a word, but it involves high computational complexity [3]. *Neural NLP* is most widely used these days, which incorporates high computational facilities to deliver better performance in comparison to rule-based and statistical NLP approaches. There are a number of models under neural NLP, such as RNN (recurrent neural network), CNN (convolutional neural network), GNN (graph neural network), transformer-based models and so on. While RNN-based models such as LSTM (long short-term memory) and GRU (gated recurrent unit) are natural, efficient in processing sequential data like speech and text and are capable of remembering histories of input but suffers from a number of issues such as gradient explosion, short-term memories and training issues [13]. CNN models engage high computational efficiency and possess the ability to learn local features. Thus, such features enable CNN to engage deep network architectures and provides better performances [14]. In order to deal with graphical data, GNN-based models are used [15]. Although these models are hard to implement for dealing with large and complex graphical data as the process of such set up and training is very costly. Again, transformer-based models are efficient in running parallel training process while processing long sequences, but lacks the ability to learn local features and information related to the position. Nowadays, combination of the aforementioned models are also preferred for their specific capabilities and better extraction performances such as CNN-LSTM networks, memory networks, RNN attention network, GCN (graph convolutional network), GCAN (graph convolutional attention network), MM (memory networks) and so on. Although, the dominance of DL and the success of neural NLP is evident in the recent times, Statistical NLP is still preferred and actively used in many healthcare studies [3].

7.3 APPLICATION OF NLP IN HEALTHCARE

Since NLP have the ability to deal with a large set of data through its different models and approaches, NLP systems have been employed in healthcare through the last few decades. The following are some of the applications of NLP in healthcare which are listed below.

7.3.1 Information Extraction and Information Retrieval

In order to find out specific set of information from a huge unstructured data especially texts in natural languages, Information extraction technology of NLP is employed [1]. On the other hand, the process of getting a set of documents in response to a set of query placed by the user is called Information retrieval, as often seen in case of using search engines like google and Bing [1, 16].

7.3.2 Developing User Interfaces through Patient Experiences and Feedback Analysis

Different NLP techniques are now available to understand and analyze the patient feedback forms, experiences and sentiments in a healthcare facility. Different models are employed to obtain meaningful data from clumsy and unclear text from patient surveys even social media comments [17, 18]. NLP approaches can learn the text to categorize feedback, analyze the sentiment and the theme of their experiences [18]. In critical care units, NLP is very useful to measure the frequency and mode of communication between doctors and patient or their families [19]. Utilization of NLP is healthcare organizations is giving valuable insights of patients' experience and those in-return generating ideas and road maps to improve the care unit and support systems and other family centered care initiatives [17, 19]. Furthermore, relying upon such data, a user interface can be developed for effective communication through computer systems as well [1].

7.3.3 Document Categorization

In order to analyze a set of documents especially containing unstructured data like textual information, it has to be examined thoroughly to identify underlying themes by application of a pre-defined set of topics. Such process is termed as document categorization [20]. Nowadays, NLP is largely used to analyze vast number of medical data, unstructured clinical notes, medical research papers and patients' experience, feedback many more health related data [21]. Again, the NLP application includes symptom classification, disease outbreak detection, and clinical decision support (CDS) systems. ML, DL are very commonly used to categorize texts using pre-processing techniques, models. It has now become crucial for handling medical terminology in medical electronic records [21]. For instances, NLP techniques have been successfully employed in breast cancer research for identifying literature, parsing pathology reports and mining electronic medical records [22].

7.3.4 Machine Translation and Text Summarization

In order to understand the text, sometimes, it has been translated to a common language that all can understand. NLP systems with translation features have the capabilities to convert textual data in a common language medium understandable by the concerned team [23]. Again, summary or a brief description of a case obtained from the original document or a set of documents provide crucial information for analysis. Moreover, such summarization process reduces the burden of going through the entire documents saving time [1].

7.3.5 Developing Predictive Models

NLP has promising applications particularly in patients' outcome prediction and communication analysis. NLP is game changer for achieving accurate classification of clinical notes of urinary inconsistence and bowel dysfunction to understand and assess patients' prostate cancer treatment outcome [24]. It has been utilized to develop predictive models based on the clinical data using rule-based models, ML and DL methods [25]. Similarly in cardiology, NLP has proven beneficial to identify patient specific diagnoses, disease severity and finally based on that predicted the clinical outcome of various cardiac conditions [26]. Thus, NLP approaches increase the effectiveness in anticipating the occurrence and probable outcome of a disease [19].

7.3.6 CDS System

CDS systems play a crucial role nowadays when the clinical data and cases are growing rapidly. In such scenario, considering the clinical data, CDS system provides suggestions on diagnosis and treatment protocols to the physicians [3]. NLP has promising applications in CDS systems. In a study, expert consensus-derived logic was incorporated into NLP to reduce endoscopy sedation type order errors [27]. In Another study, an NLP algorithm was developed and used to identify thromboembolism from electronic healthcare records [28]. NLP is gaining popularity and more and more applications of NLP are emerging like Name Entity Recognition, Summarization, and many other CDS Tools [29]. Still, there are many challenges in processing clinical notes of many diseases, temporal extraction understanding of clinical events, etc. Sidelining the limitations NLP continues to incorporate to advance clinical applications as we shift towards ML methods [30].

7.4 CASE STUDIES

NLP has revolutionized healthcare services by organizing, analyzing, and utilizing unstructured text data. There are a number of examples where NLP has been yield successful results when incorporated to healthcare system. MedlEE (Medical Language Extraction and Encoding) System, was designed by the Department of Bioinformatics, University of Columbia and Medical School, to extract data from the patients reports and encode those automatically. It has shown accurate results with 83% recall rate and precision of 89% when applied on patients suspected for tuberculosis, stroke, breast cancer and pneumonia [31]. Another example of NLP system is Ctakes (Clinical Text Analysis and Knowledge Extraction System), comes under Apache license, is an open-source NLP engine. It was developed back in 2006 with openNLP toolkit and Unstructured Information Management Architecture framework. It includes components like negation detector, drug mention annotator, phrasal chunker, part of speech tagger, boundary detector and so on [1].

7.5 CHALLENGES OF NLP IN HEALTHCARE

Although NLP has changed the way the healthcare system has dealt with the humongous volume of data, it is still evolving and often face challenges while processing unstructured data. Some of the challenges have been listed below.

7.5.1 Ethical and Legal Concerns

Although, NLP in healthcare offers huge potential but at the same time it raises ethical concerns regarding bias. NLP is capable of analyzing clinical notes to identify at-risk patients, but existing NLP models show bias related to religion, race, gender, nationality, sexuality and age. At the same time, privately funded and owned AI or ML algorithms used in healthcare facilities must be scrutinized. Because, these biases can lead to health inequalities even structural racism in biomedical research. Thus, it is noteworthy to state the importance of the ethical considerations, bias interrogating matrices, identify sensitive patients' attributes and balancing individuals' autonomy in NLP data utilization. So cross-disciplinary collaboration is must between doctors, other health professionals and data scientists. Again, NLP deals with humongous data, which involves personal information of a patient, which must remain protected within a separate database to avoid nuisances [4].

7.5.2 Exponential Growth of Vocabularies with Ambiguity

Due to constant accumulation of unstructured data every day, it is getting difficult to keep a track of every vocabulary used day by day. In other words, a specific word may have different connotation considering different clinical scenario. For example, the word mass refers to breast mass in case of breast cancer, while in case of chest's radiological report it means mass is present in the lungs. Besides, the same problem may arise with abbreviations which are used widely throughout the medical reports and clinical case reports. For instances, acronym APC may have multiple meaning such as Advanced Pancreatic Cancer, AlloPhyCocyanin, Activated Protein C and so on. Due to such ambiguity, analyzing clinical texts become hard and thus required complex modelling to deal with such issues [1, 31]. Furthermore, ambiguity in the meaning of several natural languages can make the scenario far more difficult to incorporate NLP in order to extract meaningful information from unstructured data.

7.5.3 Presence of Errors

While recording data related to case histories and clinical condition, various types of errors might happen which can distort the meaning of a data. Errors like spelling errors make it hard to analyze using NLP [23]. The term hypertension can have

CHAPTER 8

Ethical Considerations in the Use of AI in Healthcare

Anamika Chauhan, Megh R. Goyal, and J.S. Aswal

8.1 INTRODUCTION

The field of healthcare is undergoing a swift transformation because to the exponential growth of artificial intelligence (AI), which presents unparalleled prospects for improving patient care, diagnosis, and treatment. A growing number of aspects of medical practice are using AI-driven tools, such as robotic surgery and predictive analytics, which promise better results and greater efficiency. By facilitating quicker and better-informed decision-making, these advances help healthcare workers in an increasingly complicated medical environment possibly save lives and maximize resource usage. Yet, ethical concerns concerning the application of AI in healthcare have also surfaced alongside these developments, posing serious issues with privacy, responsibility, equity, and the possibility of bias in algorithmic decision-making. The reliance on vast amounts of patient data for AI models poses significant risks concerning confidentiality and data security. As Siau and Wang note, *AI systems in healthcare carry great potential but require thoughtful consideration of ethical issues to prevent harm and ensure equitable access to care* [1]. These ethical concerns necessitate a careful examination of how AI is used in healthcare to protect patients' rights and well-being, ensuring that technological progress does not come at the expense of fundamental ethical principles. The primary objective of this chapter is "to explore and analyze the ethical considerations surrounding the use of AI in healthcare. This chapter aims to provide a comprehensive understanding of the key ethical challenges, such as privacy, fairness, accountability, and bias, that arise from the integration of AI technologies in medical practice". By examining these issues, the chapter seeks to offer insights into how healthcare professionals, policymakers, and AI developers can work together to implement AI solutions that enhance patient care while upholding ethical standards. Ultimately, the goal is to establish a framework for responsible AI use in healthcare, ensuring that technological advancements are balanced with patient rights, equitable access, and trust in AI systems. In this chapter, we will explore the critical ethical issues associated with the deployment of AI in healthcare, addressing concerns related to privacy, bias, accountability, and access to AI-driven care. Through a detailed analysis, we will assess how these ethical challenges can be mitigated and how a balanced approach can ensure that AI continues to advance healthcare without compromising ethical standards (Figure 8.1).

The application of AI in healthcare includes robotics, natural language processing, machine learning algorithms, and other AI technologies that mimic human cognition in the analysis of intricate medical data. By processing and interpreting enormous volumes of data at a size and speed that surpasses human skills, AI systems are intended to complement the abilities of healthcare professionals. AI is being utilized in healthcare across multiple domains, including robotic surgery, personalized medicine, predictive analytics, diagnostic imaging, and drug development. For instance, using historical data, AI can help uncover patterns in medical imaging to identify diseases like cancer early on or forecast patient outcomes [3]. AI helps healthcare professionals make better decisions by providing recommendations based on thorough data analysis. These recommendations can result in more precise diagnoses and more effective treatment regimens [4].

DOI: 10.1201/9781003533351-8

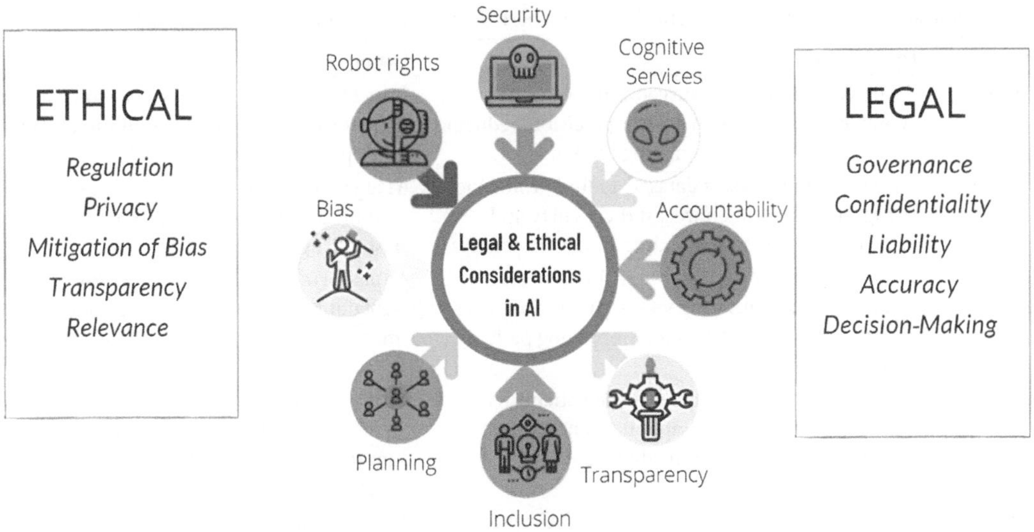

FIGURE 8.1 The use of AI in healthcare [2].

Moreover, AI in healthcare leverages deep learning and neural networks to develop predictive models that can analyze genetic information and other patient-specific data to provide tailored treatment strategies. This personalized approach to medicine has revolutionized patient care by reducing trial-and-error in treatments and improving efficiency [5]. As AI continues to evolve, its potential to transform healthcare is immense, from streamlining administrative tasks to advancing precision medicine. However, the implementation of AI in healthcare requires careful consideration of ethical issues and technological challenges to ensure that it benefits patients and healthcare providers alike (Table 8.1).

While integrating AI into healthcare systems has the potential to be revolutionary, there are important ethical issues that need to be resolved to guarantee a responsible, safe, and equitable rollout. Since AI systems primarily rely on sizable datasets containing sensitive patient data, data security and privacy are critical. In the absence of appropriate security measures, AI has the potential to escalate the likelihood of data breaches, jeopardizing patient privacy and undermining patient-provider trust [11]. Furthermore, it can be difficult to get patients' informed permission due to the extensive usage of AI in healthcare because many people may not completely comprehend how their data is being used or how judgments influenced by AI are made [12]. This lack of openness could have unforeseen repercussions and compromise the moral precepts of patient empowerment and autonomy.

Beyond privacy concerns, AI systems can also perpetuate or even exacerbate biases present in the data they are trained on, leading to unfair treatment outcomes, particularly for minority and vulnerable populations. Biases embedded in algorithms may result

TABLE 8.1 AI Applications in Healthcare (Diagnosis, Treatment, Patient Management, etc.)

AI Application	Description	Examples
Diagnosis [4, 5]	AI assists in the early detection and diagnosis of diseases through data analysis, imaging interpretation, and pattern recognition.	AI-based diagnostic tools for detecting cancer, cardiovascular diseases, and diabetic retinopathy. AI algorithms used in radiology to detect tumors and anomalies.
Predictive Analytics [3, 6]	AI uses patient data to predict health outcomes and risks, enabling proactive management of conditions before they become severe.	Predicting disease outbreaks, patient readmission, or complications from surgeries based on historical data and health patterns.
Personalized Medicine	"AI enables more precise and efficient therapies by customizing treatments for each patient based on their genetic composition, lifestyle, and medical history".	AI algorithms used to design personalized treatment plans for cancer patients, including drug prescriptions and therapy regimens.

(Continued)

TABLE 8.1 AI Applications in Healthcare (Diagnosis, Treatment, Patient Management, etc.) *(Continued)*

AI Application	Description	Examples
Robotic Surgery [3, 7]	AI-driven robots assist surgeons in performing complex surgeries with greater precision, reducing recovery time and minimizing human error.	"The Da Vinci Surgical System, which enhances the surgeon's ability to perform minimally invasive surgeries with precision and accuracy".
Drug Discovery and Development [8, 9]	"By evaluating massive datasets to find promising drug candidates and forecast their safety and efficacy, AI speeds up the drug discovery process".	AI algorithms used to identify drug interactions, screen chemical compounds, and simulate clinical trials for faster drug development.
Patient Management and Monitoring [3, 4]	AI supports healthcare providers in managing patient data, monitoring vital signs, and alerting clinicians to potential issues, improving patient care and workflow.	AI platforms handle patient records and appointment scheduling, while wearable technology driven by AI monitors vital signs such as blood pressure and heart rate.
Virtual Health Assistants [5, 10]	AI-based virtual assistants provide patients with healthcare information, appointment reminders, and help with medication adherence.	Chatbots and voice assistants offering health advice, symptom checkers, and reminders for medication, improving patient engagement and adherence to treatment plans.
Medical Imaging Analysis [4]	AI interprets medical images such as "X-rays, MRIs, and CT scans to detect diseases, identify anomalies, and assist radiologists in making more accurate diagnoses".	AI tools that detect early-stage tumors, fractures, or abnormalities in medical images with higher precision and speed than traditional methods.
Clinical Decision Support [3, 6]	AI helps clinicians in decision-making by analyzing patient data and providing recommendations based on evidence and best practices.	AI systems integrated into electronic health records (EHRs) that offer treatment suggestions or flag potential drug interactions.
Mental Health and Behavioral Support [9, 10]	AI tools monitor and assess mental health conditions through voice, text, and facial expression analysis, offering personalized interventions and support.	AI-based tools for detecting early signs of depression or anxiety by analyzing voice patterns or online interactions, and suggesting personalized treatment plans.

in disproportionate healthcare recommendations or misdiagnoses, further exacerbating existing health disparities [13]. Furthermore, questions of accountability arise when AI-driven healthcare decisions result in harm. Determining who is responsible—whether it be the developers, the healthcare providers, or the AI system itself—is complex and raises legal and moral questions about liability [12].

8.2 PRIVACY AND DATA SECURITY

A fundamental tenet of healthcare is patient confidentiality, which guarantees that private health information is shielded from unwanted access and exposure. Patients and healthcare professionals can build confidence by protecting patient data because people are more willing to divulge private information when they feel that it will be handled carefully and discreetly. In addition to adhering to legal requirements like "the Health Insurance Portability and Accountability Act (HIPAA) in the US", which imposes stringent restrictions on the use and sharing of patient data, confidentiality also supports the ethical obligation of healthcare practitioners [14]. Beyond ethical duties, confidentiality is crucial because violations of patient privacy can result in identity theft, discrimination, and a decline in public confidence in healthcare organizations [15]. In the age of AI, where large volumes of patient data are examined to develop predictive models and improve treatment outcomes, keeping anonymity becomes even more crucial. Healthcare providers and AI developers must put strong security measures in place to secure patient information because AI systems depend largely on data to operate, which increases the potential of data breaches [16]. In AI-driven healthcare, maintaining patient anonymity is essential to maintaining the integrity of the healthcare system overall as well as individual protection (Table 8.2).

Enacted in 1996, the HIPAA is a key piece of legislation that sets strict guidelines for the security of patient data in the US healthcare system. With the growing use of digital technologies and electronic health records (EHRs) for patient data management and storage, HIPAA was created to guarantee that personal health information is kept private and safe. The Privacy Rule and the Security Rule, two of the act's many sections, are the most pertinent to data privacy. Protected health information (PHI) belonging to patients is handled and safeguarded by healthcare

TABLE 8.2 Challenges in Ensuring Data Privacy with AI-driven Tools

Challenge	Description
Data Breaches [17]	AI systems in healthcare are often cloud-based, involving multiple stakeholders, which increases the risk of hacking or unauthorized access to sensitive health data.
Data Volume and Variety [18]	AI systems process vast amounts of diverse health data, including EHRs, imaging, and genetic data, amplifying the complexity of ensuring comprehensive privacy protection.
Data Sharing across Institutions [19]	AI requires large datasets for training, often involving sharing patient data between different institutions, which increases the risk of privacy violations.
Re-identification Risks	Although AI systems use de-identified data, advanced algorithms can combine datasets and re-identify individuals, compromising privacy.
Bias in Data Handling [20]	Different regions and institutions have varying standards for data protection, leading to inconsistent handling of patient data and potential biases in privacy safeguards.
Inadequate Privacy-Preserving Mechanisms [17]	Current privacy-preserving techniques, such as encryption and anonymization, may not be sufficient to protect against AI's powerful data aggregation and analysis capabilities.
Lack of Regulatory Oversight [18]	Existing regulations may not fully address the unique privacy challenges posed by AI in healthcare, leading to gaps in data protection and privacy compliance.

providers, insurers, and other covered entities by the HIPAA Privacy Rule. Significant control over personal data is given to patients by this rule, including the ability to access, modify, and agree to the sharing of their health information [14]. Additionally, HIPAA places restrictions on the use and distribution of PHI without the consent of the patient, permitting data sharing only in specific circumstances, such as those involving treatment, payment, or healthcare operations.

On the other hand, the HIPAA Security Rule focuses on the technological measures required to prevent unauthorized access, alteration, or destruction of electronic PHI (ePHI). To guarantee the privacy, availability, and integrity of ePHI, healthcare organizations are required by this rule to put in place administrative, physical, and technical safeguards [21]. Because AI systems process enormous volumes of data, adhering to HIPAA regulations in the context of AI in healthcare can be extremely difficult. Healthcare organizations utilizing AI must ensure that any data processed by these systems adheres to HIPAA's stringent security measures, including encryption, secure data transmission, and access control protocols [18]. As AI systems often involve data sharing across institutions for model training and improvement, HIPAA regulations require that data-sharing agreements comply with the law, ensuring that PHI is adequately protected throughout its lifecycle. In an era of rapidly advancing AI technologies, HIPAA remains a crucial legal framework that aims to balance the benefits of digital innovation with the need for robust patient privacy protections.

The integration of AI in healthcare brings about significant benefits, but it also introduces the potential for data breaches and misuse of sensitive health information. AI systems require vast amounts of data to function effectively, often involving EHRs, genetic information, and patient demographics. This reliance on extensive data increases the risk of unauthorized access, as healthcare data is highly attractive to cybercriminals due to its value on the black market [22]. Breaches in AI systems can lead to the exposure of confidential patient data, which may result in identity theft, insurance fraud, and loss of trust between patients and healthcare providers. Moreover, as AI systems often involve the use of cloud computing and external data storage, the data may be vulnerable to attacks during transmission or storage if encryption and security protocols are insufficiently robust [16].

Beyond breaches, there is also the risk of misuse of AI systems in healthcare, particularly when data is shared across institutions or with third-party developers. Misuse can occur if the data is used for purposes beyond patient care, such as commercial exploitation without the patient's consent, or if AI algorithms are trained on biased or incomplete datasets, leading to discriminatory treatment recommendations [18]. In the absence of strict regulatory frameworks, healthcare providers and AI developers might not implement the necessary safeguards, increasing the risk of data being used unethically. Furthermore, AI systems that are improperly configured or not regularly audited may inadvertently disclose sensitive information, exacerbating privacy concerns [23]. Given these risks, healthcare organizations must prioritize data security by implementing comprehensive security measures and maintaining transparency with patients about how their data is used and protected in AI-driven healthcare systems.

8.3 BIAS AND FAIRNESS

AI systems that exhibit systematic and unjust discrimination as a result of existing social biases reflected in the data used to train their models are said to exhibit algorithmic bias. Based on variables like color, gender, or socioeconomic position, these biases might produce skewed results that disproportionately affect particular groups. In healthcare, this issue is particularly concerning as it can lead to unequal access to care, misdiagnosis, and the perpetuation of health disparities. One of the primary sources of bias in AI systems is the data itself. If the data used to train an AI model is not representative of all demographic groups or contains historical biases, the model will likely replicate those biases in its predictions or decisions [24]. This can occur in various ways, such as underrepresenting minority groups in training datasets or prioritizing certain outcomes that favor specific demographics.

Furthermore, algorithmic bias can be introduced through the design and development of AI systems. Developers may unintentionally select features or design the AI model in ways that prioritize efficiency or accuracy for certain populations while neglecting others. For instance, many machine learning models are developed using historical healthcare data, which may not accurately reflect the needs of underrepresented groups. In addition, the "black box" nature of many AI models, especially deep learning algorithms, makes it difficult to understand how decisions are being made and where bias may be introduced. This opacity can result in biased outputs that go undetected, making it challenging to address or mitigate the bias [13]. Thus, it is essential to develop strategies for identifying and mitigating bias during the development and deployment of AI systems in healthcare.

Several case studies have highlighted the detrimental effects of algorithmic bias in AI systems, particularly in the healthcare sector. A notable example is the racial bias uncovered in an algorithm used by the US healthcare system to allocate resources to high-risk patients. Researchers found that the algorithm was less likely to recommend Black patients for advanced care than White patients, despite both having similar medical needs. This discrepancy occurred because the algorithm used healthcare spending as a proxy for healthcare needs, and due to socioeconomic factors, Black patients tend to have lower healthcare expenditures [25]. As a result, the AI system effectively reinforced existing disparities in healthcare access and exacerbated inequities in treatment. This case illustrates how seemingly neutral algorithms can perpetuate structural biases when trained on biased or incomplete data.

Gender bias is another area where AI systems have demonstrated discriminatory outcomes. For instance, AI models used to predict heart disease often underdiagnose women due to their reliance on data and symptoms predominantly associated with men. Since much of the historical research on heart disease focused on male patients, AI algorithms trained on these datasets tend to overlook symptoms that present differently in women. Consequently, women may be at higher risk of receiving inaccurate or delayed diagnoses, potentially leading to worse health outcomes [26]. These examples underscore the need for diverse and representative datasets, as well as proactive efforts to ensure fairness in AI systems. Without addressing these biases, AI risks perpetuating and exacerbating existing inequalities in healthcare.

8.4 ACCOUNTABILITY AND RESPONSIBILITY

Accountability and responsibility are critical ethical concerns in the use of AI in healthcare. As AI systems become more integrated into decision-making processes, determining who is responsible when things go wrong becomes increasingly complex. Traditionally, healthcare providers have been accountable for medical decisions and patient outcomes, but the introduction of AI adds another layer of complexity. When an AI system misdiagnoses a patient or recommends an inappropriate treatment, questions arise about whether responsibility lies with the healthcare provider who used the AI system, the developers who designed the AI, or the institution that deployed it [12]. This diffusion of responsibility creates challenges in assigning legal and ethical accountability, particularly when the AI system is a "black box", meaning its decision-making processes are not fully transparent or understandable, even to the experts using it.

One significant concern is that clinicians may become overly reliant on AI recommendations, potentially abdicating their professional responsibility to exercise independent judgment. This could lead to a shift in accountability from healthcare providers to AI systems, raising questions about liability. For example, if an AI system makes an erroneous recommendation that a clinician follows, the line between human and machine responsibility blurs [27]. There is the issue of legal frameworks lagging behind technological advancements, as current laws may not adequately address the complexities of accountability in AI-driven healthcare. To address these challenges, robust guidelines and regulations must be developed to ensure that accountability structures are clear, and that both AI developers and healthcare providers maintain responsibility for patient safety and care. Transparency in AI systems and clear legal definitions of responsibility will be essential for ensuring ethical AI integration in healthcare.

8.5 EQUITY AND ACCESS TO AI-DRIVEN HEALTHCARE

Equity and access are pressing concerns in the implementation of AI-driven healthcare systems, as these technologies hold the potential to both alleviate and exacerbate existing healthcare disparities. While AI can improve diagnosis, treatment, and patient outcomes, its benefits are not equally accessible to all populations. One major issue is the "digital divide", where underserved communities, particularly those in rural or low-income areas, may lack the technological infrastructure or resources necessary to benefit from AI advancements [19]. Furthermore, healthcare institutions in resource-poor regions may struggle to afford and implement AI systems, leaving these populations with limited access to AI-driven innovations that could significantly improve healthcare outcomes. This disparity creates an ethical concern where AI systems may reinforce existing inequalities, providing better care for those who already have access to high-quality healthcare while leaving vulnerable populations behind [17].

Concerns have been raised over whether AI systems are made with diversity in mind, in addition to infrastructure obstacles. Datasets that primarily represent the healthcare requirements and results of wealthier, mostly Western, populations are used to train many AI models. Because of this, these models might not work as well for underrepresented groups, like women, people with unusual diseases, or members of racial minorities [25]. The disparity in healthcare equity may be widened further by skewed results and diminished AI system effectiveness for particular populations as a result of this lack of representativeness. Improving the inclusivity of data used to train AI models and making sure that AI technologies are implemented in ways that support equitable access across all socioeconomic and geographic contexts are both necessary to address these problems. To guarantee that AI-driven healthcare systems promote equity rather than maintain gaps, policies targeted at closing the digital divide must be paired with moral standards for the creation and application of AI.

8.6 AI IN END-OF-LIFE CARE

The use of AI in end-of-life care presents both significant opportunities and profound ethical challenges. AI technologies have the potential to assist healthcare professionals in making critical decisions regarding palliative care, pain management, and life-sustaining interventions. AI-driven predictive models can analyze patient data to estimate life expectancy or the likely progression of terminal illnesses, helping clinicians and families make informed decisions about end-of-life options [28]. Additionally, AI can support personalized care by tailoring treatment plans to the unique needs and preferences of each patient, ensuring that they receive appropriate comfort measures during their final days [29]. These innovations can improve the quality of life for terminally ill patients and help alleviate the emotional and logistical burdens on their families.

However, the use of AI in such sensitive and personal situations raises significant ethical concerns. One issue is the potential for AI to undermine the humanistic aspects of end-of-life care, where compassion, empathy, and emotional support are paramount. Relying on AI algorithms to guide decisions about whether to continue or withdraw life-sustaining treatment may lead to situations where patients and their families feel disconnected from the care process, or where AI-driven decisions conflict with the personal values and wishes of the patient [30]. The accuracy of AI predictions in end-of-life care remains a concern, as incorrect or overly deterministic predictions could prematurely limit treatment options or lead to decisions that hasten death. The use of AI in this context necessitates careful ethical oversight and a commitment to maintaining a patient-centered approach, where AI serves as a tool to enhance, rather than replace, human judgment and compassion.

8.7 REGULATION AND GOVERNANCE OF AI IN HEALTHCARE

AI in healthcare must be governed and regulated in order for these technologies to be applied in a way that is morally, legally, and efficiently. Comprehensive legal and regulatory frameworks that handle the particular difficulties presented by AI are becoming more and more necessary as AI is incorporated into healthcare decision-making. Guidelines for the development and implementation of AI-driven medical devices and data protection have started to be established by current regulatory frameworks, such as "the General Data Protection Regulation (GDPR) of the European Union and the Food and Drug Administration (FDA) of the United States" [31]. Regulating accountability, safety, and ethical use is lacking in many cases, and these laws frequently lag behind the quick speed of AI progress. For instance, the FDA has unveiled a framework that is still in the process of being developed for AI-based medical devices. It centers on post-market surveillance and real-time monitoring in order to handle the dynamic learning capabilities of AI systems. This is unlike how regulation has always been done, which usually focuses on stationary medical equipment [3].

The governance of AI in healthcare also needs to address issues of transparency, accountability, and discrimination. Many AI systems operate as "black boxes", which makes it difficult for regulators, physicians, and patients to understand how decisions are made. This lack of transparency raises concerns about accountability when AI systems make mistakes or generate biased findings [27]. Furthermore, national and regional laws pertaining to patient safety, data privacy, and technology use complicate governance because AI development is an international undertaking. Therefore, uniform rules for the ethical use of AI in healthcare must be provided through international cooperation. Initiatives like the World Health Organization's guidelines on the ethical governance of AI in health are crucial to developing a regulatory framework that balances innovation with safeguarding patients' rights and welfare [32]. To fully utilize AI in healthcare, robust, adaptable governance structures that develop alongside the technology are essential.

8.8 CONCLUSION

To conclude, the ethical implications of using AI in healthcare are vast and multifaceted. While AI offers transformative potential in improving diagnostic accuracy, personalized treatment, and overall patient care, it also introduces significant ethical challenges that must be addressed. These include privacy concerns related to the handling of sensitive patient data, the risk of algorithmic bias, questions of accountability in AI-driven decisions, and unequal access to AI technologies across different socioeconomic groups. The central ethical concern revolves around the preservation of patient privacy and confidentiality. AI systems rely on large datasets, often containing highly sensitive health information, which raises the risk of data breaches and unauthorized access. Without robust security measures and strict regulatory frameworks, such as compliance with HIPAA, patient data could be misused or exposed to malicious actors, undermining trust in AI-driven healthcare solutions.

The issue of algorithmic bias in AI systems poses a serious risk to equitable healthcare delivery. When trained on biased or non-representative datasets, AI models can perpetuate disparities in healthcare outcomes, particularly for marginalized or underrepresented populations. Addressing these biases through better data inclusivity and fairness-focused AI development is essential for ensuring that AI promotes rather than undermines health equity. Accountability is another critical challenge in AI healthcare applications. When AI systems influence or make clinical decisions, determining who is responsible when errors occur becomes complex. Healthcare providers, developers, and institutions must share responsibility for ensuring that AI is used ethically and that clinicians remain actively involved in the decision-making process. Lastly, access to AI-driven healthcare remains uneven, with technological advancements primarily benefiting those with access to quality care. Bridging the digital divide and ensuring that AI technologies are available to underserved communities is essential for preventing further disparities in healthcare outcomes. Hence, while AI has the potential to revolutionize healthcare, its integration must be carefully managed to ensure that ethical principles are upheld. Policymakers, healthcare professionals, and AI developers must collaborate to create a framework that promotes transparency, fairness, and equity, ensuring that AI-driven innovations enhance rather than compromise patient care.

REFERENCES

1. Siau, K., & Wang, W. (2020). Artificial intelligence (AI) ethics: Ethics of AI and ethical AI. *Journal of Database Management*, *31*, 74–87.
2. Morley, J., & Floridi, L. (2020). An ethically mindful approach to AI for health care. *The Lancet*, *395*(10220), 254–255.
3. Jiang, F., Jiang, Y., Zhi, H., Dong, Y., Li, H., Ma, S., Wang, Y., Dong, Q., Shen, H., & Wang, Y. (2017). Artificial intelligence in healthcare: Past, present and future. *Stroke and Vascular Neurology*, *2*, 230–243. https://doi.org/10.1136/svn-2017-000101
4. Esteva, A., Robicquet, A., Ramsundar, B., Kuleshov, V., DePristo, M., Chou, K., Cui, C., Corrado, G., Thrun, S., & Dean, J. (2019). A guide to deep learning in healthcare. *Nature Medicine*, *25*, 24–29.
5. Topol, E. J. (2019). High-performance medicine: The convergence of human and artificial intelligence. *Nature Medicine*, *25*, 44–56.
6. Miotto, R., Li, L., Kidd, B. A., & Dudley, J. T. (2016). Deep patient: An unsupervised representation to predict the future of patients from the electronic health records. *Scientific Reports*, *6*(1), 1–10.
7. Silvera-Tawil, D. (2024). Robotics in healthcare: A survey. *SN Computer Science*, *5*(1), 189.
8. Paul, D., Sanap, G., Shenoy, S., Kalyane, D., Kalia, K., & Tekade, R. K. (2021). Artificial intelligence in drug discovery and development. *Drug Discovery Today*, *26*(1), 80–93.
9. Xiong, H., Chen, H., Xu, L., Liu, H., Fan, L., Tang, Q., & Cho, H. (2022). A survey of data element perspective: Application of artificial intelligence in health big data. *Frontiers in Neuroscience*, *16*, 1031732.
10. Bickmore, T., Schulman, D., & Yin, L. (2010). Maintaining engagement in long-term interventions with relational agents. *Applied Artificial Intelligence*, *24*(6), 648–666.
11. Char, D. S., Shah, N. H., & Magnus, D. (2018). Implementing machine learning in health care - addressing ethical challenges. *The New England Journal of Medicine*, *378*(11), 981–983.
12. Morley, J., Machado, C. C. V., Burr, C., Cowls, J., Joshi, I., Taddeo, M., & Floridi, L. (2020). The ethics of AI in health care: A mapping review. *Social Science & Medicine (1982)*, *260*, 113172.
13. Rajkomar, A., Hardt, M., Howell, M. D., Corrado, G., & Chin, M. H. (2018). Ensuring fairness in machine learning to advance health equity. *Annals of Internal Medicine*, *169*(12), 866–872.
14. Agich, G. J. (1994). Privacy and medical record information. *Journal of the American Medical Informatics Association*, *1*(4), 323–324.
15. Al Kuwaiti, A., Nazer, K., Al-Reedy, A., Al-Shehri, S., Al-Muhanna, A., Subbarayalu, A. V., Al Muhanna, D., & Al-Muhanna, F. A. (2023). A review of the role of artificial intelligence in healthcare. *Journal of Personalized Medicine*, *13*(6), 951.
16. Murdoch, B. (2021). Privacy and artificial intelligence: Challenges for protecting health information in a new era. *BMC Medical Ethics*, *22*(1), 122.
17. Bajwa, J., Munir, U., Nori, A., & Williams, B. (2021). Artificial intelligence in healthcare: Transforming the practice of medicine. *Future Healthcare Journal*, *8*(2), e188–e194.
18. Chada, B. V., & Summers, L. (2022). AI in the NHS: A framework for adoption. *Future Healthcare Journal*, *9*(3), 313–316.
19. Davenport, T., & Kalakota, R. (2019). The potential for artificial intelligence in healthcare. *Future Healthcare Journal*, *6*(2), 94–98.
20. Khalid, N., Qayyum, A., Bilal, M., Al-Fuqaha, A., & Qadir, J. (2023). Privacy-preserving artificial intelligence in healthcare: Techniques and applications. *Computers in Biology and Medicine*, *158*, 106848.
21. Sewell, J. P. (2019). *Informatics and Nursing: Opportunities and Challenges*. Wolters Kluwer Health.
22. Cremer, F., Sheehan, B., Fortmann, M., Kia, A. N., Mullins, M., Murphy, F., & Materne, S. (2022). Cyber risk and cybersecurity: A systematic review of data availability. *The Geneva Papers on Risk and Insurance. Issues and Practice*, *47*(3), 698–736.
23. Goodman, K. W., & Meslin, E. M. (2014). *Ethics, Information Technology, and Public Health: Duties and Challenges in Computational Epidemiology* (pp. 191–209). London: Springer.
24. Mehrabi, N., Morstatter, F., Saxena, N., Lerman, K., & Galstyan, A. (2021). A survey on bias and fairness in machine learning. *ACM Computing Surveys*, *54*, 1–35.
25. Obermeyer, Z., Powers, B., Vogeli, C., & Mullainathan, S. (2019). Dissecting racial bias in an algorithm used to manage the health of populations. *Science (New York, N.Y.)*, *366*(6464), 447–453.
26. Buolamwini, J., & Gebru, T. (2018). Gender shades: Intersectional accuracy disparities in commercial gender classification. *Proceedings of Machine Learning Research*, *81*, 77–91.
27. Floridi, L., Cowls, J., King, T. C., & Taddeo, M. (2021). How to design AI for social good: Seven essential factors. *Ethics, Governance, and Policies in Artificial Intelligence*, 125–151.

28. De Panfilis, L., Peruselli, C., Tanzi, S., & Botrugno, C. (2023). AI-based clinical decision-making systems in palliative medicine: Ethical challenges. *BMJ Supportive & Palliative Care*, *13*(2), 183–189.
29. Peruselli, C., De Panfilis, L., Gobber, G., Melo, M., & Tanzi, S. (2020). Intelligenza Artificiale e cure palliative: opportunità e limiti [Artificial intelligence and palliative care: Opportunities and limitations. *Recenti Progressi in Medicina*, *111*(11), 639–645.
30. Farhud, D. D., & Zokaei, S. (2021). Ethical issues of artificial intelligence in medicine and healthcare. *Iranian Journal of Public Health*, *50*(11), i–v.
31. Gerke, S., Minssen, T., & Cohen, G. (2020). Ethical and legal challenges of artificial intelligence-driven healthcare. *Artificial Intelligence in Healthcare*, 295–336.
32. World Health Organization (2021). *Ethics and Governance of Artificial Intelligence for Health: WHO Guidance: Executive Summary*. World Health Organization.

CHAPTER 9

Regulatory Framework for AI in Smart Healthcare

Preeti and Sheetal

9.1 INTRODUCTION

The use of artificial intelligence (AI) in healthcare is both vast and significantly promising, holding tremendous potential to revolutionize patient care, streamline clinical work flow, and improve healthcare outcomes, while also presenting opportunities for innovation and growth. "AI is highly potential to revolutionize healthcare, but its adoption also raises significant concerns. These include unethical data collection practices, heightened cybersecurity risks, and the potential to perpetuate and amplify existing biases or spread misinformation, ultimately compromising patient trust and outcomes. Addressing these challenges is crucial to harnessing the benefits of AI in healthcare while ensuring safety, equity, and transparency." So, we need a unified regulatory framework that strikes a balance between giving space for innovation and establishing strong robust regulatory framework, allowing us to harness the potential of AI in healthcare while ensuring safety accuracy and precessions through clear guidelines and standards. Reports of data breaches and security lapses in AI-powered healthcare platforms have raised concerns about patient data protection. Experts in AI healthcare recommend a rigorous development approach, akin to pharmaceuticals and medical devices, involving comprehensive trials and testing before public deployment. They advocate for stringent evaluation and regulation to ensure AI systems meet high standards for safety, efficacy, and reliability. This necessitates a robust ethical and cybersecurity framework to mitigate risks and protect sensitive patient data. Effective regulation should prioritize transparency, accountability, and human oversight, safeguarding against biases and errors. By adopting this structured approach, AI in healthcare can realize its potential to improve patient outcomes while maintaining trust and confidence [1].

9.2 REGULATORY BODIES

Regulatory bodies of AI in healthcare confirm that there must be safe, efficient, as well as ethical use of AI. The World Health Organization (WHO) has expressed some important problems regarding the application of AI in healthcare. The WHO is deeply concerned about the increasing use of AI in healthcare, citing safety and ethics worries. In response, WHO published outline of six areas for regulation of AI in healthcare which stresses on transparency and documentation, risk management, validation of data quality, and privacy [1]. Over many years every guideline and research paper on AI and digital health have given an evolving landscape to AI technologies in healthcare. As concerns surrounding AI in healthcare mount, stakeholders worldwide are taking proactive steps to establish robust regulations, ensuring its safe and effective integration into healthcare systems.

9.2.1 International Regulatory Bodies

9.2.1.1 IMDRF (International Medical Device Regulator Forum)

The IMDRF is regarded across-the-board convention of therapeutic gadget supervisory offices, joining together remote agents to cluster and harmonize directorial approaches, cultivating a bound together outline for medical device oversight by facilitating collaboration and consensus-building, IMDRF aims

DOI: 10.1201/9781003533351-9

9.3.3 Complexity of AI Algorithm

Due to complex functioning and less knowledge among healthcare professional they are not able to understand as well as explain treatment process to patient properly. For algorithms that continuously learn and improve, with features that may evolve beyond the initially approved performance, appropriate policies and regulatory guidelines are essential to ensure proper management and oversight. [13]. So, it comes with challenge while forming regulatory framework.

9.3.4 Liability

Liability issue is also the main concern, for example, in case any error occurs in clinical settings due to use of AI resources then who will be responsible [13]. Hence, the regulation regarding medical malpractice and line between responsibility of AI and clinical comes with a challenge.

9.4 SUMMARY

AI usage in healthcare is still in initial phase which requires standardized guidelines, regulatory bodies, and training of healthcare workers. The proper handling of patient's sensitive data, consent, and privacy need to be maintained. With the strong foundation of AI in healthcare can bring a revolutionary change. Till date regulatory bodies present are either not unified or there is general lack of standard guidelines we need. In many countries, organizations are working together to build a strong base above which AI can function smoothly for improving healthcare outcomes. It is imperative that we establish a robust, unified, and harmonized regulatory framework to enable the safety and effective integration of AI in healthcare, enabling us to harness its full potential and achieve optimal outcomes for patient and healthcare systems worldwide.

REFERENCES

1. WHO. (2023). *WHO Outlines Considerations for Regulation of Artificial Intelligence for Health.* Retrieved from https://www.who.int/news/item/19-10-2023-who-outlines-considerations-for-regulation-of-artificial-intelligence-for-health
2. Government of Singapore. (2024). *International Medical Device Regulator Forum (IMDRF).* Retrieved from https://www.hsa.gov.sg/medical-devices/international-collaboration-medical-devices/imdrf
3. WHO. (2024). *Ethics and Governance of Artificial Intelligence for Health: Guidance on Large Multi-Modal Models.* Geneva: WHO, 1–5.
4. FDA. (2019). *Proposed Regulatory Framework for Modifications to Artificial Intelligence/Machine Learning-Based Software as a Medical Device.* Silver Spring, MD: FDA.
5. U.S. Food and Drug Administration. (2021). *Artificial Intelligence/Machine Learning (AI/ML)-Based Software as a Medical Device (SaMD) Action Plan.* Retrieved from https://www.fda.gov/media/145022/download
6. The EU Newsletter. (2021). *The Artificial Intelligence Act.* Retrieved from https://artificialintelligenceact.eu/
7. Regulatory Horizons Council. (2022). *The regulation of artificial intelligence as a medical device* (Report PE 638.391). London, UK: Regulatory Horizons Council.
8. European Commission. (2019). *Ethics Guidelines for Trustworthy AI: Shaping Europe's Digital Future.* Retrieved from https://digital-strategy.ec.europa.eu/en/library/ethics-guidelines-trustworthy-ai
9. Song, X., Hu, M., Li, B., Zhang, K., Zhang, X., & Wang, L. (2022). Advancing medical device regulatory reforms for innovation, translation and industry development in China. *Journal of Orthopaedic Translation, 37,* 89–93.
10. Rao Vardhana, V. M. (2023). *Ethical Guidelines for Application of Artificial Intelligence in Biomedical Research and Healthcare* (pp. 9–30). New Delhi: Indian Council of Medical Research, DHR-ICMR Artificial Intelligence Cell.
11. Char, D. S., Shah, N. H., & Magnus, D. (2018). Implementing machine learning in health care: Addressing ethical challenges. *New England Journal of Medicine, 378,* 981–983. Retrieved from https://pmc.ncbi.nlm.nih.gov/articles/PMC5962261/
12. NICE. (2024). *Evidence Standards Framework (ESF) for Digital Health Technologies.* Retrieved from https://www.nice.org.uk/about/what-we-do/our-programmes/evidence-standards-framework-for-digital-health-technologies
13. Reddy, S., Fox, J., & Purohit, M. P. (2020). Artificial intelligence-enabled healthcare delivery. *Journal of the American Medical Informatics Association.* Retrieved from https://pubmed.ncbi.nlm.nih.gov/31682262/
14. World Health Organization. (2021, June 28). WHO issues first global report on Artificial Intelligence (AI) in health and six guiding principles for its design and use. https://www.who.int/news/item/28-06-2021-who-issues-first-global-report-on-ai-in-health-and-six-guiding-principles-for-its-design-and-use

CHAPTER 10

Patient Engagement through Artificial Intelligence-Driven Solutions

Arindam Ghosh, Aditya Anand, Deepak Kumar, Sonalika Roy, and Asitava Deb Roy

10.1 INTRODUCTION

Within the realm of intelligent healthcare systems, the active involvement of patients has emerged as a crucial factor in enhancing healthcare results and patient contentment [1]. Conventional methods of involvement, although somewhat effective, often lack the ability to be expanded and tailored to individual patients, resulting in a standardized approach that does not accommodate the specific requirements of each patient. By incorporating artificial intelligence (AI) into smart health systems, there is a significant opportunity to improve patient engagement by overcoming these limitations. AI-powered solutions facilitate ongoing, tailored, and preemptive engagements with patients, leading to a substantial enhancement in their healthcare encounter [2].

This chapter explores the impact of AI-driven solutions on patient engagement. It examines the various ways in which AI is used, the advantages and difficulties associated with its implementation, and its potential for future development in the context of smart health systems. AI-driven chatbots and virtual assistants provide patients with 24/7 access to information and support, significantly improving their experience by providing immediate responses and minimizing waiting periods [3]. Predictive analytics employs patient data to identify patients at risk of non-adherence to treatment regimens or possible health deterioration. This enables healthcare providers to intervene promptly and deliver tailored assistance. AI-driven personalized health education delivers customized information to patients based on their unique health profiles, empowering them to make informed decisions about their healthcare [4].

Remote monitoring technologies and telehealth integration facilitate continuous patient engagement by consistently tracking vital signs, symptoms, and medication compliance in real-time [5, 6]. These tools facilitate the ongoing connection between patients and their healthcare providers, even when they are physically separated, enabling prompt interventions and decreasing the necessity for emergency medical attention [7]. The chapter also explores the various advantages of AI-powered patient engagement, such as improved customization, scalability, ongoing involvement, cost efficiency, and optimized resource allocation [8].

Nevertheless, the execution of AI-powered solutions is not devoid of obstacles. The delicate nature of medical records makes data privacy and security paramount [9, 10]. Maintaining patient trust requires strict adherence to regulations and the implementation of strong data protection measures [9]. Ensuring equitable care for all patients also requires addressing ethical considerations, such as the need to avoid biases in AI algorithms [11]. Moreover, the process of incorporating AI solutions into current healthcare systems can be intricate and necessitates meticulous strategizing and implementation [12]. For these technologies to be widely used, it is essential to build

DOI: 10.1201/9781003533351-10

trust and acceptability among healthcare providers and patients [13].

The capabilities of AI-driven patient engagement solutions are anticipated to be enhanced in the future by advancements in AI technologies like deep learning and natural language processing (NLP) [14, 15]. In order to give individualized suggestions and treatments, wearable devices can continuously monitor health variables, providing real-time data that can be analyzed by AI [16, 17, 18]. Technology developers, healthcare professionals, and patients can work together using co-design and collaborative techniques to create AI-driven solutions that are personalized to each user's needs and preferences [18]. This approach guarantees higher acceptance and effectiveness of these solutions.

AI-powered patient engagement is positioned to have a crucial impact on the future of intelligent health systems, providing a higher level of customized, streamlined, and patient-focused healthcare [19]. Healthcare providers can utilize AI technologies to provide ongoing, personalized assistance to patients, thereby increasing their contentment and enhancing health results [20]. Tackling the challenges associated with data privacy, ethics, integration, and acceptance is crucial for the successful implementation of these solutions and the promotion of efficient and fair healthcare delivery [21].

10.2 THE EVOLUTION OF PATIENT ENGAGEMENT IN SMART HEALTH SYSTEMS

The evolution of patient engagement in smart health systems has transitioned from traditional methods like face-to-face consultations and printed materials to advanced AI-driven solutions [19]. Early digital portals offered limited interaction, while AI now enables personalized, scalable, and continuous engagement [19]. AI-powered tools such as chatbots, predictive analytics, and remote monitoring have revolutionized patient care, providing timely support and proactive interventions [21]. This shift has enhanced patient satisfaction, improved health outcomes, and optimized healthcare resource utilization [23].

10.2.1 Traditional Methods of Patient Engagement

Traditional patient engagement strategies, such as face-to-face appointments and consultations, offer a personal touch and foster trust between patients and providers. AI-driven solutions build on this foundation by extending engagement beyond the clinic through personalized digital interactions, virtual assistants, and predictive insights that keep patients continuously supported and connected. These approaches have been crucial in laying the groundwork for a strong relationship between patients and providers and in making sure patients have the resources they need to communicate openly and honestly. Phone consultations extend this interaction beyond the clinic, offering patients a way to seek advice and follow-up care from the comfort of their homes.

Printed educational materials, such as brochures and pamphlets, have been widely used to inform patients about their conditions, treatments, and healthy lifestyle choices. Patients may now view their medical information, schedule appointments, and even connect with their doctors online thanks to simple digital interfaces. These approaches have worked to a certain degree, but they don't always deliver what's needed: consistent, tailored, and scalable engagement. Rather than being specific to each patient's needs, the information given is frequently generalized. There may be lapses in care and chances for prompt interventions due to a lack of constant contact. More complex methods are required to meet the rising demand for patient-centered care and the increasing difficulty of healthcare delivery. The increasing prevalence of chronic diseases and the complexity of healthcare systems highlight the critical need for patient engagement strategies that can evolve with patients' changing requirements and offer continuous assistance. Although they have their uses, traditional approaches have limitations when it comes to providing individualized treatment to huge populations. This disparity highlights the need to include cutting-edge tech like AI to improve patient involvement and fulfill the demands of contemporary healthcare [24].

10.2.2 Emergence of AI in Healthcare

AI technologies, including machine learning, NLP, and predictive analytics, are transforming patient engagement [25]. In intelligent healthcare systems, AI enhances the examination of extensive datasets, allowing for proactive, individualized, and prompt interactions unattainable by conventional methods [26, 27]. Machine learning algorithms can examine patient data to discern patterns and forecast outcomes, facilitating early intervention and personalized care plans [28]. Predictive analytics can identify patients at risk of medication non-adherence or potential health deterioration, enabling healthcare providers to implement preemptive measures [29]. NLP facilitates the creation of advanced chatbots and virtual assistants capable of interacting with patients in a human-like fashion. These AI-driven tools can address patient enquiries, deliver health education, and perform initial evaluations based on patient feedback [15]. This enhances access to healthcare information and guarantees that patients obtain reliable and precise guidance.

The capability of AI to analyze extensive datasets in real-time facilitates ongoing monitoring and interaction [30]. Wearable devices and remote monitoring tools produce a continuous flow of health data, which AI can analyze to identify anomalies and deliver real-time feedback to patients and healthcare providers [16, 31]. This ongoing engagement aids in the management of chronic conditions, tracking recovery progress, and ensuring compliance with treatment protocols.

Moreover, AI-driven solutions can customize patient engagement by adapting information and interventions to the specific needs of each patient. By evaluating factors such as medical history, lifestyle, and preferences, AI can provide tailored health education and recommendations, thereby improving patient adherence and satisfaction [32].

The incorporation of AI in healthcare enhances traditional patient engagement methods and creates new opportunities for providing high-quality, personalized, and efficient care [19]. As AI technologies advance, their utilization in patient engagement is anticipated to become increasingly sophisticated, enhancing health outcomes and revolutionizing the patient experience within intelligent health systems. This is summarized in Table 10.1.

TABLE 10.1 Contrast of Traditional vs. AI-Powered Patient Engagement Methods

Aspect	Traditional Methods	AI-Powered Methods
Scalability	Limited	High
Personalization	Generalized	Highly personalized
Accessibility	Limited to office hours and physical presence	24/7 accessibility
Continuity of Engagement	Episodic	Continuous
Resource Utilization	High	Optimized
Cost-Effectiveness	Moderate to high	High
Patient Satisfaction	Variable	Generally higher

10.3 AI-DRIVEN SOLUTIONS FOR PATIENT ENGAGEMENT

AI-driven solutions for patient engagement encompass personalized health education customized to individual requirements and preferences. AI-driven chatbots and virtual assistants offer 24/7 support and instantaneous replies to patient enquiries [19]. Predictive analytics enable proactive care by identifying health risks and prompting early interventions [4]. Remote monitoring tools, integrated with telehealth, ensure continuous engagement and timely feedback [5]. These solutions enhance patient satisfaction, improve health outcomes, and optimize healthcare resource utilization.

10.3.1 AI-Powered Chatbots and Virtual Assistants

AI-powered chatbots and virtual assistants facilitate continuous engagement, delivering prompt responses to enquiries, appointment scheduling, medication reminders, and fundamental health information [3]. These tools are engineered to engage seamlessly with patients, improving their experience and guaranteeing support whenever required. AI chatbots employ NLP to comprehend and address patient enquiries. They offer comprehensive information regarding medications, symptoms, and treatment alternatives, rendering them an invaluable resource for patients in search of prompt answers [33]. A patient may enquire with a chatbot regarding the side effects of a prescribed medication and receive an immediate, precise response, alleviating anxiety and uncertainty.

Furthermore, virtual assistants can be integrated with electronic health records (EHRs) to deliver personalized recommendations based on a patient's medical history, thereby ensuring relevance and precision [34]. For example, if a patient has a history of hypertension, a virtual assistant can provide customized guidance on managing their condition, including suggesting particular lifestyle modifications or notifying them of impending check-ups. This integration enhances care personalization and ensures that the advice provided is grounded in a thorough understanding of the patient's health status. AI-powered chatbots and virtual assistants enhance patient engagement and satisfaction by offering 24/7 access to information and support, eliminating the need for patients to wait for office hours to resolve their concerns [35].

10.3.2 Predictive Analytics for Proactive Engagement

Predictive analytics utilize patient data to anticipate health risks and identify individuals susceptible to non-adherence to treatment regimens. This proactive strategy allows healthcare providers to intervene promptly, delivering customized support and resources to patients requiring assistance, thereby enhancing adherence and health outcomes [4]. Predictive models can identify patients with chronic conditions at risk of exacerbations. By examining patterns in health data, including variations in vital signs or discrepancies in medication adherence, these models can generate alerts for healthcare providers to intervene proactively, potentially averting hospitalizations [17].

This degree of proactive care can markedly improve patient engagement and overall health outcomes. A patient with chronic obstructive pulmonary disease (COPD) may be observed for alterations in their respiratory patterns. Should the predictive model identify early indicators of a potential exacerbation, it may alert a healthcare provider to contact the patient with tailored instructions or modifications to their treatment regimen, thus averting a more serious health crisis [36]. This timely intervention not only helps in managing the patient's condition more effectively but also reassures the patient that their health is being closely monitored, fostering a sense of continuous care and engagement.

10.3.3 Personalized Health Education

AI algorithms analyze individual patient data to deliver customized health education materials. By tailoring content to the specific needs, preferences, and conditions of each patient, AI enhances the relevance and effectiveness of educational efforts, empowering patients to make informed health decisions [18]. Personalized education may manifest in various formats, such as tailored videos, articles, and interactive modules. A diabetic patient may receive customized dietary guidance and exercise suggestions tailored to their lifestyle and health metrics. This focused strategy guarantees that patients obtain the information most relevant to their health, enhancing their engagement and compliance with treatment regimens [37].

Additionally, AI-driven platforms can adapt educational content over time as they learn more about the patient's behavior and preferences. If a patient frequently searches for information on managing stress, the platform might prioritize providing resources on mental health and stress management techniques. This adaptive personalization maintains the patient's engagement and motivation to adhere to their treatment regimen. Moreover, interactive educational tools can evaluate patient understanding via quizzes or feedback systems, enabling healthcare providers to pinpoint areas requiring further assistance. AI-driven personalized education provides pertinent and comprehensible health information, enabling patients to actively manage their health, resulting in improved health outcomes and enhanced patient satisfaction [38].

10.3.4 Remote Monitoring and Telehealth Integration

AI-powered remote monitoring instruments assess vital signs, symptoms, and medication compliance in real-time. When combined with telehealth services, these tools facilitate ongoing patient engagement and prompt interventions, especially for chronic disease management [5, 6]. This integration guarantees that patients maintain connectivity with their healthcare providers, even remotely. Remote monitoring devices, including wearable sensors and smart home technology, gather continuous health data. AI algorithms examine this data to identify anomalies and trends, notifying healthcare providers of potential issues [16, 31]. This ongoing surveillance facilitates prompt intervention, diminishing the necessity for emergency treatment and improving patient outcomes.

A patient with cardiovascular disease may utilize a device that tracks their heart rate and activity levels. Should the AI system identify anomalies, such as an abrupt elevation in heart rate absent of related physical exertion, it can autonomously alert both the healthcare provider and the patient.

The provider can utilize telehealth platforms to perform a virtual consultation, evaluate the patient's condition, and modify their treatment plan as needed. The seamless integration of remote monitoring and telehealth guarantees timely and effective management of chronic conditions while offering patients reassurance through continuous health monitoring [13].

Additionally, remote monitoring tools can assess medication adherence by prompting patients to take

TABLE 10.2 Key AI Technologies Used in Patient Engagement

AI Technology	Application in Patient Engagement	Benefits
Natural Language Processing (NLP)	Chatbots, virtual assistants	Enhanced interaction quality, instant responses
Machine Learning (ML)	Predictive analytics, Personalized health education	Accurate predictions, tailored recommendations
Computer Vision	Remote monitoring, diagnostic support	Automated data analysis, real-time health monitoring
Robotics	Assistive care, automated physical therapy	Improved patient support, consistent care delivery

their medications and recording their compliance. This information may be disseminated to healthcare providers, enabling them to monitor patients who are noncompliant with their prescribed regimens. By promptly addressing non-adherence, healthcare providers can avert complications and enhance patient outcomes. The incorporation of AI-powered remote monitoring and telehealth services signifies a substantial progression in patient engagement, providing ongoing, tailored, and proactive care that adjusts to the changing requirements of patients, as outlined in Table 10.2.

10.4 BENEFITS OF AI-DRIVEN PATIENT ENGAGEMENT IN SMART HEALTH SYSTEMS

AI-driven patient engagement in smart health systems offers numerous benefits, including enhanced personalization and tailored care, leading to higher patient satisfaction [6]. It provides scalability and accessibility, ensuring consistent support for large and diverse populations. Continuous and proactive engagement improves health outcomes by enabling timely interventions. Additionally, AI reduces healthcare costs and optimizes resource utilization by automating routine tasks and enhancing efficiency [39].

10.4.1 Enhanced Personalization and Patient Experience

AI-driven solutions provide exceptional personalization, customizing interactions and interventions to the distinct circumstances of each patient [40]. This degree of customization improves patient satisfaction and engagement, as patients perceive themselves as understood and supported. AI systems evaluate numerous data points, encompassing medical history, lifestyle factors, and behavioral patterns, to formulate personalized care plans [41]. An AI platform may evaluate a patient's historical treatment responses and propose adjustments to enhance effectiveness. This meticulous approach guarantees that patients obtain the most suitable and efficacious treatments for their particular ailments.

Personalization encompasses more than merely educational content. AI can tailor communication styles and engagement strategies according to patient preferences and behaviors. For instance, certain patients may prefer to receive information through text messages, whereas others might favor email or application notifications. By accommodating these preferences, AI improves the overall patient experience. Furthermore, AI can modify the frequency and tone of communications according to patient feedback, guaranteeing that interactions are pertinent and captivating. A patient who regularly enquires about exercise regimens may receive tailored fitness advice and motivational communications to promote commitment to their physical activity objectives. This dynamic personalization cultivates continuous support and engagement, resulting in increased patient satisfaction and improved health outcomes.

10.4.2 Scalability and Accessibility

AI tools can simultaneously engage extensive patient populations, rendering them highly scalable. They also improve accessibility, offering assistance to patients in remote or underserved regions, thus diminishing disparities in healthcare access. Scalability is crucial for managing extensive patient populations and guaranteeing that all individuals obtain the requisite support. AI-driven solutions can concurrently manage thousands of patient interactions, ensuring consistent and high-quality engagement without burdening healthcare providers.

During a public health crisis, AI can rapidly and efficiently disseminate information to a vast audience. Chatbots can address enquiries regarding symptoms, preventive strategies, and treatment alternatives, offering prompt assistance and alleviating the strain on healthcare hotlines and emergency services [3]. This capability is particularly advantageous in rural or underserved regions where healthcare resources

TABLE 10.4 Difficulties and Considerations Regarding AI-Driven Solutions Implementation

Challenge	Description	Mitigation Strategy
Privacy and Security in Data	Dealing with private patient records	Implementing robust encryption and security measures
Ethical Concerns	Avoiding biases and ensuring equitable care	Using diverse datasets, regular fairness audits
System Integration	Seamlessly integrating with existing healthcare systems	Investing in interoperable systems and integration strategies
Acceptance by Users	Building trust and ensuring usability	Educating and training patients and providers

experience with AI tools, and highlight how these solutions can enhance, rather than replace, human interaction. Demonstrating successful case studies and sharing positive patient and provider experiences can also foster acceptance and trust. Real-world examples of how AI-driven solutions have improved patient outcomes and streamlined healthcare processes can help alleviate fears and highlight the potential benefits of these technologies.

Furthermore, including providers and patients in the process of development and execution will help them to feel more responsible for the technology and trustworthy in it. Healthcare companies can inspire the acceptance of AI-driven solutions and guarantee their successful integration into patient care by addressing these issues and laying a strong basis of trust and transparency.

Table 10.4 emphasizes key components in the problems and considerations in using AI-driven solutions.

10.6 FUTURE DIRECTIONS

Advanced machine learning algorithms will be included into future directions in AI-driven patient involvement to further personalize treatment and predict health issues with more accuracy [61]. Enhanced interoperability between AI tools and existing healthcare systems will streamline workflows and improve data sharing. Innovations in wearable technology and Internet of Things (IoT) devices will enable more comprehensive remote monitoring. Ethical AI development will focus on reducing biases and ensuring equitable care. Overall, these advancements will continue to enhance patient engagement, improve health outcomes, and optimize resource use in smart health systems.

10.6.1 Advancements in AI Technologies

Deep learning and natural language understanding among other continuous developments in AI technologies will improve the capacity of AI-driven patient involvement solutions [62]. More accurate forecasts, better personalizing, and more straightforward patient interactions resulting from these developments will help to define them.

NLP developments, for instance, allow AI chatbots to interact with patients in more natural and context-aware manner, so enhancing the general interaction quality. Analogous developments in machine learning can improve predictive analytics, so improving risk assessments and tailored treatments.

10.6.2 Integration with Emerging Technologies

The convergence of AI with other developing technologies including wearable devices, the IoT, and blockchain will open fresh chances for patient involvement and [69]. Wearable devices can continuously monitor health indicators and generate real-time data, which AI can analyze to deliver personalized recommendations and targeted interventions.

By means of a distributed and tamper-proof method of storing medical records, blockchain technology can improve patient privacy and data security. Blockchain combined with AI guarantees transparent and safe data handling, so addressing many of the privacy and security issues related to solutions driven by artificial intelligence [69].

10.6.3 Collaborative and Co-design Approaches

Effective application of AI-driven patient engagement solutions will call for patient, healthcare provider, and technological developer cooperation. By co-designing these solutions with feedback from all stakeholders, one can guarantee they satisfy consumer needs and preferences, so increasing acceptance and efficacy [27].

Including patients in the design process will help to reveal their needs and preferences, so guaranteeing that AI-driven solutions are relevant and user-friendly. Likewise, including healthcare professionals guarantees that these instruments are useful and easily included into clinical procedures as shown in Table 10.5.

TABLE 10.5 Future Directions for AI in Patient Engagement

Future Direction	Description	Potential Impact
Advancements in AI Technologies	Improved NLP, machine learning, and deep learning	More accurate predictions, better patient interactions
Integration with Emerging Technologies	Combining AI with IoT, wearables, and blockchain	Enhanced data security, real-time health monitoring
Collaborative Approaches	Co-designing solutions with input from all stakeholders	Higher acceptance and effectiveness of AI solutions

10.7 CONCLUSIONS

Solutions driven by AI have great power to transform patient involvement inside smart health systems, so changing the way healthcare is experienced and provided. Healthcare providers can produce more customized and responsive interactions going beyond conventional models of treatment by leveraging advanced AI technologies. AI-enabled personalized engagement lets one better grasp the unique needs, preferences, and behaviors of every patient, so guiding the delivery of tailored interventions, reminders, and support. These systems can guarantee that patients get constant, real-time feedback and direction, so enhancing their adherence to treatment plans and hence their general health outcomes.

Furthermore, AI can improve scalability, allowing healthcare systems to interact and manage an increasing patient count without sacrificing efficiency or quality. AI-powered virtual assistants and chatbots, for example, can schedule appointments, respond instantly to patient questions, and offer health education, so freeing up human resources for more challenging clinical work. Predictive analytics helps to further a preventive and patient-centered approach of treatment by identifying at-risk patients and proactively addressing possible health issues.

Still, the effective acceptance of these creative ideas depends on tackling several important problems. Data privacy and security must first be given top priority since patients have to believe that their private information is handled sensibly. To prevent aggravating already existing healthcare inequalities, ethical issues including reducing algorithmic bias and guaranteeing fair access to AI-driven tools must be taken under great attention. Maximizing the value of AI technologies without disturbance depends also on their seamless integration into current healthcare systems and processes. Overcoming opposition to change also depends on encouraging acceptance among patients and healthcare professionals by means of education and open communication.

As AI develops, it will become increasingly important in determining patient involvement going forward. As machine learning, NLP, and predictive modeling continue to advance, AI systems will grow ever more complex and adept at providing hyper-personalized experiences. These developments could raise patient satisfaction, increase the effectiveness of treatments, and help to support more wise use of resources. In the end, AI will improve patient involvement and help to bring about a paradigm change toward more effective, efficient, and really patient-centered healthcare delivery.

10.8 SUMMARY

AI is transforming patient engagement within healthcare, offering personalized, scalable, and responsive solutions that improve patient satisfaction and outcomes. AI lets healthcare providers customize treatments and simplify care delivery by using technologies including chatbots, predictive analytics, and machine learning. The role of AI is particularly significant in predictive care, ensuring early detection of health risks, and in automating routine tasks to enhance efficiency.

However, successful implementation requires addressing challenges such as data privacy, ethical considerations, integration with existing systems, and fostering acceptance among stakeholders. Overcoming these hurdles is crucial to leveraging AI's potential fully. As advancements continue, AI will further redefine patient-centered healthcare, ensuring more dynamic and effective engagement tailored to individual needs while optimizing healthcare system resources.

BIBLIOGRAPHY

1. Vicente, M. A., Fernández, C., Guilabert, M., Carrillo, I., Martín-Delgado, J., Mira, J. J., & Prometeo Working Group (2022). Patient engagement using telemedicine in primary care during COVID-19 pandemic: A trial study. *International Journal of Environmental Research and Public Health*, *19*(22), 14682.
2. Byrne, M. D. (2020). Telehealth and the COVID-19 pandemic. *Journal of Perianesthesia Nursing*, *35*(5), 548–551.

3. Kramer, A. A. (2019). A continuously updated predictive analytics model for the timely detection of critically ill patients with a high risk of mortality. *Medical Research Archives*, *7*(1), 1–15.
4. Dixon, D., Sattar, H., Moros, N., et al. (2024). Unveiling the influence of AI predictive analytics on patient outcomes: A comprehensive narrative review. *Cureus*, *16*(5), e59954.
5. Vikas, V., Saisri, G., Meghana, T. S., Harshini, A. S., & Kaveri, G. (2023). Web security audit and penetration testing: Identifying vulnerabilities and strengthening website security. *International Journal for Research in Applied Science and Engineering Technology*, *11*(7), 794–805.
6. Williams, G., & Fahy, N. (2019). Building and maintaining public trust to support the secondary use of personal health data. *Eurohealth*, *25*(2), 7–10.
7. Carrasco Ramírez, J. G. (2024). AI in healthcare: Revolutionizing patient care with predictive analytics and decision support systems. *Journal of Artificial Intelligence and General Science (JAIGS)*.
8. Ghane, G., Ghiyasvandian, S., Chekeni, A. M., Karimi, R. (2024). Revolutionizing nursing education and care: The role of artificial intelligence in nursing. *Nurse & Author Editor*, e12057. https://doi.org/10.1111/nae2.12057
9. Norori, N., Hu, Q., Aellen, F. M., Faraci, F. D., & Tzovara, A. (2021). Addressing bias in big data and AI for healthcare: A call for open science. *Patterns*, *2*(10), 100347.
10. Babu, A., & Boddu, S. B. (2024). BERT-based medical chatbot: Enhancing healthcare communication through natural language understanding. *Exploratory Research in Clinical and Social Pharmacy*, *13*, 100419. https://doi.org/10.1016/j.rcsop.2024.100419
11. Donia, J., & Shaw, J. A. (2021). Co-design and ethical artificial intelligence for health: An agenda for critical research and practice. *Big Data & Society*, *8*(2).
12 Clark, M., & Bailey, S. (2024, January 22). Chatbots in healthcare: Connecting patients to information. *Canadian Journal of Health Technologies*, *4*(1), 2024. https://doi.org/10.51731/cjht.2024.818.Full-text:https://canjhealthtechnol.ca/index.php/cjht/article/view/EH0122
13. Zou, J., & Schiebinger, L. (2021). Ensuring that biomedical AI benefits diverse populations. *EBioMedicine*, *67*, 103358.
14. Olyanasab, A., & Annabestani, M. (2024). Leveraging machine learning for personalized wearable biomedical devices: A review. *Journal of Personalized Medicine*, *14*(2), 203.
15. Vora, L. K., Gholap, A. D., Jetha, K., Thakur, R. R. S., Solanki, H. K., & Chavda, V. P. (2023). Artificial intelligence in pharmaceutical technology and drug delivery design. *Pharmaceutics*, *15*(7), 1916.
16. Bohr, A., & Memarzadeh, K. (2020). The rise of artificial intelligence in healthcare applications. In A. Bohr & K. Memarzadeh (Eds.), *Artificial Intelligence in Healthcare* (pp. 25–60).
17. Nazer, L. H., Zatarah, R., Waldrip, S., Ke, J. X. C., Moukheiber, M., & Khanna, A. K. (2023). Bias in artificial intelligence algorithms and recommendations for mitigation. *PLOS Digital Health*, *2*(6), e0000278.
18. Francisca Chibugo Udegbe, O., Ebulue, O. R., Ebulue, C. C., & Ekesiobi, C. S. (2024). AI's impact on personalized medicine: Tailoring treatments for improved health outcomes. *Engineering Science & Technology Journal*, *5*(4), 1386–1394.
19. Robinson, C. L., D'souza, R. S., Yazdi, C., Diejomaoh, E. M., Schatman, M. E., Emerick, T. D., & Orhurhu, V. (2024). Reviewing the potential role of artificial intelligence in delivering personalized and interactive pain medicine education for chronic pain patients. *Journal of Pain Research*, *17*, 923–929.
20. Puaschunder, J. M., & Feierabend, D. (2019). Artificial intelligence in the healthcare sector. *CompSciRN: Other Machine Learning (Topic)*.
21. Walker, R. C., Tong, A., Howard, K., & Palmer, S. C. (2019). Patient expectations and experiences of remote monitoring for chronic diseases: Systematic review and thematic synthesis of qualitative studies. *International Journal of Medical Informatics*, *124*, 78–85.
22. Chand, N., Bhattacharyya, S., & Sarkar, A. (2021). A novel encryption technique to protect patient health information electronically using Playfair Cipher 15 by 14 matrix. In *Advances in Medical Physics and Healthcare Engineering* (pp. 423–431). Singapore: Springer.
23. Almasri, I. A. (2024). Almasri I. The Power of Artificial Intelligence for Improved Patient Outcomes, Ethical Practices and Overcoming Challenges. IgMin Res. 2024 Jul;2(7):585–588.
24. Triantafillou, G., & Minas, M. Y. (2020). Future of the artificial intelligence in daily health applications. *The European Journal of Social & Behavioural Sciences*, *29*(3), 145–153. https://doi.org/10.15405/ejsbs.278
25. Ferrick, A. M., Raj, S. R., Deneke, T., et al. (2023). 2023 HRS/EHRA/APHRS/LAHRS expert consensus statement on practical management of the remote device clinic. *Journal of Arrhythmia*, *39*(3), 250–302.
26. Badawy, M. M. (2023). Integrating artificial intelligence and big data into smart healthcare systems: A comprehensive review of current practices and future directions. *Artificial Intelligence Evolution*, *4*(2), 133–135.
27. Banerjee, S., Alsop, P., Jones, L., & Cardinal, R. N. (2022). Patient and public involvement to build trust in artificial intelligence: A framework, tools, and case studies. *Patterns (N Y)*, *3*(6),100506. https://doi.org/10.1016/j.patter.2022.100506
28. Danso, G., Daniel, K. K., & Quansah, M. (2024, January–February). Assessing security mechanisms for protecting patient health information at the Berekum Holy Family Hospital. *International Journal for Multidisciplinary Research*, *6*(1).

29. Ruan, S. J. (2019). Intelligent systems for smart healthcare: Leveraging information for better well-being. *IEEE Consumer Electronics Magazine*, *8*(2), 71–71.
30. Porbundarwala, T., Lalji. S., Patil, H., & Dongardive, J. (2023). Impact of AI in personalized health monitoring using wearable devices. *IJFMR*, *5*(5).
31. Jadczyk, T., Wojakowski, W., Tendera, M., Henry, T. D., Egnaczyk, G., & Shreenivas, S. (2021). Artificial intelligence can improve patient management at the time of a pandemic: The role of voice technology. *Journal of Medical Internet Research*, *23*(5), e22959.
32. Perri-Moore, S., Kapsandoy, S., Doyon, K., Hill, B., Archer, M., Shane-McWhorter, L., Bray, B. E., & Zeng-Treitler, Q. (2016). Automated alerts and reminders targeting patients: A review of the literature. *Patient Education and Counseling*, *99*(6), 953–959.
33. Xie, Y., Lu, L., Gao, F., et al. (2021). Integration of artificial intelligence, blockchain, and wearable technology for chronic disease management: A new paradigm in smart healthcare. *Current Medical Science*, *41*, 1123–1133.
34. Abul-Husn, N. S., & Kenny, E. E. (2019). Personalized medicine and the power of electronic health records. *Cell*, *177*, 58–69.
35. Dandotiya, N., Khatri, P., & Dandotiya, A. S. (2020). Digital auditing: A technique to ensure security. International Journal of Scientific Research in Computer Science, Engineering and Information Technology, 6(3), 371–379
36. Maleki Varnosfaderani, S., & Forouzanfar, M. (2024). The role of AI in hospitals and clinics: Transforming healthcare in the 21st century. *Bioengineering*, *11*(4), 337.
37. Alfano, M., Kellett, J., Lenzitti, B., & Helfert, M. (2023). *Intelligent provision of tailored, easily understood, and trusted health information for patient empowerment.* In F. Cabitza, A. Fred, & H. Gamboa (Eds.), *Proceedings of the 16th International Joint Conference on Biomedical Engineering Systems and Technologies (BIOSTEC 2023) – HEALTHINF* (Vol. 5, pp. 384–391). Setúbal, Portugal: SCITEPRESS. ISBN 978-989-758-631-6; ISSN 2184-4305.
38. Udegbe, F. C., Ebulue, O. R., Ebulue, C. C., & Ekesiobi, C. S. (2024). The role of artificial intelligence in healthcare: A systematic review of applications and challenges. *International Medical Science Research Journal*, *4*(4), 500–508.
39. Srinivasaiah, B. (2024). Transformative wearables: How AI and ML are shaping healthcare innovations. *International Journal of Science and Research (IJSR)*, *13*(4), 152–156.
40. Harishbhai Tilala, M., Kumar Chenchala, P., Choppadandi, A., Kaur, J., Naguri, S., Saoji, R., & Devaguptapu, B. (2024). Ethical considerations in the use of artificial intelligence and machine learning in healthcare: A comprehensive review. *Cureus*, *16*(6), e62443.
41. Sundar, R., Choudhury, Z., Chiranjivi, M., Parasa, G., Ravuri, P., Sivaram, M., Subramanian, B., Muppavaram, K., & Challa, V. L. (2024). Future directions of artificial intelligence integration: Managing strategies and opportunities. *Journal of Intelligent & Fuzzy Systems*, *46*, 1–14.
42. Chen, Y., Clayton, E. W., Novak, L. L., Anders, S., & Malin, B. (2023). Human- centered design to address biases in artificial intelligence. *Journal of Medical Internet Research*, *25*, e43251.
43. Rathore, Y. K., Mishra Chaturvedi, V., Sujay Madhukar, K., Karwande, V. S., Rokade, A. H., & Nagargoje, Y. (2023). Patient engagement and satisfaction in AI-enhanced healthcare management. In *2023 International Conference on Artificial Intelligence for Innovations in Healthcare Industries (ICAIIHI).*
44. Khan, B., Fatima, H., Qureshi, A., Kumar, S., Hanan, A., Hussain, J., & Abdullah, S. (2023, February). Drawbacks of artificial intelligence and their potential solutions in the healthcare sector. *Biomed Mater Devices*, *8*, 1–8. Epub ahead of print. PMID: 36785697; PMCID: PMC9908503.
45. Schwartz, K. A., & Britton, B. (2011). Use of telehealth to improve chronic disease management. *North Carolina Medical Journal*, *72*(3), 216–218.
46. Bekbolatova, M., Mayer, J., Ong, C. W., & Toma, M. (2024). Transformative potential of AI in healthcare: Definitions, applications, and navigating the ethical landscape and public perspectives. *Healthcare*, *12*(2), 125.
47. Mocydlarz-Adamcewicz, M. (2021). Effective communication between hospital staff and patients in compliance with personal data protection regulations. *Reports of Practical Oncology and Radiotherapy*, *26*(6), 833–838.
48. Chandwani, A., Shrivastava, K., Sadashiv, S., Saha, I., & Mishra, P., P. (2023, April). Virtual assistant for appointment booking. In *2023 IEEE 8th International Conference for Convergence in Technology (I2CT)* (pp. 1–5). Piscataway: IEEE.
49. Mennella, C., Maniscalco, U., De Pietro, G., & Esposito, M. (2024). Ethical and regulatory challenges of AI technologies in healthcare: A narrative review. *Heliyon*, *10*(4), e26297.
50. Singh, K. (2023). Artificial intelligence & cloud in healthcare: Analyzing challenges and solutions within regulatory boundaries. *SSRG International Journal of Computer Science and Engineering*, *10*(9), 1–9.
51. Nathaniel Kumar, S. P., & Mangam, T. V. (2024). AI-driven natural language processing in healthcare: Transforming patient-provider communication. *Indian Journal of Pharmacy Practice*, *17*(1), 21–26.
52. Michard, F., & Teboul, J. (2019). Predictive analytics: Beyond the buzz. *Annals of Intensive Care*, *9*(1), 1–10.
53. Chang, A. (2020). The role of artificial intelligence in digital health. In S. Wulfovich & A. Meyers (Eds.), *Digital Health Entrepreneurship* (pp. 25–40).

Health Informatics. Cham: Springer. http://dx.doi.org/10.1007/978-3-030-12719-0_7
54. DeCamp, M., & Lindvall, C. (2023). Mitigating bias in AI at the point of care. *Science, 381*(6534), 150–152.
55. Dave, M., & Patel, N. (2023). Artificial intelligence in healthcare and education. *British Dental Journal, 234*(5), 761–764.
56. Wubineh, B. Z., Deriba, F. G., & Woldeyohannis, M. M. (2024). Exploring the opportunities and challenges of implementing artificial intelligence in healthcare: A systematic literature review. *Urologic Oncology: Seminars and Original Investigations, 42*(3), 48–56.
57. Parul, B., & Deep, M. D. (2024). Revolutionizing healthcare platforms: The impact of AI on patient engagement and treatment efficacy. *International Journal of Science and Research (IJSR), 13*(2), 273–280.
58. Pahune, S. (2023). How does AI help in rural development in the healthcare domain: A short survey. *International Journal for Research in Applied Science and Engineering Technology, 11*(6), 4184–4191.
59. Comito, C., Falcone, D., & Forestiero, A. (2020). Current trends and practices in smart health monitoring and clinical decision support. In *2020 IEEE International Conference on Bioinformatics and Biomedicine (BIBM)* (pp. 2577–2584). Piscataway, NJ: IEEE
60. Hernandez, I., & Zhang, Y. (2017). Using predictive analytics and big data to optimize pharmaceutical outcomes. *American Journal of Health-System Pharmacy, 74*(18), 1494–1500.
61. Chavez-Cano, A. M. (2023, December). Artificial intelligence applied to telemedicine: Opportunities for healthcare delivery in rural areas. *LatIA, 1*(3), Article 3. https://doi.org/10.62486/latia20233
62. Oyeniyi, J. (2024). The role of AI and mobile apps in patient-centric healthcare delivery. *World Journal of Advanced Research and Reviews, 22*(1), 1897–1907.

CHAPTER 11

Artificial Intelligence in Personalized Medicine

Mohammad Al Diab Al Azzawi

11.1 INTRODUCTION

Personalized medicine is replacing the traditional approach of conventional medical treatment and transitioning toward a more tailored and individualized method [1]. This revolution laid down the assertion that every individual is different, in terms of genetic endowment, environment, and behavioral predispositions to health and disease [1]. Personalized medicine tries to adapt healthcare treatments to these individual features, hence boosting therapeutic efficacy while limiting potential adverse events [2]. Central to this change is the quick expansion of artificial intelligence (AI), which has emerged as a crucial element in hastening the implementation of personalized medicine [3]. AI's capacity to analyze vast volumes of data, identify precise patterns, and construct predictive models is dramatically transforming the way we approach medical care [3]. AI provides for a greater knowledge of each patient's unique health profile by merging data from multiple sources, such as genetic information, electronic health records (EHRs), and medical imaging [4]. AI technology is currently integrated into many components of tailored medicine. Machine learning (ML) and deep learning (DL) algorithms, for example, increase sickness risk projections by analyzing complex information and discovering subtle correlations that earlier techniques may ignore [4]. Additionally, these technologies play a significant part in generating tailored treatment regimens that adjust drugs to the individual needs of each patient. Furthermore, AI impacts medication development by expediting the identification of promising drug candidates and predicting patient reactions to innovative medications [5]. This not only speeds the drug development process, but it also raises the probability that new medications may be useful for particular patient populations [5]. AI is shifting healthcare from reactive to preventive care, models detect pre-symptomatic risk signals, enabling earlier evaluation and treatment that may avoid progression and improve outcomes [6]. Furthermore, AI-powered solutions, such as wearable health devices, enable patients to actively manage their health by offering real-time feedback and personalized recommendations. As AI improves, it provides both potential benefits and challenges [7].

This chapter aims to review the use of AI within the field of personalized medicine and the ways in which AI technologies help to augment diagnostic precision, treatment strategies, and patients' outcomes due to the customization of medical processes.

11.2 AI TECHNOLOGIES IN PERSONALIZED MEDICINE

AI technologies in personalized medicine have many aspects, for example, ML and DL, natural language processing (NLP), clinical data extraction, literature review and evidence synthesis, patient-provider communication, genomics and AI, and imaging and AI.

11.2.1 Machine Learning and Deep Learning

Leading AI technologies in personalized medicine include ML and DL. Without overtly programming, these technologies may learn from data and provide predictions based on that learning. In healthcare especially, where the capacity to examine vast amounts of

DOI: 10.1201/9781003533351-11

data and identify trends may result in more accurate diagnosis and tailored treatment programs, this capacity is quite important [8]. In personalized medicine, ML algorithms examine a range of data sources including genetic information, EHRs, imaging data, even patient-reported outcomes. ML algorithms analyze different data sources such as genetic/omics data, EHRs, and medical imaging to identify relationships between disease risk and genetic markers that may not be visible with conventional analytical methods [9]. ML algorithms, for instance, can use a patient's genetic composition, family history, and lifestyle choices to forecast their chance of acquiring a given illness [10]. Early intervention made possible by this predictive capacity can be rather important for controlling chronic diseases or stopping the course of a sickness [10]. Conversely, DL a subfield of ML involves neural networks with several layers capable of modeling intricate relationships inside data [11]. Processing and understanding massive, unstructured datasets such as medical imaging or genetic sequences where DL is especially useful is rather difficult [12]. DL models may examine medical images in the framework of personalized medicine to identify early illness indicators, including minute changes in tissue structure that might point to the beginning of cancer [13]. Analyzing genomic data using these models can also help to uncover genetic variations linked to certain illnesses or project patient response to various therapies [10]. Personalized medicine's application of ML and DL transcends diagnosis and therapy planning, by locating possible medication candidates more likely to be beneficial for particular patient groupings, these tools are also helping to maximize drug development [14]. This strategy not only speeds up the medication development process but also raises the possibility that new medicines may be safe and successful for a greater spectrum of patients.

11.2.2 Natural Language Processing

NLP is a subfield of AI allowing computers to comprehend and analyze human discourse. NLP is enabling insightful analysis of unstructured data sources including clinical notes, patient records, and medical literature in the context of healthcare. These unstructured data sources contain rich information that when analyzed for example with NLP can enhance healthcare decision-making and support personalized care as well as public-health practice [15, 16].

11.2.2.1 Clinical Data Extraction

One of the main uses of NLP in customized medicine is the relevant information extraction from patient records and clinical notes. These notes frequently provide thorough explanations of a patient's symptoms, treatment background, and reaction to past therapy. Usually, this material is disorganized and difficult to access for study. By use of structured data derived from these records, NLP algorithms may discover important information pieces and translate them into guide for treatment decisions [17].

11.2.2.2 Literature Review and Evidence Synthesis

Another crucial use of NLP in customized medicine is the evidence synthesis from medical literature. Unprecedented rates of medical research publication make it challenging for healthcare professionals to remain current with the most recent results. NLP can rapidly locate pertinent research, summarize their results, and underline important insights, thereby automating the process of literary review. In customized medicine, where treatment decisions frequently depend on the most recent research, this capacity is very important [18].

11.2.2.3 Patient-Provider Communication

NLP is also helping to enhance patient-provider contact by means of its applications. AI-powered chatbots, for instance, can utilize NLP to grasp patient questions and respond personally depending on the patient's medical background and present state of health. Additionally, useful patient-reported outcomes may be gathered by these chatbots, which can be tracked therapy efficacy and modified treatment plans as necessary. Using NLP can help healthcare professionals to better grasp the particular situation of the patient, so enabling more individualized and successful treatment [18].

11.2.3 Genomics and AI

Personalized medicine has a revolutionary development in integration of AI with genetics [19]. Studying an individual's whole DNA allows one to get important understanding of hereditary elements influencing health and illness [20]. Analyzing the huge and complicated genetic data is difficult without AI, though. By spotting variations associated to illnesses like breast cancer, Alzheimer's, or diabetes, which helps to enable early preventative actions, AI systems shine in genetic risk prediction [21]. In pharmacogenomics, AI improves pharmacogenomics by using genetic variability to forecast individual medication reactions, therefore improving treatment and reducing negative effects. AI-driven ML algorithms utilize genomic and clinical data to analyze massive datasets, enabling the identification of crucial gene-drug interactions and possible biomarkers. This facilitates individualized treatment programs, enabling doctors to choose the most efficacious drugs and doses customized to the specific genetic makeup of each patient [22].

11.2.4 Imaging and AI

Modern healthcare is based on medical imaging, which offers vital understanding of the structure and function of the body [23]. Medical image interpretation has always depended on radiologists' and other professionals' experience. But in image analysis, AI is increasingly supplementing and, in some circumstances, surpassing human ability, resulting in earlier and more precise diagnosis [24]. AI algorithms are being applied in radiology to examine X-ray, MRIs, CT scan, and other imaging modalities. Often finding early indicators of disease that could be missed by human observers, these algorithms can detect anomalies with great degree of accuracy. On chest CT images, for instance, AI has been applied to identify tiny lung nodules that could point to early-stage lung cancer [18]. Early disease detection allows AI to help guarantee that patients get therapy when it is most likely to be beneficial [25]. Beyond conventional imaging, sophisticated methods include functional imaging and molecular imaging are also using AI [27]. AI can improve functional imaging which gauges physiological activity inside the body by offering real-time analysis of how diseases are compromising biological processes [27]. Targeting treatments efficacy in real-time may be tracked via molecular imaging, which views biological events at the molecular level: by means of their analysis of these intricate pictures, AI systems can evaluate the efficacy of a therapy and enable necessary corrections [28]. Treatment planning is being tailored depending on imaging data using AI as well. AI may, for instance, examine imaging data in radiation treatment for cancer to identify the exact size, shape, and position of a tumor, therefore enabling more specific targeting of radiation. This accuracy boosts the treatment's efficacy and lessens harm to nearby healthy tissue [29]. By simulating various treatment situations, AI can also assist doctors in selecting the optimal course of treatment for any particular patient [29]. By means of AI integration with medical imaging, healthcare practitioners are improving their capacity to diagnosis and treat illnesses more precisely, hence improving patient outcomes [30].

11.3 APPLICATIONS OF AI IN PERSONALIZED MEDICINE

Applications of AI in personalized medicine can be used in different ways like in, oncology, cardiovascular disease management, and neurological disorders. A summary is illustrated in Table 11.1.

11.3.1 Precision Oncology

Personalized medicine has led the way in oncology [31]. A complicated situation, cancer affects every patient with a different genetic and molecular profile [32]. This variety means that treatments meant for one patient might not work for another. Precision oncology seeks to personalized cancer treatment to the particular traits of every patient's tumor; AI is helping to make this feasible [29]. Analyzing tumor genetic and molecular profiles using AI helps to pinpoint the particular mutations causing cancer spread [21]. Once these mutations are found, AI can match the tumor profile with the most probable efficient, focused treatments [33]. AI may, for instance, find breast cancer patients who are likely to react to HER2 inhibitors a kind of medication aimed at a particular protein overexpressed in some cancers [34]. This focused strategy reduces adverse effects connected with more general chemotherapy treatments in addition to raising the likelihood of treatment success. Immunotherapy a treatment using the body's immune system to combat cancer is another field where AI is having a major influence [35]. By examining tumor and immune system data, AI can forecast which patients would most likely respond to immunotherapy [35]. Immunotherapy can have major negative effects and is not successful for every patient; hence, this is very crucial. AI can assist to minimize needless treatments by spotting patients who are ideal for immunotherapy, thereby concentrating on those most likely to benefit [35]. Monitoring patients throughout treatment, AI is also collecting data from imaging, blood tests, and other sources to

TABLE 11.1 Applications of AI in Personalized Medicine

Aspect	Traditional Approach	AI-Enhanced Personalized Medicine	Key Benefits of AI
Diagnosis	• Relies on manual interpretation of medical data by healthcare professionals.	• AI analyzes large datasets (e.g., imaging, genetic data) for faster, more accurate diagnoses.	• Improved accuracy and early detection.
Treatment Planning	• Standardized treatment plans based on population averages.	• AI tailors treatment plans to individual patient profiles (genetics, lifestyle, medical history).	• Personalized, targeted therapies.
Cancer Treatment	• Often uses a one-size-fits-all approach to chemotherapy and radiation.	• AI predicts patient response to therapies and adjusts treatment in real-time.	• Avoids unnecessary treatments, reduces side effects.
Cardiovascular Disease Management	• Treatment based on generalized risk factors and clinical guidelines.	• AI predicts individual risk and suggests personalized preventive measures and treatments.	• Preventive care, optimized treatment for chronic conditions.
Neurological Disorders	• Diagnosis and treatment are challenging due to complex disease interactions.	• AI detects early signs through data analysis and personalizes treatment plans.	• Early intervention, tailored therapy for conditions like epilepsy.
Real-Time Monitoring and Adaptation	• Limited real-time data analysis; adjustments are based on periodic assessments.	• AI continuously monitors patient data and suggests real-time adjustments.	• Dynamic treatment adjustments, better patient outcomes.
Data Privacy and Security	• Data is stored in centralized systems, potentially vulnerable to breaches.	• AI systems can incorporate advanced encryption and blockchain for secure data storage.	• Enhanced data protection and patient trust.
Bias and Fairness	• Potential for bias in clinical decisions due to human judgment and nondiverse data.	• AI trained on diverse datasets to reduce bias and improve equity in care.	• Fairer, more equitable healthcare delivery.
Regulatory Oversight	• Traditional medical tools are regulated by well-established frameworks.	• AI tools are still developing regulatory guidelines, requiring ongoing updates.	• Need for adaptive regulation, ensuring safe AI use.
Ethical Considerations	• Ethical concerns managed through standard medical protocols and guidelines.	• AI introduces new ethical challenges, requiring careful consideration of patient-centered care and equity.	• Potential for more ethically sound, patient-focused care.
Future Potential	• Limited by current technologies and traditional methodologies.	• AI promises ongoing innovation, integration with IoT, blockchain, quantum computing.	• Future-ready healthcare with advanced capabilities.

evaluate the effectiveness of a medication. If the treatment fails to produce the intended results, AI can recommend modifications to the treatment schedule, like dose changes or alternative therapy substitution [36]. This real-time adaptation guarantees that patients get the best treatment available.

11.3.2 Cardiovascular Disease Management

Cardiovascular disease (CVD) is the leading cause of mortality worldwide [37], CVDs need a personalized strategy in their management. By greatly improving diagnosis, treatment, and CVD management, AI is reducing patient and healthcare system burden. By examining a mix of genetic, lifestyle, and clinical elements, AI algorithms are being applied more and more to forecast a person's risk of CVDs [38]. For example, AI may analyze data from wearable devices such as fitness trackers and heart rate monitors to find trends suggesting a higher risk of heart disease, enabling early preventive actions such as lifestyle or drug changes [39]. By means of analysis of a patient's medical history, genetic information, and reactions to past therapies, AI also helps to create customized treatment regimens by suggesting the most appropriate drugs and doses for any particular situation. AI-powered monitoring systems also provide continual insights into the health state of patients with chronic CVDs like heart failure, seeing early indicators of deterioration like changes in heart rate or blood pressure, and thus allowing quick treatments [40, 41]. By recommending real-time changes to drug schedules or lifestyle advice based on current data, AI can help to further maximize disease management [7]. By offering more individualized, proactive treatment in the field of CVD management, AI is overall enhancing patient outcomes [40].

11.3.3 Neurological Disorders

Because of the complex interaction of genetic, environmental, and behavioral elements involved in neurological diseases such as Alzheimer's disease, Parkinson's disease, and epilepsy, diagnosis and treatment present unique challenges [42, 43]. Through improved early detection, diagnosis, and tailored treatment plans, AI is significantly overcoming these obstacles. AI's early-stage neurological disease detection is one of its most exciting uses in neurology [44, 45].

AI can detect early symptoms of diseases such as Alzheimer's disease by examining a broad range of data, including genetic information, brain imaging, and cognitive tests, enabling prompt therapies that can halt disease development and improve patient quality of life [46, 47]. AI is also transforming individualized therapy for neurological conditions. For example, AI may examine electroencephalograms (EEGs) in epilepsy to identify seizure trends, therefore enabling customized treatment regimens, including either the investigation of neurostimulation treatments or optimal pharmaceutical doses [48, 49]. In Parkinson's disease, too, AI may use wearable device data to track symptoms and offer real-time therapy recommendation changes, thereby improving patient care [48]. For neurological patients, AI-powered tools are also being included in cognitive-behavioral treatment (CBT), providing personalized therapy sessions depending on specific symptoms and advancement. These instruments not only improve the outcomes of mental health issues, but they also enable continuous therapy monitoring, and therapy modification ensures that treatment remains successful [50]. Generally, the use of AI in neurology entails improving the diagnosis and treatment of neurological diseases, resulting in better patient outcomes and more tailored therapy.

11.4 CHALLENGES AND ETHICAL CONSIDERATIONS

Although AI might revolutionize tailored treatment, its use in healthcare presents major ethical questions and difficulties. Dealing with these issues guarantees that AI technologies are created and applied in ways that give patient safety, equity, and confidence first priority. Personalized medicine incorporation of AI raises significant concerns about data privacy, security, prejudice, justice, regulatory hurdles, and ethical usage. Large datasets, including medical records and genetic data, necessitate robust security measures such as modern encryption methods and safe storage systems such as blockchain to prevent illegal access and breaches. Maintaining patient privacy also requires obtaining informed consent, patients should be clearly informed about how their data will be used and given the opportunity to cancel data sharing if they wish at any time. AI algorithms are only as good as the data they are trained on; hence, if the data is biased, it will help to widen healthcare inequalities. AI models must be trained on varied and representative datasets and designed with openness to guarantee that decision-making procedures are intelligible and challengeable in order to help to offset this. The rapid pace at which AI is developing also poses regulatory challenges, as current systems find it difficult to match emerging technology. Before clearance, regulatory authorities have to make sure AI products go through thorough testing and are under constant observation to preserve safety and efficiency in practical environments. Regardless of socioeconomic level or region, ethically AI should be used to improve patient-centered care, supporting rather than substituting the patient-provider interaction, and guaranteeing fair access to the advantages of AI-driven therapies. Realizing the full potential of AI in customized medicine and protecting patient rights and therefore advancing justice in healthcare depend on addressing these few difficulties.

11.5 FUTURE DIRECTIONS

The future of AI in personalized medicine is promising and has lots of opportunities for advancement. AI technologies will progressively be increasingly included in healthcare as they develop, offering patients more individualized and efficient treatments. Advances in AI algorithms, integration with other technologies, patient empowerment, and multidisciplinary cooperation will fundamentally influence customized medicine going forward. Advanced ML and DL models will enable AI algorithms, which are ever more sophisticated, to examine ever more complicated information from many sources, including genetic, imaging, and lifestyle data. This will allow even more exact forecasts and tailored therapy advice.

38. Almansouri, N. E., Awe, M., Rajavelu, S., Jahnavi, K., Shastry, R., Hasan, A., Hasan, H., Lakkimsetti, M., AlAbbasi, R. K., Gutiérrez, B. C., & Haider, A. (2024). Early diagnosis of cardiovascular diseases in the era of artificial intelligence: An in-depth review. Cureus, 16(3). https://doi.org/10.7759/CUREUS.55869
39. Javaid, A., Zghyer, F., Kim, C., Spaulding, E. M., Isakadze, N., Ding, J., Kargillis, D., Gao, Y., Rahman, F., Brown, D. E., Saria, S., Martin, S. S., Kramer, C. M., Blumenthal, R. S., & Marvel, F. A. (2022). Medicine 2032: The future of cardiovascular disease prevention with machine learning and digital health technology. American Journal of Preventive Cardiology, 12, 100379. https://doi.org/10.1016/J.AJPC.2022.100379
40. Armoundas, A. A., Narayan, S. M., Arnett, D. K., Spector-Bagdady, K., Bennett, D. A., Anthony Celi, L., Friedman, P. A., Gollob, M. H., Hall, J. L., Kwitek, A. E., Lett, E., Menon, B. K., Sheehan, K. A., & Al-Zaiti, S. S. (2024). Use of artificial intelligence in improving outcomes in heart disease: A scientific statement from the. American Heart Association. Circulation, 149(14), E1028–E1050. https://doi.org/10.1161/CIR.0000000000001201
41. Singh, M., Kumar, A., Khanna, N. N., Laird, J. R., Nicolaides, A., Faa, G., Johri, A. M., Mantella, L. E., Fernandes, J. F. E., Teji, J. S., Singh, N., Fouda, M. M., Singh, R., Sharma, A., Kitas, G., Rathore, V., Singh, I. M., Tadepalli, K., Al-Maini, M., … Suri, J. S. (2024). Artificial intelligence for cardiovascular disease risk assessment in personalised framework: A scoping review. EClinicalMedicine, 73, 102660. https://doi.org/10.1016/J.ECLINM.2024.102660
42. Ayeni, E. A., Aldossary, A. M., Ayejoto, D. A., Gbadegesin, L. A., Alshehri, A. A., Alfassam, H. A., Afewerky, H. K., Almughem, F. A., Bello, S. M., & Tawfik, E. A. (2022). Neurodegenerative diseases: Implications of environmental and climatic influences on neurotransmitters and neuronal hormones activities. International Journal of Environmental Research and Public Health, 19(19), 12495. https://doi.org/10.3390/IJERPH191912495
43. Wang, H., Yang, F., Zhang, S., Xin, R., & Sun, Y. (2021). Genetic and environmental factors in Alzheimer's and Parkinson's diseases and promising therapeutic intervention via fecal microbiota transplantation. npj Parkinson's Disease, 7(1). https://doi.org/10.1038/S41531-021-00213-7
44. Kalani, M., & Anjankar, A. (2024). Revolutionizing neurology: The role of artificial intelligence in advancing diagnosis and treatment. Cureus, 16(6). https://doi.org/10.7759/CUREUS.61706
45. Vinny, P. W., Vishnu, V. Y., & Padma Srivastava, M. V. (2021). Artificial intelligence shaping the future of neurology practice. Medical Journal, Armed Forces India, 77(3), 276. https://doi.org/10.1016/J.MJAFI.2021.06.003
46. How AI Can Help Spot Early Risk Factors for Alzheimer's Disease | UC San Francisco. (n.d.). Retrieved August 22, 2024, from https://www.ucsf.edu/news/2024/02/427131/how-ai-can-help-spot-early-risk-factors-alzheimers-disease
47. Vrahatis, A. G., Skolariki, K., Krokidis, M. G., Lazaros, K., Exarchos, T. P., & Vlamos, P. (2023). Revolutionizing the early detection of Alzheimer's disease through non-invasive biomarkers: The role of artificial intelligence and deep learning. Sensors (Basel, Switzerland), 23(9). https://doi.org/10.3390/S23094184
48. Knight, A., Gschwind, T., Galer, P., Worrell, G. A., Litt, B., Soltesz, I., & Beniczky, S. (2023). Artificial intelligence in epilepsy phenotyping. Epilepsia. https://doi.org/10.1111/EPI.17833
49. Roy, S., Kiral, I., Mirmomeni, M., Mummert, T., Braz, A., Tsay, J., Tang, J., Asif, U., Schaffter, T., Ahsen, M. E., Iwamori, T., Yanagisawa, H., Poonawala, H., Madan, P., Qin, Y., Picone, J., Obeid, I., Marques, B. D. A., Maetschke, S., … Harrer, S. (2021). Evaluation of artificial intelligence systems for assisting neurologists with fast and accurate annotations of scalp electroencephalography data. EBioMedicine, 66. https://doi.org/10.1016/J.EBIOM.2021.103275
50. Khalid, U., Naeem, M., Stasolla, F., Syed, M., Abbas, M., & Coronato, A. (2024). Impact of AI-powered solutions in rehabilitation process: Recent improvements and future trends. International Journal of General Medicine, 17, 943–969. https://doi.org/10.2147/ijgm.s453903

CHAPTER 12

Robotics in Healthcare

Surgical and Patient Care Automation

Rahma Mahmoud Saad, Muhammad Islampanah, Husam Alddin Maher Alkajah, Gergis Goseph Altalab, Ayah Abdulgadir, Ahmed Tharwat Al-Namrouti, and Abdelrahman Abdelshafi

12.1 INTRODUCTION

Robotics in healthcare represents a transformative integration of advanced technologies with medical practices, enhancing precision and outcomes and alleviating the burden of healthcare workers [1]. In the context of healthcare, robotics refers to the application of robotic systems to assist in surgical procedures, patient rehabilitation, telemedicine, and hospital automation [1–4].

The concept of robot-assisted surgery dates to 1967. The US Department of Defense and other stakeholders created the first completely operational surgically robot to decrease wartime fatalities [5]. Using the Puma 560, an industrial robot arm, Kwoh carried out the first robotic surgery in 1985 for a CT-guided brain biopsy. First employed in neurosurgery as a stereotactic frame, Westinghouse Limited later deemed it unsafe for surgical purposes due to its industrial design, thus prohibiting its surgical use [6]. Later, milestones such as the introduction of AESOP 1000 in 1994 and the da Vinci Surgical System in 2000 marked significant advancements, enabling surgeons to perform complex procedures with enhanced precision through minimally invasive techniques [7]. Over time, the area has broadened to include a diverse range of applications, from rehabilitation robots aiding in patient recovery to telepresence robots facilitating remote consultations [2, 4].

Robotics has revolutionized modern medicine, impacting surgical procedures and patient care [1, 8]. Traditional surgical methods, often limited by human fatigue, have been supplanted by robotic systems that offer unparalleled accuracy and precision [9, 10]. These systems not only reduce the margin of error but also enhance patient outcomes by minimizing trauma and accelerating recovery times [3]. In patient care, robotics provide information and communication functions, send an emergency call, record fall events, lift and carry objects such as oxygen equipment and reduce anxiety [11]. The review aims to investigate the development, application, and impact of healthcare robots in surgery and patient care. Also, it provides a complete understanding of the role of these robots in enhancing healthcare delivery and patient outcomes by assessing technology improvements as well as the possible benefits and issues they may present.

12.2 CLASSIFICATION OF HEALTHCARE ROBOTS

Robotics has a wide range of applications in the healthcare industry. Healthcare robots are frequently classified according to their functions or applications,

DOI: 10.1201/9781003533351-12

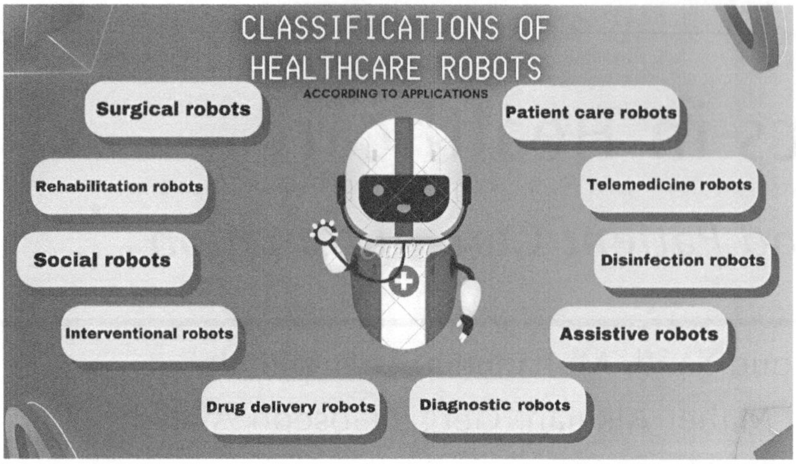

FIGURE 12.1 Different classifications of healthcare robots depending on their functions.

with the most prevalent classification being their specialized applications and functions as the following [12, 13] (Figure 12.1):

• Surgical robots	• Robots for diagnosis and imaging
• Rehabilitation robots	• Telemedicine robots
• Social robots/companion	• Drug delivery robots
• Radiotherapy/interventional robots	• Disinfection robots
• Assistive robots	• Patient care robots

In this chapter, we focus on two most important robots' applications nowadays: surgical and patient care robots.

12.3 OVERVIEW OF SURGICAL ROBOTS

The word "robot" originated at the beginning of the 20th century and was derived from the Czech word "robota," which means "labor." Surgical procedures have recently benefited from the application of robotics and computer assistance based on telepresence and virtual reality concepts. The use of robots in surgery has been around for about 35 years, and during the last 20 years, thanks to the development of cutting-edge technologies, it has grown significantly [14]. These systems primarily consist of robotic arms equipped with surgical instruments, a surgeon-controlled console, and a high-definition vision system that provides a magnified, three-dimensional (3D) view of the surgical field [15].

The study of designing, building, operating, and deploying robots to carry out specific physical activities for people with variable degrees of autonomy is known as robotics. Robots are typically employed for tasks that need a lot of repetition or pose a serious risk to human safety [13]. Additionally, the Robot Institute of America established the first scientific definition of a robot in 1972, which was the latest scientific definition available as it was described as "A multifunctional, reprogrammable manipulator designed to perform a variety of tasks by utilizing a variety of preprogrammed motions to move components, materials, equipment, or specialized devices" [16].

12.4 TYPES OF SURGICAL ROBOTIC SYSTEMS

12.4.1 Master-Slave Systems

Master-slave systems represent a widely used form of surgical robots [17]. These systems function by replicating the surgeon's movements, providing enhanced accuracy and dexterity [18]. This setup minimizes tremors and allows for minimally invasive procedures with improved precision [19].

12.4.2 Supervisory-Controlled Systems

Supervisory-controlled systems operate under the guidance of a surgeon but perform specific tasks autonomously once programmed [20]. These systems are preprogrammed with detailed instructions and can execute tasks such as radiotherapy, suturing, or needle placement procedures, with high precision [20–22].

12.4.3 Shared-Control Systems

Shared-control systems facilitate a collaborative approach between the surgeon and the robot [23, 24]. In these systems, both the surgeon and the robot share control over the surgical system [23]. The robot assists by doing preplanned tasks, while the surgeon guides the procedure. These systems offer the same comparable accuracy and efficiency as the traditional robot-aided procedures, while being more comfortable and less tiring [25]. Examples include systems used in regrasping the needle using dual-arm movements and laser ablation [24, 26].

12.4.4 Autonomous Robotic Systems

Autonomous robotic systems represent a frontier in surgical technology, where robots perform surgical tasks with minimal human intervention [27]. These systems are designed to carry out procedures based on pre-set algorithms and real-time data analysis [28–30]. Current examples, such as the Smart Tissue Autonomous Robot (STAR), demonstrate promising capabilities in suturing and soft tissue manipulation [31].

12.4.5 Tele-surgical Systems

Tele-surgical systems enable surgeons to perform operations remotely, overcoming geographical barriers [32]. These systems are particularly beneficial in providing expert surgical care to remote or underserved areas [32, 33]. The Raven system exemplifies tele-surgery capabilities, allowing surgeons to operate on patients from distant locations using high-speed internet connections and advanced robotic interfaces [32, 34]. These systems are crucial in situations where immediate surgical intervention is needed but local expertise is unavailable, thereby expanding the reach of specialized surgical care [35].

12.4.6 Hybrid Systems

Hybrid systems integrate various robotic technologies to enhance surgical capabilities [36]. These systems combine features of master-slave, shared-control, and autonomous systems to provide a versatile surgical platform [37–39]. By leveraging the strengths of different robotic approaches, hybrid systems offer unique applications and improved outcomes [40].

12.5 COMPONENTS AND FEATURES OF SURGICAL ROBOTS

The more advancements in robotic surgical systems the more variability of its components and features it gained to function in further surgical procedures. Here are a few of the most widely used surgical robotic systems used nowadays (Table 12.1).

TABLE 12.1 Components and Features of Surgical Robots

Surgical System	da Vinci [41, 42]	Ion [43, 44]	Mako [45]	NAVIO [46]
Components	1. Patient cart 2. Surgeon console 3. Vision cart	1. Controller and system cart 2. PlanPoint software and Planning Laptop 3. Tools: • Fully articulating, flexible catheter • Peripheral vision probe • Flexion biopsy needles	1. 3D CT-based planning 2. AccuStop hepatic technology 3. Insightful data analytics	1. Robotic-assisted hand piece 2. Software applications 3. CT-free registration and planning
Features	• Surgeons control the instruments from a console • High-definition 3D view of the patient's anatomy • Real-time translation of hand movements to guide instruments	• Precise navigation: using an ultrathin, ultra maneuverable catheter, physicians can reach small lung lesions with more precision, stability, and reach • Shape-sensing technology: provide precise location, shape, and orientation information throughout navigation and	• 3D CT-based planning: surgeons use detailed 3D images of a patient's anatomy to plan the procedure accurately • AccuStop™ haptic technology: this feature allows	• Customized robotics-aided technologies: The NAVIO system combines a simple CT-free registration method, patient-specific planning, and handheld robots. This technology allows for precise implant planning and bone preparation

(Continued)

TABLE 12.1 Components and Features of Surgical Robots *(Continued)*

Surgical System	da Vinci [41, 42]	Ion [43, 44]	Mako [45]	NAVIO [46]
	• Supports 3D high-definition vision system • Highly magnified views • Operates through small incisions • Precise, human-like movements	biopsy. It measures the catheter's shape hundreds of times per second • Integrated imaging: integrates Siemens Healthineers' Cios Spin mobile imaging system. 3D scans performed by Cios Spin are automatically sent to Ion, allowing clinicians to refine navigation and verify tool-in-lesion using the touchscreen • Biopsy planning: Ion PlanPoint software uses 3D airway trees generated from CT scans to identify targeted pulmonary nodules and automatically create a path to the nodule	precise bone resections, reduces soft tissue damage, and promotes bone preservation during surgery • Insightful data analytics: Mako provides a platform to track surgical metrics (e.g., time, complications) and engage with patients throughout their care	• Individualized patient plans: Each NAVIO procedure is based on the patient's unique anatomy, ensuring personalized treatment • Soft-tissue Balancing: the system helps doctors create implant schemes that locate components and balance soft tissue. Achieving proper soft-tissue balance is crucial for successful knee replacement • Multiple control modes: The handheld instrumentation offers various control modes to assist surgeons during bone preparation for implantation
Models	da Vinci Si, X, Xi, and SP1	–	–	NAVIO TKA (total knee arthroplasty), NAVIO PFJ (patellofemoral joint), and NAVIOUKA (unicompartmental knee arthroplasty)

12.6 APPLICATION OF ROBOTICS IN DIFFERENT SURGICAL FIELDS

Robotic application has invaded the surgical field either in diagnosis, minimal procedures, or even in complicated surgeries. Robotic approaches showed their efficacy in diagnosis, as a robotic-assisted bronchoscopy model showed its efficacy in examining peripheral pulmonary nodules (PPNs) and even enabled us to take clear biopsies [47]. The standard bronchoscopy showed sensitivity of 34% in lesions less than 2 cm and 63% in more than 2 cm lesions [48]. The Ion™ Endoluminal System (Sunnyvale, CA: Intuitive Surgical, Inc.) used in this procedure showed its favorability due to the ultra-maneuverable and ultrathin catheters which mean size is about 3.4 mm, the Ion system showed accessibility in over 96.6% of cases [49]. The surgical system was further tested in two separate populations in China and the United States. The diagnostic yield was relatively high to advanced bronchoscopy techniques where it ranges between 79.3% and 88.1% [49–52]. Although it showed stability and high diagnostic yield to PPNs, most of the physicians used this technology and their teams were unfamiliar with this procedure on real-life cases, therefore, the outcome data were gathered at an early stage of the training procedure. Nearly all the physicians who practiced this technique were highly experienced pulmonologists and thoracic surgeons in diagnostic field so the data couldn't be guaranteed to be matched with the general field physicians. Robotic-assisted approaches in thoracic surgery were not just minimized to the diagnostic aspect. It interfered in most thoracic surgery cases like: thoracic outlet syndrome, chest wall resection, tracheobronchomalacia, airway and sleeve lung surgery, lobectomy after neoadjuvant therapy, complex segmentectomy, giant paraoesophageal hernia repair, esophagectomy and esophageal enucleation, mediastinal masses and thymectomy and lung transplantation. In the matter of chest wall resection which is one of the major open surgeries conducted to excise a part or all the soft tissue, cartilage, sternum, and/or ribs in the chest wall to manage chest wall tumors, infections, necrosis, trauma and parts affected by lung cancer invasion [53]. It was based on the technique used in robotic first rib resection where it also proved its superiority over traditional infraclavicular, transaxillary, or supraclavicular approaches in providing better visualization for the surrounding anatomy which decreased the incidence of neurovascular bundle or incomplete resections of the rib [54]. A study conducted by Verm et al. in their institution (6 cases) and from the national cancer database (NCDB) (96 cases) which found that robotic-assisted resections showed its feasibility and safety in excising chest wall tumors along the follow-up period for 90 days [55]. As long as in chest wall tumors, robotic-assisted surgeries expanded across all the surgical oncology field. A comparative study conducted by Niklas et al. in 2016 showed the beneficial pros of robotic-assisted surgical system in prostatectomy over the conventional open surgery. The study results favored robotic-assisted prostatectomy in regard of positive safety margins, hospitalization, intraoperative transfusion, and duration of catheter ($P < 0.0001$). Although, the longer operative time in robotic-assisted prostatectomy (184.4 minutes) compared to open surgery (128.0 min), this is mostly due to that surgeons are in the early phase of the learning curve of the procedure [56]. Also, a recent study assessing the efficacy of a common surgery conducted in pancreatic cancer which is pancreaticoduodenectomy using a robotic-assisted surgical system compared to the standard open surgery. Robotic-assisted approach showed its superiority in length of hospital stay (12.0 (9.0–16.0) days) and estimated blood loss (190 mL) over open approach (260.0 mL) and (14.5 (11.0–19.0) days), respectively ($P < 0.0001$). However, main complications like delayed gastric emptying, bile leakage, clinically relevant postoperative pancreatic fistula (CR-POPF), postoperative pancreatectomy hemorrhage (PPH), reoperation, 90-day mortality rates were not significant between both groups [57]. Yet, robotic-assisted pancreaticoduodenectomy reduced operation time was not comparable even with laparoscopic approach. The comparative study with the laparoscopic approach conducted by Zong et al. in 2022 revealed a significant shorter operation time with a mean time of 5.97 hours vs. 5.42 hours in laparoscopic approach $P < 0.05$ [58]. However as much as the operative time important as long as the complexity of robotic-assisted approach, operative time is significantly related to the experience of the surgeon conducting the surgery. For the importance of this outcome Tai Huang et al. in 2023 in China studied the relation between the number of surgeries conducted by a single surgeon over 6 years between 2016 and 2022 to relate between the accumulative gained experience of operating 160 consecutive cases with his operative time in robotic-assisted pancreaticoduodenectomy. The cumulative sum analysis of both resection and reconstruction steps showed significant decline in their time after a number of cases. The resection time declined significantly after 30 cases reaching a median operative time of 195 minutes while the reconstruction time curve showed its inflection point after 45 cases reaching a median of 120 minutes [59] (Figure 12.2). The mean number of operations or maneuvers needed to excel in robotic-assisted approach surgeries is relevant to the starting point of the surgeon as it varies from one surgeon to another and whether the surgeon has experience in endoscopic approach or not. The robotic surgical system did not just exclusively evolve within surgical oncology field; it also developed in other surgical fields like joint replacement surgeries, gynecologic surgeries, and anomalies correction. Specialized surgical

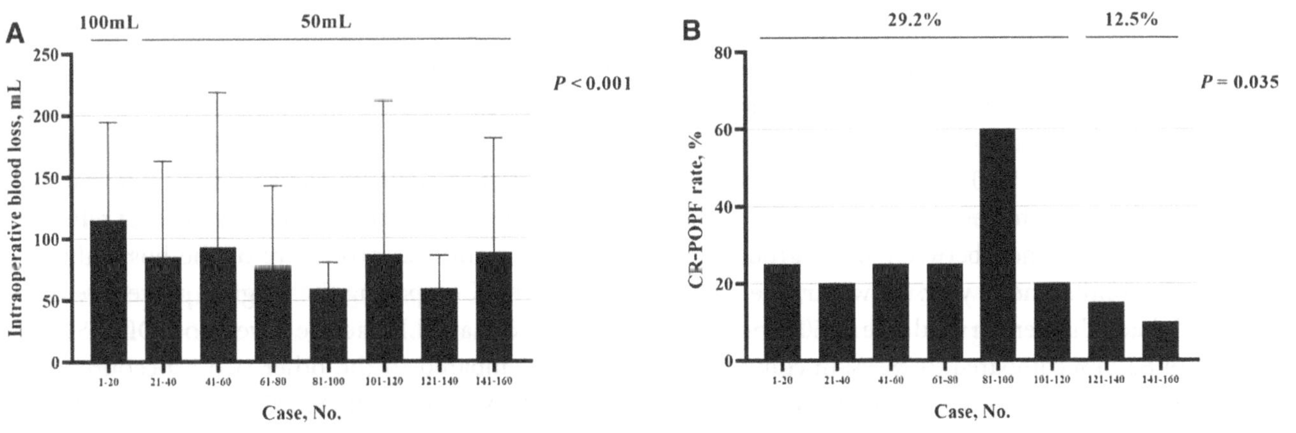

FIGURE 12.2 The learning curve of robotic-assisted pancreaticoduodenectomy [44].

haptic devices, which enable precise tissue manipulation by receiving data from distant sensors and relaying it back to the surgeon. Additionally, intraperitoneal surgical robots are being refined for less invasive surgical applications and enhanced intraoperative surgical triangulation [79]. Since robots can execute repetitive tasks for extended periods without degrading performance, there is a decreased chance of errors resulting from fatigue, which could lead to more consistent patient results [80]. Surgical robotic technology has had a dramatic impact on procedures by automating and teleoperating robotic platforms, surgeons will benefit from more compatible equipment and devices with greater dexterity. Furthermore, if surgical robotics continues to progress, it might become feasible to carry out operations in remote, challenging-to-reach locations. For example, tele-remotely operated platforms are being developed to enable the surgeon to be geographically distant from the patient [79].

Additionally, robotic guidance prevents radiation exposure for the patient, the surgeon, and other operating room staff [80]. Robot-assisted training systems are poised to have a growing and important role in the training and education of surgeons as these devices will offer prospective surgeons with immersive and authentic simulations, accurately reproducing diverse surgical settings to enhance surgeons' confidence and skills by providing guidance and hand-on experiences [15]. Last, considering the successful use of more automation in other industries, robotic surgical devices with the ability to detect tissue types, suture automatically, and perform autonomous procedures might be developed [79].

12.9 COMPLICATIONS OF SURGICAL ROBOTS

We must clarify robotic-assisted surgery complications to determine the surgical outcomes in perioperative circumstances. When compared to laparoscopic treatments, robotic procedures have been reported to need longer operating periods, which might result in consequences such as visual impairments, heart problems, endotracheal tube displacement, and nerve damage [81]. Authors have classified robotic surgery complications into two sides:

1. Related to patient positioning.
2. Related to robot malfunction.

12.9.1 Complications Related to Patient Positioning

Longer operating periods have been reported for robotic operations in comparison to laparoscopic procedures, which might result in nerve damage, ocular difficulties, endotracheal tube displacement, and cardiac issues [81]. The whole issue in patient complication is Trendelenburg's position, since it is not a physiological state for the human body, it may have a variety of hemodynamic effects. The principal effects of such a position are ocular, neurological, and musculoskeletal complications. The main impact of body position on the eyes is a raising in intraocular pressure, an increase in central venous pressure, an increase in blood flow toward the head, and a decrease in cardiac output. Furthermore, individuals who were already susceptible to raised intraocular pressure before the procedure may be more susceptible to developing glaucoma because of the increased intraocular pressure and extended Trendelenburg's position duration [82]. Increased intraocular pressure, corneal abrasions, subcutaneous ecchymosis, and periorbital edema are among the primary ocular consequences [83, 84]. Numerous case reports and case series claimed varying degrees of blindness, visual field loss, keratopathy, and conjunctival edema [82, 85]. The incidence of ocular complications rises with neck robot-assisted surgery such as robotically assisted endoscopic thyroidectomies and large vessel surgery like the jugular veins during the neck's CO_2 insufflation might increase intraocular pressure [82]. A large percentage of the reported complications were related to neurologic issues, which could be attributed to both positional and procedural variables. The most frequently reported problems are mild head contusion and damage to the peripheral nerve system caused by improper posture. Compression, stretching, and ischemia are among the potential mechanisms underlying nerve damage. Long-term compression can cause demyelination and destruction of Schwann cells, which can result in irreversible nerve damage [86, 87]. Following robotic-assisted surgery in gynecological cases, postoperative lower extremities rhabdomyolysis is an uncommon but potentially serious complication that seems to be linked to morbid obesity, hypothermia, and prolonged lithotomy for more than 5 h, considerable preoperative medical comorbidities, and potentially extended operative time [88, 89]. Pulmonary complications described in the reviews were not unique to robotic-assisted surgery and in the end, surgeons observed very few heart-related issues related to robotic-assisted surgery [86].

12.9.2 Complications Related to Robot Malfunction

Despite the advantages, more robotic-assisted surgery procedures carry the risk of mechanical malfunctions, which can cause delays in surgery, cases to be abandoned, open or laparoscopic procedures to be used instead, and patient harm [90]. Even though system failures have a greater serious malfunction rate, most faults seem to be caused by instrument mistakes. Lower-volume centers may lack the ability to easily obtain spare tools or resources to swiftly switch robots, and these error types could result in surgery delays and replacement expenses [91]. Pierce et al. reported that the majority of robotic malfunctions, 76.4%, were discovered to be instrument failures. The most frequent ones included the ProGrasp forceps that accounted for 21.0% of the total, whereas the bipolar Maryland forceps accounted for 15.3% and the monopolar-curved scissors accounted for 11.3% [92]. There were numerous instances of mechanical malfunctions with the da Vinci system, such as console malfunctions, robotic arm malfunctions, optic system malfunctions as well as software incompatibility and metal fatigue/break of the console handpiece [93]. A few Surgeons had reported robotic damage to body organs. The majority of organ damage instances are linked to unintended thermal or cold injury from the instrument as well as failure to detect anatomy, particularly in the presence of adhesion or tumor infiltration [87]. Despite that, mechanical failure or malfunction as well as open or laparoscopic conversion in various fields is infrequent [93].

12.10 OUTCOMES OF SURGICAL ROBOTS

When compared to traditional laparoscopy and open surgery, surgeons and industry reported better results because of several technological advancements, including increased motion scaling, tremor filtration, improved ergonomics in response to a more pleasant user interface, improved wrist motion flexibility, and improved vision due to the 3D interface [79]. Compared to laparotomy, there were fewer perioperative complications, and a shorter stay in hospital [87]. The results of Anderson et al.'s study indicate that, in comparison to open procedures, robotic procedures were associated with lower mortality rates, lower length of stay for all procedures but valvuloplasty, and higher overall costs for most procedures except cardiac procedures (where valvuloplasty had no difference in total charges and CABG had lower rates) [94]. However, other gynecological procedures had a somewhat greater death rate when carried out by robots. When robotic assistance is used in procedures that were previously conducted as open procedures, the results are most improved. Despite procedures already done laparoscopically, improvement in results is less evident [94]. Robotic surgery has been linked to shorter LOS, warm ischemia durations, projected blood loss, and operative times during partial nephrectomy as compared to pure laparoscopic partial nephrectomy [95, 96]. Despite that, there were no differences in 30-day mortality and readmission between the robotic and laparoscopic approaches to gastric bypass surgery, the robotic approach also required longer operating times [81]. Furthermore, an examination of the Surgery Accreditation and Quality Improvement Program database comparing robotic and laparoscopic sleeve gastrectomy revealed that robotic surgery has a greater rate of leaks, surgical site infections, and longer operating times [97]. Surgical robots for rectal cancer are associated with shorter hospital stays, less blood loss, reduced conversion rates, and quicker operating times [98]. The conversion rate for robotic surgery used to treat rectal cancer was lower. The positive outcomes seem to have been influenced by many advantageous features of the robotic system, which were especially noticeable during lower rectal cancer surgery, compared to laparoscopic surgeries. Nevertheless, in the event of major, perhaps fatal intraoperative instances (such as uncontrollable bleeding), the time required to undock the robot and convert to an open laparotomy could have catastrophic consequences [99]. Prior studies revealed that the anticipated blood loss during procedures was comparable between laparoscopic and robotic surgery [100], while some research indicated that the robotic technique resulted in a much smaller volume of blood loss than the laparoscopic approach [101]. Meta-analyses revealed that there was no discernible difference between robotic and traditional laparoscopic rectal cancer surgery in terms of the frequency of overall problems in short-term healthcare as they require continuous attention to their lifespan, support, and updates [114].

12.15.1 High Cost

Robots and automation are very expensive which in turn will increase the cost of robotic surgeries than manual or laparoscopic surgeries as we mentioned before [113, 121], also the cost of installation, maintenance, software updates, and training of healthcare providers to adapt these new technologies makes them more expensive which is challenging smaller healthcare facilities [15, 113].

12.15.2 Adaptability

It might be challenging and time consuming to adapt to new things especially in case of emergency as happened during COVID-19; the automated environment was unfamiliar to those who operate in the system [113, 121].

12.15.3 Training Requirements

Proficiency in robotic surgery requires extensive time and effort investment along with specialized training. To efficiently operate robotic systems, traverse the surgical console, improve patient care, and accurately control robotic equipment, surgeons must learn the necessary abilities and adapt to new roles [15, 122]. Surgeons must get continual education and training to stay current with emerging technologies and procedures [15].

12.15.4 Job Displacement

Certain healthcare positions might be replaced by robotics, necessitating worker transfers and reskilling [114]. Robotics adoption will result in higher unemployment [113].

12.15.5 Access Disparities

The integration of robotics in healthcare worsens disparities in care [121]. Also, the high cost exacerbates these disparities in access to benefits of advanced surgical options specifically for small, underfunded facilities [15, 123].

12.15.6 Intraoperative Challenges

Robotic surgery involves several obstacles, including dealing with unforeseen intraoperative complications such as robotic arm, vision, or sensors failures and communication concerns, which necessitates extensive debugging and contingency planning training for surgeons and their teams [15, 113]. There is also an increased risk of infection if robotic systems are not adequately sanitized between surgeries, thus rigorous attention to cleaning standards is required [15]. Furthermore, the lack of comprehensive haptic input in robotic systems might impair a surgeon's ability to judge tissue qualities, forcing dependence on visual and aural cues, with continuing research targeted at enhancing such feedback [15, 113].

12.16 ETHICAL AND LEGAL CONSIDERATION

The use of artificial intelligence (AI) and automation in the realm of surgery presents several ethical dilemmas that need meticulous examination. It is essential to find the correct equilibrium between the benefits of automation and the ethical standards that form the foundation of medical practice. This is necessary in order to build public confidence and guarantee the appropriate utilization of robotic surgery [15]. The following are the primary factors to consider while implementing robots in healthcare.

12.16.1 Transparency and Informed Consent

Healthcare practitioners have a moral obligation to inform patients about the hazards, benefits, and consequences of using robots and how their data will be used by robots to take their consent [114, 124].

12.16.2 Privacy and Data Security

Robotic systems gather, retain, and examine patient data, giving rise to issues around privacy, permission, and potential misuse; so stricter data protection regulations, clear permission procedures, and other measures are needed to address these issues and guarantee patient anonymity [114, 124].

12.16.3 Accountability

To properly handle ethical concerns, it is crucial to create precise legislative frameworks that specify obligations and liabilities when it comes to the use of robotics in healthcare [114]. The ability of AI systems to diagnose and treat patients without errors is inevitable, and determining who is accountable for these errors may prove difficult [125].

12.16.4 Equity in Access and Treatment

Robotics technology has the potential to exacerbate health disparities by limiting access to advanced treatments for some groups thus ethical guidelines should include addressing price and accessibility, as well as ensuring that all demographic groups benefit equally from technological advancements [114].

Regulatory agencies are continually formulating guidelines and standards to assure patient safety and equity. These regulations are essential for minimizing risks and establishing a framework for accountability, ensuring that robotic surgery adheres to the highest standards of safety and quality [125].

12.17 FUTURE DIRECTIONS

The implantation of healthcare robotics in patient monitoring and management has significantly raised patient care time provided by healthcare professionals [12]. Development of microbots (tiny robots) in the future will reduce recovery time as they can navigate and do surgical procedures within the human body without need of surgical incisions so the study of microbots that can repair the body is highly recommended [115]. Due to the convergence of precision and minimally invasive treatments, patients will have access to safer, more effective, and more successful therapies; as new advancements emerge, this is projected to revolutionize the surgical practice environment [114]. Also, the use of soft robotics in minimally invasive surgery will facilitate agility and improve tissue injury [114]. In robotic surgical training complex procedures will be simulated using virtual and augmented reality allowing healthcare practitioners to employ AR and VR in telemedicine encounters to strengthen their virtual presence [114]. Furthermore, humanoid robots will help with everyday tasks and patient emotional support by offering social interaction and reducing loneliness [114]. Also, further development in AI algorithms to be designed to examine patient data in real-time and provide the surgeon with precise recommendations during the operation [15]. Nanorobotics, which operates on the cellular and molecular levels, is a promising topic in surgical research. These small robots have the potential to transform illness detection and therapy by targeting and tackling diseases with unprecedented accuracy [15]. Advancements in data analytics, computer vision, machine learning, and AI will enhance the abilities of medical robots, allowing them to perform tasks more efficiently, accurately, and independently. Further research is needed to greatly improve the efficiency of these robotic and automated systems, making them safe and reliable enough to replace the human-driven healthcare system. Additionally, exploring new areas for practical application is encouraged [115].

12.18 SUMMARY

The more the robotic systems involved in medical procedures, the more diversity and fitting of the robotic components to the complexity of the procedures. Most of robotic systems mainly surgical robots, processing units, and surgical console offer improved control, stability, and precision, making operations easier. It also offers ergonomic benefits, such as joystick-operated consoles and haptic devices. Robotic surgery can be performed in remote locations, reducing radiation exposure. Robotics has significantly impacted the surgical field, particularly in diagnosis, minimal procedures, and complicated surgeries. It became a crucial part of the healthcare system, but their high cost and novelty has limited their usage in certain healthcare centers and countries which led to minimum accessibility to most physicians in the globe and the current data collected is defined to their early training process and at the available centers. Still, the current evidence of the robotic-assisted procedures has shown feasibility, safety, and efficacy over the conventional approaches in some aspects. Most of the concerned outcomes found to be enhanced significantly with further surgeon's experience and training.

Despite the challenges of robotic systems represented in robotic errors, patient positioning, and instrument malfunction, robotic surgery has shown better results compared to traditional laparoscopy and open surgery, with shorter hospital stays, less blood loss, reduced conversion rates, and quicker operating time.

The use of AI and automation in surgery raises ethical dilemmas. Healthcare practitioners must balance the benefits of automation with ethical standards to ensure public confidence and appropriate use. Factors to consider include transparency, privacy, accountability, and equity in access and treatment. Regulatory

human use of a new robotic-assisted fiber optic sensing navigation system for small peripheral pulmonary nodules. *Respiration*, *98*(2), 142–150.
50. Hammad Altaq, H., Parmar, M., Syed Hussain, T., Salim, D. J., & Chaudry, F. A. (2023). The use of robotic-assisted bronchoscopy in the diagnostic evaluation of peripheral pulmonary lesions: A paradigm shift. *Diagnostics*, *13*(6), 1049.
51. Kalchiem-Dekel, O., Connolly, J. G., Lin, I. H., Husta, B. C., Adusumilli, P. S., Beattie, J. A., … Bott, M. J. (2022). Shape-sensing robotic-assisted bronchoscopy in the diagnosis of pulmonary parenchymal lesions. *Chest*, *161*(2), 572–582.
52. Xie, F., Zhang, Q., Mu, C., Zhang, Q., Yang, H., Mao, J., & Sun, J. (2024). Shape-sensing robotic-assisted bronchoscopy (SS-RAB) in sampling peripheral pulmonary nodules: A prospective, multicenter clinical feasibility study in China. *Journal of Bronchology & Interventional Pulmonology*, *31*(4), e0981.
53. Purnell, S., Odeh, A., Freeman, R., Raad, W., Servais, E., & Abdelsattar, Z. (2024). National trends, safety, and effectiveness of minimally invasive concomitant chest wall resection for locally advanced lung cancer. *JTCVS Open*, 19, 311–324.
54. Martinez, B. D., Wiegand, C. S., Evans, P., Gerhardinger, A., & Mendez, J. (2005). Computer-assisted instrumentation during endoscopic transaxillary first rib resection for thoracic outlet syndrome: A safe alternate approach. *Vascular*, *13*(6), 327–335.
55. Verm, R. A., Vigneswaran, W. T., Lin, A., Zywiciel, J., Freeman, R., & Abdelsattar, Z. M. (2023). Robotic chest wall resection for primary benign chest wall tumors and locally advanced lung cancer: An institutional case series and national report. *Journal of Thoracic Disease*, *15*(9), 4849.
56. Niklas, C., Saar, M., Berg, B., Steiner, K., Janssen, M., Siemer, S., … Ohlmann, C. H. (2016). da Vinci and open radical prostatectomy: Comparison of clinical outcomes and analysis of insurance costs. *Urologia Internationalis*, *96*(3), 287–294.
57. Liu, Q., Zhao, Z., Zhang, X., Wang, W., Han, B., Chen, X., & Liu, R. (2023). Perioperative and oncological outcomes of robotic versus open pancreaticoduodenectomy in low-risk surgical candidates: A multicenter propensity score-matched study. *Annals of Surgery*, *277*(4), e864–e871.
58. Zong, K., Luo, K., Chen, K., Ye, J., Liu, W., & Zhai, W. (2022). A comparative study of robotics and laparoscopic in minimally invasive pancreatoduodenectomy: A single-center experience. *Frontiers in Oncology*, *12*, 960241.
59. Huang, X. T., Wang, X. Y., Xie, J. Z., Cai, J. P., Chen, W., Chen, L. H., & Yin, X. Y. (2023). Learning curves of resection and reconstruction procedures in robotic-assisted pancreatoduodenectomy by a single surgeon: A retrospective cohort study of 160 consecutive cases. *Gastroenterology Report*, *11*, goad042.
60. Ma, N., Sun, P., Xin, P., Zhong, S., Xie, J., & Xiao, L. (2024). Comparison of the efficacy and safety of MAKO robot-assisted total knee arthroplasty versus conventional manual total knee arthroplasty in uncomplicated unilateral total knee arthroplasty a single-centre retrospective analysis. *International Orthopaedics*, 1–8.
61. Hoznek, A., Zaki, S. K., Samadi, D. B., Salomon, L., Lobontiu, A., Lang, P., & Abbou, C. C. (2002). Robotic assisted kidney transplantation: An initial experience. *The Journal of Urology*, *167*(4), 1604–1606.
62. Patil, A., Ganpule, A., Singh, A., Agrawal, A., Patel, P., Shete, N., & Desai, M. (2023). Robot-assisted versus conventional open kidney transplantation: A propensity matched comparison with median follow-up of 5 years. *American Journal of Clinical and Experimental Urology*, *11*(2), 168.
63. Emerson, D., Catarino, P., Rampolla, R., Chikwe, J., & Megna, D. (2024). Robotic-assisted lung transplantation: First in man. *The Journal of Heart and Lung Transplantation*, *43*(1), 158–161.
64. Krishnan, A., Kasinpila, P., Wang, H., Ruaengsri, C., Shudo, Y., Jackson, E., & Woo, Y. J. (2023). First-in-human beating-heart transplant. *JTCVS Techniques*, *19*, 80–85.
65. Troisi, R. I., Cho, H. D., Giglio, M. C., Rhu, J., Cho, J. Y., Sasaki, K., … Kim, K. H. (2024). Robotic and laparoscopic right lobe living donation compared to the open approach: A multicenter study on 1194 donor hepatectomies. *Liver Transplantation*, *30*(5), 484–492.
66. Chatterjee, S., Das, S., Ganguly, K., & Mandal, D. (2024). Advancements in robotic surgery: Innovations, challenges and future prospects. *Journal of Robotic Surgery*, *18*(1), 28.
67. Barbash, G. I., & Glied, S. A. (2010). New technology and health care costs—the case of robot-assisted surgery. *New England Journal of Medicine*, *363*(8), 701–704.
68. Koh, Y. X., Zhao, Y., Tan, I. E. H., Tan, H. L., Chua, D. W., Loh, W. L., … Goh, B. K. P. (2024). Comparative cost-effectiveness of open, laparoscopic, and robotic liver resection: A systematic review and network meta-analysis. *Surgery*, 176(1), 11–23.
69. Petrochenkov, E., Bencini, G., Martinino, A., Lian, A., Olazar, J., Akshelyan, S., … Tzvetanov, I. (2024). Analyzing the impact of CIT on the largest reported cohort of robotic kidney transplantation from the deceased donors. *Transplantation Direct*, *10*(9), e1671.
70. Zhang, J., Ndou, W. S., Ng, N., Gaston, P., Simpson, P. M., Macpherson, G. J., & Clement, N. D. (2022). Robotic-arm assisted total knee arthroplasty is associated with improved accuracy and patient reported outcomes: A systematic review and meta-analysis. *Knee Surgery, Sports Traumatology, Arthroscopy*, *30*(8), 2677–2695.

71. Zhang, J., Feng, Q., Huang, Y., Ouyang, L., & Luo, F. (2022). Updated evaluation of robotic-and video-assisted thoracoscopic lobectomy or segmentectomy for lung cancer: A systematic review and meta-analysis. *Frontiers in Oncology, 12*, 853530.
72. Mwesigwa, N. W., & Tentzeris, V. (2024). A comparison of short-term outcomes of robotic-assisted thoracic surgery versus video-assisted thoracic surgery following lung cancer surgery at a tertiary hospital in the United Kingdom: A propensity-matched analysis. *Cureus, 16*(7), e64298.
73. Patel, V. R. (Ed.). (2012). *Robotic Urologic Surgery*. Springer Science & Business Media.
74. Cypko, M. A., Timmermann, L., Sauer, I. M., & Müller-Birn, C. (2022). Towards human-robotic collaboration: Observing teamwork of experienced surgeons in robotic- assisted surgery. In *Proceedings of Mensch und Computer 2022*. Association for Computing Machinery, (pp. 566–571).
75. Kristensen, S. E., Mosgaard, B. J., Rosendahl, M., Dalsgaard, T., Bjørn, S. F., Frøding, L. P., … Lajer, H. (2017). Robot-assisted surgery in gynecological oncology: Current status and controversies on patient benefits, cost and surgeon conditions–a systematic review. *Acta Obstetricia et Gynecologica Scandinavica, 96*(3), 274–285.
76. Abdelrahman, A. M., Lowndes, B., Rand, C., Shah, H., El-Sherif, N., Prokop, L., & Hallbeck, M. S. (2017). Impact of robotic surgery versus laparoscopic surgery on surgeon musculoskeletal symptoms and workload: A systematic review and meta-analysis. *Surgical Endoscopy and Other Interventional Techniques, 31*(1), S324.
77. Lee, G. I., Lee, M. R., Clanton, T., Sutton, E., Park, A. E., & Marohn, M. R. (2014). Comparative assessment of physical and cognitive ergonomics associated with robotic and traditional laparoscopic surgeries. *Surgical Endoscopy, 28*, 456–465.
78. Park, L. S., Pan, F., Steffens, D., Young, J., & Hong, J. (2021). Are surgeons working smarter or harder? A systematic review comparing the physical and mental demands of robotic and laparoscopic or open surgery. *World Journal of Surgery, 45*(7), 2066–2080.
79. Peters, B. S., Armijo, P. R., Krause, C., Choudhury, S. A., & Oleynikov, D. (2018). Review of emerging surgical robotic technology. *Surgical Endoscopy, 32*, 1636–1655.
80. Hernigou, P., Lustig, S., & Caton, J. (2023). Artificial intelligence and robots like us (surgeons) for people like you (patients): Toward a new human–robot-surgery shared experience. What is the moral and legal status of robots and surgeons in the operating room? *International Orthopaedics, 47*(2), 289–294.
81. Jara, R. D., Guerrón, A. D., & Portenier, D. (2020). Complications of robotic surgery. *Surgical Clinics, 100*(2), 461–468.
82. Gkegkes, I. D., Karydis, A., Tyritzis, S. I., & Iavazzo, C. (2015). Ocular complications in robotic surgery. *The International Journal of Medical Robotics and Computer Assisted Surgery, 11*(3), 269–274.
83. Mangham, M. (2016). Positioning of the anaesthetised patient during robotically assisted laparoscopic surgery: Perioperative staff experiences. *Journal of Perioperative Practice, 26*(3), 50–52.
84. Oksar, M., Akbulut, Z., Ocal, H., Balbay, M. D., & Kanbak, O. (2014). Anesthetic considerations for robotic cystectomy: A prospective study. *Revista Brasileira de Anestesiologia, 64*, 109–115.
85. Gainsburg, D. M. (2012). Anesthetic concerns for robotic-assisted laparoscopic radical prostatectomy. *Minerva Anestesiologica, 78*(5), 596.
86. Maerz, D. A., Beck, L. N., Sim, A. J., & Gainsburg, D. M. (2017). Complications of robotic-assisted laparoscopic surgery distant from the surgical site. *British Journal of Anaesthesia, 118*(4), 492–503.
87. Tse, K. Y., Ngan, H. Y. S., & Lim, P. C. (2017). Robot-assisted gynaecological cancer surgery—Complications and prevention. *Best Practice & Research Clinical Obstetrics & Gynaecology, 45*, 94–106.
88. Mattei, A., Di Pierro, G. B., Rafeld, V., Konrad, C., Beutler, J., & Danuser, H. (2013). Positioning injury, rhabdomyolysis, and serum creatine kinase-concentration course in patients undergoing robot-assisted radical prostatectomy and extended pelvic lymph node dissection. *Journal of Endourology, 27*(1), 45–51.
89. Terry, R. S., Gerke, T., Mason, J. B., Sorensen, M. D., Joseph, J. P., Dahm, P., & Su, L. M. (2015). Postoperative rhabdomyolysis following robotic renal and adrenal surgery: A cautionary tale of compounding risk factors. *Journal of Robotic Surgery, 9*, 195–200.
90. Lavery, H. J., Thaly, R., Albala, D., Ahlering, T., Shalhav, A., Lee, D., … Patel, V. R. (2008). Robotic equipment malfunction during robotic prostatectomy: A multi-institutional study. *Journal of Endourology, 22*(9), 2165–2168.
91. Nayyar, R., & Gupta, N. P. (2010). Critical appraisal of technical problems with robotic urological surgery. *BJU International, 105*(12), 1710–1713.
92. Pierce, H., Roberts, B., Scherr, D., Barbieri, C., Anger, J., McClure, T., & Chughtai, B. (2021). Patient injuries and malfunctions associated with robotic prostatectomy: Review of the manufacturer and user facility device experience database. *Journal of Robotic Surgery, 15*, 179–185.
93. Kim, W. T., Ham, W. S., Jeong, W., Song, H. J., Rha, K. H., & Choi, Y. D. (2009). Failure and malfunction of da Vinci surgical systems during various robotic surgeries: Experience from six departments at a single institute. *Urology, 74*(6), 1234–1237.
94. Anderson, J. E., Chang, D. C., Parsons, J. K., & Talamini, M. A. (2012). The first national examination of outcomes and trends in robotic surgery in

the United States. *Journal of the American College of Surgeons*, *215*(1), 107–114.

95. Anderson, J. E., Kellogg Parsons, J., Chang, D. C., & Talamini, M. A. (2012). Hospital costs and length of stay related to robot-assisted versus open radical and partial nephrectomy for kidney cancer in the USA. *Journal of Robotic Surgery*, *6*, 19–22.
96. Pierorazio, P. M., Patel, H. D., Feng, T., Yohannan, J., Hyams, E. S., & Allaf, M. E. (2011). Robotic-assisted versus traditional laparoscopic partial nephrectomy: Comparison of outcomes and evaluation of learning curve. *Urology*, *78*(4), 813–819.
97. Fazl Alizadeh, R., Li, S., Inaba, C. S., Dinicu, A. I., Hinojosa, M. W., Smith, B. R., … Nguyen, N. T. (2019). Robotic versus laparoscopic sleeve gastrectomy: A MBSAQIP analysis. *Surgical Endoscopy*, *33*, 917–922.
98. del Gutiérrez Delgado, M. P., Mera Velasco, S., Turino Luque, J. D., González Poveda, I., Ruiz López, M., & Santoyo Santoyo, J. (2021). Outcomes of robotic-assisted vs conventional laparoscopic surgery among patients undergoing resection for rectal cancer: An observational single hospital study of 300 cases. *Journal of Robotic Surgery*, 16(1), 179–187.
99. Nozawa, H., & Watanabe, T. (2017). Robotic surgery for rectal cancer. *Asian Journal of Endoscopic Surgery*, *10*(4), 364–371.
100. Xiong, B., Ma, L., Zhang, C., & Cheng, Y. (2014). Robotic versus laparoscopic total mesorectal excision for rectal cancer: A meta-analysis. *Journal of Surgical Research*, *188*(2), 404–414.
101. Erguner, I., Aytac, E., Boler, D. E., Atalar, B., Baca, B., Karahasanoglu, T., … Uras, C. (2013). What have we gained by performing robotic rectal resection? Evaluation of 64 consecutive patients who underwent laparoscopic or robotic low anterior resection for rectal adenocarcinoma. *Surgical Laparoscopy Endoscopy & Percutaneous Techniques*, *23*(3), 316–319.
102. Scarpinata, R., & Aly, E. H. (2013). Does robotic rectal cancer surgery offer improved early postoperative outcomes? *Diseases of the Colon & Rectum*, *56*(2), 253–262.
103. Trastulli, S., Farinella, E., Cirocchi, R., Cavaliere, D., Avenia, N., Sciannameo, F., & Boselli, C. (2012). Robotic resection compared with laparoscopic rectal resection for cancer: Systematic review and meta-analysis of short-term outcome. *Colorectal Disease*, *14*(4), e134–e156.
104. Kamali, D., Omar, K., Imam, S. Z., Jha, A., Reddy, A., & Jha, M. (2017). Patient quality of life and short-term surgical outcomes between robotic and laparoscopic anterior resection for adenocarcinoma of the rectum. *Techniques in Coloproctology*, *21*, 355–361.
105. World Health Organization. (2023). Artificial intelligence for healthy aging: A literature review. In *2023 10th International Conference on ICT for Smart Society (ICISS)* (pp. 1–7). IEEE.
106. Rich, P. B., & Adams, S. D. (2014). Health care: Economic impact of caring for geriatric patients. *The Surgical Clinics of North America*, *95*(1), 11–21.
107. Mahlknecht, P., Kiechl, S., Bloem, B. R., Willeit, J., Scherfler, C., Gasperi, A., & Seppi, K. (2013). Prevalence and burden of gait disorders in elderly men and women aged 60–97 years: A population-based study. *PloS One*, *8*(7), e69627.
108. Bor, J., Cohen, G. H., & Galea, S. (2017). Population health in an era of rising income inequality: USA, 1980–2015. *The Lancet*, *389*(10077), 1475–1490.
109. Bouwhuis, D. G. (2016). Current use and possibilities of robots in care. *Gerontechnology*, *15*(4), 198–208.
110. Teng, R., Ding, Y., & See, K. C. (2022). Use of robots in critical care: Systematic review. *Journal of Medical Internet Research*, *24*(5), e33380.
111. Abou Allaban, A., Wang, M., & Padır, T. (2020). A systematic review of robotics research in support of in-home care for older adults. *Information*, *11*(2), 75.
112. Sather, R., Soufineyestani, M., Imtiaz, N., & Khan, A. (2021). Assistive robots designed for eldery care and caregivers. *International Journal of Robotics and Control*, *3*(1), 1.
113. Javaid, M., Haleem, A., Singh, R. P., Rab, S., Suman, R., & Kumar, L. (2025). Utilization of robotics for healthcare: A scoping review. *Journal of Industrial Integration and Management*, 10(01), 43–65.
114. Kaluvakuri, S. (2022). Revolutionizing healthcare: The impact of robotics on health services. *Malaysian Journal of Medical and Biological Research*, *9*(2), 41–50.
115. Nwazor, N. O., & Orakwue, S. I. (2023). Enhancing healthcare through automation and robotics. In *Modernity in Health and Disease Diagnosis: The Account from STEM Women* (pp. 59–67). Springer Nature Switzerland.
116. Denecke, K., & Baudoin, C. R. (2022). A review of artificial intelligence and robotics in transformed health ecosystems. *Frontiers in Medicine*, *9*, 795957.
117. Stasevych, M., & Zvarych, V. (2023). Innovative robotic technologies and artificial intelligence in pharmacy and medicine: Paving the way for the future of health care—a review. *Big Data and Cognitive Computing*, *7*(3), 147.
118. Orejana, J. R., MacDonald, B. A., Ahn, H. S., Peri, K., & Broadbent, E. (2015). Healthcare robots in homes of rural older adults. In *Social Robotics: 7th International Conference, ICSR 2015, Paris, France, October 26-30, 2015, Proceedings 7* (pp. 512–521). Springer International Publishing.
119. Huang, R., Li, H., Suomi, R., Li, C., & Peltoniemi, T. (2023). Intelligent physical robots in health care: Systematic literature review. *Journal of Medical Internet Research*, *25*, e39786.
120. Farid, F., Bello, A., Ahamed, F., & Hossain, F. (2023). The roles of AI technologies in reducing hospital readmission for chronic diseases: A comprehensive analysis.

Preprints. https://doi.org/10.20944/preprints202307.1000.v1

121. Chauhan, A. (2021). Robotics and automation: The rescuers of COVID era. *Artificial Intelligence for COVID, 19*, 119–151.
122. Elendu, C., Amaechi, D. C., Elendu, T. C., Jingwa, K. A., Okoye, O. K., Okah, M. J., … Alimi, H. A. (2023). Ethical implications of AI and robotics in healthcare: A review. *Medicine, 102*(50), e36671.
123. Weerarathna, I. N., Raymond, D., & Luharia, A. (2023). Human-robot collaboration for healthcare: A narrative review. *Cureus, 15*(11).
124. Ness, S., Xuan, T. R., & Oguntibeju, O. O. (2024). Influence of AI: Robotics in healthcare. *Asian Journal of Research in Computer Science, 17*(5), 222–237.
125. Davenport, T., & Kalakota, R. (2019). The potential for artificial intelligence in healthcare. *Future Healthcare Journal, 6*(2), 94–98.

CHAPTER 13

Telemedicine and Remote Patient Monitoring

Transforming Healthcare Delivery

Awez A. Sikkalgar, Kalpesh R. Yeole, Naina Chaurasiya, Sagar A. More, Devendra K. Jadhav, and Zubair Saghir Ahmed Shaikh

13.1 INTRODUCTION

Telemedicine and RPM have emerged as transformative forces in healthcare delivery, particularly after the COVID-19 pandemic. These technologies leverage information and communication technology to facilitate healthcare services over distances, thereby enhancing access to medical care for patients who may otherwise face barriers due to geographical, physical, or economic constraints. The historical roots of telemedicine can be traced back to the 1980s when it was primarily utilized in disaster medicine and rural healthcare settings. However, the rapid advancement of technology, including the internet, mobile devices, and artificial intelligence (AI), has significantly expanded the scope and effectiveness of telemedicine, making it a vital component of modern healthcare systems (1, 2).

Historically, telemedicine began with simple telephone consultations and has evolved into sophisticated systems that integrate various technologies for remote monitoring and patient engagement. The development of implantable cardiac devices, such as pacemakers and defibrillators, has particularly benefited from RPM, allowing for continuous monitoring of patient's heart health without requiring frequent in-clinic visits (3). Studies have demonstrated that RPM can significantly reduce hospital visits and healthcare costs while maintaining or even improving patient outcomes (4). For instance, the Home Guide Registry confirmed that continuous remote monitoring of cardiac implantable electronic devices (CIEDs) can be effectively integrated into standard clinical practice, reducing patient visits by nearly 50%.

The importance of telemedicine and RPM in healthcare cannot be overstated. These modalities have improved patient outcomes, particularly for individuals with chronic conditions. For instance, studies indicate that telemedicine can effectively manage chronic diseases, enhance patient satisfaction, and reduce healthcare costs by minimizing the need for in-person visits (5). Moreover, telemedicine has proven invaluable for older adults and those residing in rural areas, where access to healthcare facilities is often limited. By offering convenience and flexibility, telemedicine facilitates more consistent management of chronic conditions, ultimately leading to improved health outcomes (5, 6). The COVID-19 pandemic catalyzed the widespread adoption of telemedicine, as healthcare systems sought to maintain continuity of care while adhering to social distancing measures. This shift has underscored the potential of telemedicine to alleviate the burden on overtaxed healthcare systems and improve access to care for vulnerable populations (7). This chapter aims to provide a comprehensive overview of the current state of telemedicine and RPM, highlighting their historical evolution, significance in contemporary healthcare, and the challenges that lie ahead. It will explore the various telemedicine applications, including virtual consultations, remote monitoring of chronic diseases, and the integration of advanced technologies such as mobile health applications (mHealth) and

DOI: 10.1201/9781003533351-13

wearable devices. Furthermore, the chapter addresses the barriers to effective telemedicine implementation, such as technological disparities, privacy concerns, and regulatory challenges (8, 9). By examining these aspects, the chapter aims to contribute to the ongoing discourse on how telemedicine can be optimized to enhance healthcare delivery and patient outcomes.

13.2 TECHNOLOGICAL FOUNDATIONS OF TELEMEDICINE AND RPM

Telemedicine and RPM have revolutionized healthcare delivery by leveraging advanced technologies to enhance patient care, accessibility, and efficiency. The technological foundations of telemedicine and RPM encompass various components, including communication platforms, wearable devices, AI, cloud computing, and security measures, all crucial for effective remote healthcare delivery. Communication platforms are integral to telemedicine, facilitating real-time interactions between healthcare providers and patients. Video conferencing tools enable virtual consultations, allowing healthcare professionals to assess and diagnose patients remotely. The rise of mHealth has further enhanced this capability, providing patients with tools to manage their health and communicate with providers seamlessly. Asynchronous telehealth, or "store-and-forward" telehealth, allows for the collection and transmission of patient data, such as electronic health records (EHRs) and images, which can be reviewed later by healthcare professionals, thus improving efficiency and accessibility (10, 11). Wearable devices and the Internet of Things (IoT) play a pivotal role in RPM by enabling continuous health monitoring. These devices collect vital signs and health metrics, which are then transmitted to healthcare providers for analysis. Integrating IoT with wearable technology allows real-time data collection and monitoring, significantly enhancing patient management, especially for chronic conditions (12, 13). For instance, smart devices can alert healthcare providers to critical patient condition changes, facilitating timely interventions (14, 15). The basic details of telemedicine are shown in Figure 13.1.

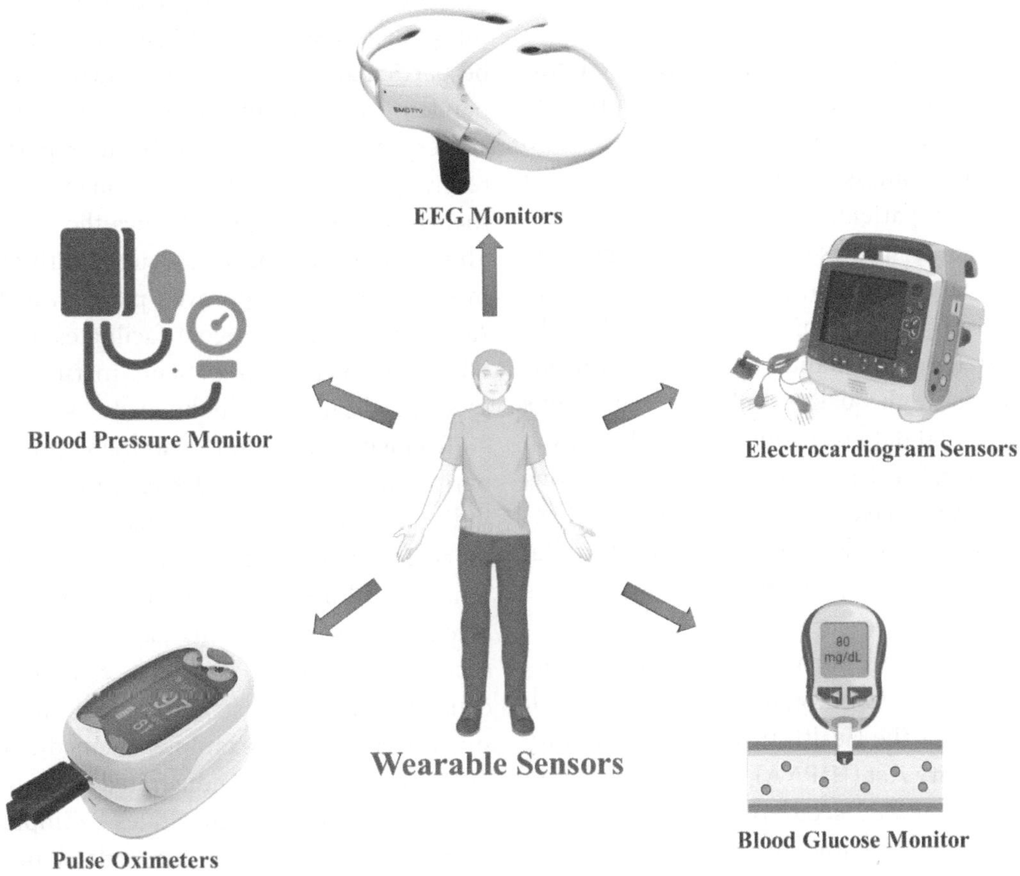

FIGURE 13.1 The application of wearable sensors (telemedicine) in healthcare.

AI and data analytics transform remote healthcare by enabling predictive analytics and personalized medicine. AI algorithms can analyze vast amounts of data collected from wearable devices and EHRs to identify trends and predict health outcomes. This capability enhances decision-making processes, allowing healthcare providers to tailor interventions based on individual patient needs (16, 17). Moreover, AI-driven tools can assist in diagnosing conditions and recommending treatment plans, thereby improving the overall quality of care (18). Cloud computing is the backbone of telemedicine and RPM, providing data storage, management, and sharing infrastructure. It enables healthcare providers to access patient data from anywhere, facilitating collaboration and continuity of care (19). The scalability of cloud solutions allows healthcare organizations to manage large volumes of data efficiently, ensuring that patient information is readily available for analysis and decision-making (20). Furthermore, cloud computing supports integrating various telemedicine services, enhancing the healthcare delivery system (21).

The successful implementation of telemedicine and RPM necessitates robust infrastructure and reliable connectivity. High-speed internet access is essential for seamless communication and data transfer between patients and healthcare providers. In rural and underserved areas, improving connectivity can significantly enhance access to telemedicine services, bridging the gap in healthcare delivery (22, 23). Bandwidth considerations are critical, as high-quality video consultations require substantial data transmission capabilities to ensure clear communication without interruptions (17). Security and privacy are paramount in telemedicine, given the sensitive nature of health data. Implementing stringent security measures, such as encryption and secure access protocols, is essential to protect patient information from unauthorized access (24, 25). Additionally, compliance with regulations such as the Health Insurance Portability and Accountability Act (HIPAA) is crucial to ensure patient data is handled securely and ethically. By prioritizing security, healthcare providers can foster patient trust, encouraging the adoption of telemedicine services (26, 27).

13.3 CLINICAL APPLICATIONS OF TELEMEDICINE AND RPM

Chronic diseases such as diabetes, hypertension, and cardiovascular diseases are prevalent globally, necessitating effective management strategies. Telemedicine facilitates continuous monitoring and management of these conditions, allowing healthcare providers to track patients' health metrics remotely. For instance, RPM technologies enable real-time data collection on blood glucose levels in diabetic patients, which can lead to timely interventions and adjustments in treatment plans. Studies have shown that patients with chronic conditions who engage in telemedicine report higher satisfaction levels and improved health outcomes than traditional care methods (28, 29). Furthermore, telemedicine reduces the need for frequent in-person visits, which can be burdensome for patients with mobility issues or those living in remote areas (30).

Telemedicine offers innovative solutions for post-operative care, and rehabilitation is critical to patient recovery. Remote monitoring allows healthcare providers to assess recovery progress without necessitating patient visits to healthcare facilities, which can be particularly advantageous during the COVID-19 pandemic (31). For instance, patients recovering from surgery can use telemedicine to report symptoms, receive guidance on rehabilitation exercises, and have their progress monitored by healthcare professionals. This approach has been associated with reduced readmission rates and improved patient satisfaction (32). Additionally, telemedicine facilitates access to rehabilitation services for patients in rural areas, where specialized care may not be readily available (33).

Telemedicine is vital in addressing healthcare disparities in rural and underserved populations. Many individuals in these areas face significant barriers to accessing healthcare, including long travel distances and limited availability of specialists (34, 35). Telemedicine bridges this gap by providing remote access to healthcare services, enabling patients to consult with specialists without the need for extensive travel (36). For example, telemedicine initiatives in rural India have demonstrated a significant reduction in maternal mortality rates by improving access to prenatal care. Furthermore, telemedicine has been shown to reduce overall healthcare costs for patients in rural settings, as it minimizes travel expenses and time away from work (37). The successful implementation of telemedicine in these communities requires ongoing investment in technology and infrastructure to ensure equitable access (38).

The integration of telemedicine into mental health services has significantly transformed care delivery. Teletherapy and remote counseling have become increasingly popular, allowing patients to receive mental health support from the comfort of their homes (39). This modality has been particularly beneficial for individuals in rural areas, where mental health resources may be scarce (40). Studies indicate that teletherapy is as effective as in-person therapy for many patients, leading to improved access to care and reduced stigma associated with seeking help. Furthermore, telemedicine facilitates ongoing support for patients with chronic mental health conditions, enabling regular check-ins and adjustments to treatment plans as needed (41).

13.4 BENEFITS OF TELEMEDICINE AND RPM

Telemedicine and RPM have emerged as transformative components of modern healthcare delivery, particularly during the COVID-19 pandemic. These technologies enhance patient care and improve healthcare efficiency, accessibility, and cost-effectiveness. This chapter discusses the multifaceted benefits of telemedicine and RPM, supported by recent research findings. One of the primary advantages of telemedicine is its ability to provide timely and accessible healthcare services. By facilitating virtual consultations, telemedicine reduces the need for in-person visits, minimizing patient travel time and associated costs. Walker et al. highlight that remote monitoring in peritoneal dialysis supports timely care and fosters patient education, which is crucial for chronic disease management (42). Similarly, the use of RPM in cardiac care has been shown to enhance patient outcomes by enabling early detection of arrhythmias and device malfunctions, leading to timely interventions (43). The COVID-19 pandemic accelerated the adoption of these technologies, demonstrating their effectiveness in maintaining continuity of care while adhering to social distancing guidelines (44).

Moreover, telemedicine and RPM significantly enhance patient engagement and self-management. In real-time, patients with remote monitoring devices can track their health metrics, such as blood pressure and heart rate. This capability empowers patients to take an active role in their healthcare, fostering a sense of autonomy and responsibility. Research by El-Rashidy et al. emphasizes that RPM systems facilitate the collection of vital signs, enabling healthcare providers to make informed decisions promptly (45). Additionally, smartphone-assisted monitoring has been shown to improve communication between patients and healthcare providers, further enhancing patient satisfaction and adherence to treatment plans (46).

The economic implications of telemedicine and RPM are also noteworthy. By reducing the frequency of in-person visits, healthcare systems can lower operational costs while reallocating resources to more critical areas of patient care. Hwang and Bae report that remote monitoring of CIEDs not only decreases unnecessary outpatient visits but also enhances patient satisfaction (47). Furthermore, integrating RPM in chronic disease management has been associated with reduced healthcare utilization, as evidenced by studies showing lower hospitalization rates among patients engaged in remote monitoring programs. This economic efficiency is particularly relevant in resource-limited settings, where the cost of care can be a significant barrier to access (48).

In addition to economic benefits, telemedicine and RPM contribute to improved healthcare outcomes. For instance, a study on remote monitoring of implantable cardioverter defibrillators (ICDs) found that it led to more rapid clinical event detection and a reduction in inappropriate shocks, which are critical for patient safety (47). Similarly, Goto et al. reported that remote monitoring in pacemaker patients provided a sense of safety comparable to traditional follow-ups while also facilitating early detection of potential issues (43). These findings underscore the potential of telemedicine and RPM to enhance clinical outcomes through proactive monitoring and timely interventions.

However, the implementation of telemedicine and RPM is not without challenges. Data privacy concerns, technological barriers, and the need for adequate training for patients and healthcare providers can hinder the effective use of these technologies. Lindsay et al. note that the lack of understanding and support for the necessary work involved in remote monitoring can impede its routine adoption in clinical practice (49). Additionally, disparities in digital literacy among patients can affect the overall effectiveness of RPM interventions, as highlighted by research indicating that younger, more educated individuals are more likely to benefit from remote monitoring (50).

Despite these challenges, the potential benefits of telemedicine and RPM are substantial. The ongoing evolution of healthcare technology and increasing acceptance of virtual care models suggest a promising future for these modalities. Integrating advanced technologies such as the IoT and AI into telemedicine and RPM systems can enhance their effectiveness and user-friendliness (51). For instance, IoT-enabled devices can provide continuous monitoring and real-time feedback, making it easier for patients to manage their health conditions effectively.

13.5 CHALLENGES AND BARRIERS

The advent of telemedicine and RPM has revolutionized healthcare delivery, particularly during the COVID-19 pandemic. However, several challenges and barriers impede its widespread adoption and effectiveness. These challenges include regulatory and legal issues, technological limitations, and patient and provider adoption barriers. One of the primary challenges in telemedicine is navigating the complex regulatory landscape. Licensing and cross-border telemedicine practices pose significant hurdles. Each jurisdiction has licensing requirements, which complicate care provision across state or national lines. During the COVID-19 pandemic, many countries relaxed regulations to facilitate telemedicine use, but these changes were often temporary and varied significantly across regions (52, 53). The lack of a standardized framework can lead to confusion among healthcare providers and patients alike, as they may be unsure about the legality of telemedicine practices in different locations.

Moreover, compliance with data protection regulations such as HIPAA in the United States and GDPR in Europe is critical yet challenging. Telemedicine involves the exchange of sensitive health information, necessitating robust security measures to protect patient data (54). The rapid implementation of telemedicine during the pandemic highlighted vulnerabilities in data security, as many healthcare providers were unprepared for the increased risk of data breaches (55). Therefore, establishing clear legal frameworks and ensuring compliance with data protection regulations are essential for fostering trust in telemedicine services.

Technological barriers are another significant challenge in the implementation of telemedicine. Connectivity issues in remote areas can severely limit access to telehealth services. Many patients in rural or underserved regions lack reliable internet access, which can hinder their ability to participate in virtual consultations (56). This digital divide exacerbates existing health disparities, as those who would benefit most from telemedicine may be the least able to access it. Additionally, device interoperability and integration challenges present obstacles to effective telemedicine. Many healthcare providers utilize different platforms and technologies, leading to fragmented care and inefficiencies (57, 58). The lack of standardized protocols for telemedicine can complicate the integration of various devices and systems, making it difficult for providers to share information seamlessly and coordinate care effectively.

Adoption barriers among patients and providers also play a crucial role in telemedicine's challenges. Digital literacy is a significant concern, particularly among older adults and individuals from lower socioeconomic backgrounds. Many patients may struggle to navigate telemedicine platforms, leading to frustration and reluctance to use these services (59, 60). Training and educational resources are essential to enhance digital literacy and ensure that patients can effectively engage with telehealth technologies (61). Furthermore, ethical concerns in virtual diagnosis can deter both patients and providers from fully embracing telemedicine. The inability to conduct physical examinations can lead to apprehensions about the quality of care delivered through virtual consultations (62, 63). Providers may feel less confident in diagnosing and treating patients remotely, impacting their willingness to adopt telemedicine practices (64). Addressing these ethical concerns through clear guidelines and best practices is vital for fostering confidence in telemedicine.

13.6 REIMBURSEMENT POLICIES AND ECONOMIC CONSIDERATIONS

The integration of telemedicine and RPM into healthcare systems has been accelerated by the COVID-19 pandemic, prompting significant changes in reimbursement policies and economic considerations. As healthcare providers increasingly adopt these technologies, understanding the financial implications and reimbursement frameworks becomes crucial for sustainable implementation. Reimbursement for telehealth services has historically been inconsistent and influenced by various factors, including state regulations, payer policies, and the services' nature. For instance, differences in reimbursement rates for video versus telephone consultations have created disparities in access to care (65). The pandemic prompted emergency legislation that temporarily relaxed restrictions on telehealth reimbursement, allowing for broader coverage by insurance companies (66). This shift has been essential for maintaining continuity of care during heightened demand for remote services (67). However, the long-term sustainability of these changes remains uncertain, as many providers express concerns about the permanence of these policies (68).

Economic considerations surrounding telemedicine extend beyond reimbursement rates. The cost-effectiveness of RPM technologies, particularly in managing chronic diseases, has been a focal point of research. Studies indicate that RPM can lead to reduced hospitalizations and lower overall healthcare costs by facilitating timely interventions and improving patient outcomes (69). For example, a systematic review highlighted the potential of RPM to enhance disease management in cardiovascular patients, suggesting that it could be a cost-effective alternative to traditional in-person visits (70). However, the economic benefits of RPM are often contingent on adequate reimbursement structures that incentivize its use (71).

Moreover, the implementation of telehealth services has revealed disparities in access and utilization, particularly among vulnerable populations. Higher telehealth utilization rates among Medicaid patients have been attributed to state-level policy changes that expanded coverage (72). This indicates that equitable reimbursement policies are essential to ensure all patient demographics can benefit from telemedicine services. Additionally, the need for ongoing education and support for both patients and providers is critical to maximize the effectiveness of RPM systems (44). The need for technological infrastructure and training also shapes the economic landscape of telemedicine. Investments in telehealth platforms and training for healthcare professionals are necessary to ensure effective service delivery (73). However, these initial costs can be a barrier for some healthcare systems, particularly smaller practices that may lack the resources to implement comprehensive telehealth solutions (74). Reimbursement policies must account for these initial investments to encourage widespread adoption.

13.7 CASE STUDIES: SUCCESSFUL IMPLEMENTATIONS AND LESSONS LEARNED

Telemedicine and RPM have emerged as transformative forces in healthcare delivery, particularly in managing chronic diseases, enhancing access to care in rural areas, and leveraging AI for predictive healthcare. This chapter explores three case studies that exemplify these advancements: Remote Monitoring for Chronic Disease Management, Telehealth in Rural Healthcare Systems, and AI-Powered Telemedicine in Predictive Healthcare.

13.7.1 Case Study 1: Remote Monitoring for Chronic Disease Management

Remote monitoring has proven to be a pivotal strategy in managing chronic diseases such as diabetes, hypertension, and chronic obstructive pulmonary disease (COPD). Studies indicate that telemedicine interventions significantly improve disease-specific outcomes, including better glycemic control in diabetic patients and enhanced blood pressure management in hypertensive individuals (75). The integration of RPM technologies allows healthcare providers to collect real-time health data from patients, facilitating timely interventions and personalized care plans (70). For instance, the inSight program demonstrated that pharmacists could effectively manage chronic diseases through telephone and remote monitoring, improving patient outcomes. Furthermore, a systematic review highlighted that RPM not only enhances patient engagement but also reduces hospital admissions and emergency visits, showcasing its efficacy in chronic disease management (76).

13.7.2 Case Study 2: Telehealth in Rural Healthcare Systems

Telehealth has been particularly beneficial in rural healthcare systems, where access to traditional healthcare services is often limited. Research indicates that telehealth interventions have led to decreased travel costs and times for patients while maintaining similar levels of patient satisfaction compared to in-person visits (77). For example, an extensive integrated healthcare system in New York utilized telemedicine to provide express care for minor complaints, effectively addressing the healthcare needs of rural populations (78). Additionally, the COVID-19 pandemic accelerated the adoption of telehealth services, which played a critical role in maintaining healthcare access for underserved communities (79). The ability to conduct remote consultations has proven essential in managing chronic conditions, particularly during public health emergencies, thereby reducing the burden on healthcare facilities (80).

13.7.3 Case Study 3: AI-Powered Telemedicine in Predictive Healthcare

The integration of AI into telemedicine represents a significant advancement in predictive healthcare. AI-enhanced remote monitoring systems facilitate continuous health monitoring and data analysis, allowing for proactive management of chronic diseases (81).

These systems can identify patterns and predict potential health crises, enabling timely interventions to prevent hospitalizations (82). For instance, wearable devices with AI capabilities can track vital signs and alert healthcare providers to anomalies, thus improving patient outcomes. Moreover, the Internet of Medical Things (IoMT) has emerged as a critical component in this domain, enabling seamless data transmission and real-time monitoring of patients' health conditions (83). Combining AI and telemedicine enhances healthcare delivery efficiency and empowers patients to manage their health actively.

13.8 CONCLUSION

Telemedicine and RPM have significantly transformed modern healthcare by improving accessibility, patient outcomes, and cost-effectiveness. This chapter has explored these evolving healthcare modalities' technological foundations, clinical applications, benefits, challenges, and economic considerations. The integration of telemedicine with communication platforms, wearable devices, AI, and cloud computing has enabled real-time patient monitoring and data-driven decision-making. These advancements have particularly benefited chronic disease management, postoperative care, rural healthcare, and mental health services. One of the most profound impacts of telemedicine and RPM is their role in improving patient engagement and self-management. By facilitating continuous health monitoring, these technologies empower patients to take an active role in their healthcare while reducing the burden on healthcare systems. Furthermore, the COVID-19 pandemic accelerated their adoption, demonstrating their potential to maintain continuity of care during crises. Despite these advantages, challenges such as regulatory barriers, data privacy concerns, digital literacy issues, and disparities in access to telehealth remain significant obstacles to widespread implementation. Addressing these challenges requires policy reforms, enhanced security measures, and investment in digital infrastructure, particularly in underserved areas.

The economic implications of telemedicine and RPM also underscore their potential to optimize healthcare costs. Studies have shown that RPM reduces hospitalizations and emergency visits, leading to substantial cost savings for patients and healthcare providers. However, sustaining these technologies requires supportive reimbursement policies and financial incentives for healthcare systems. The future of telemedicine lies in integrating AI, the IoMT, and predictive analytics to enhance personalized medicine and early disease detection. In conclusion, telemedicine and RPM represent a paradigm shift in healthcare delivery. While challenges persist, these technologies' continued advancement and strategic implementation can bridge healthcare disparities, improve patient outcomes, and shape the future of global healthcare systems.

REFERENCES

1. Vudathaneni VKP, Lanke RB, Mudaliyar MC, Movva KV, Kalluri LM, Boyapati R, et al. The impact of telemedicine and remote patient monitoring on healthcare delivery: A comprehensive evaluation. Cureus [Internet]. 2024 [cited 2025 Feb 26];16(3). Available from: https://www.cureus.com/articles/227384-the-impact-of-telemedicine-and-remote-patient-monitoring-on-healthcare-delivery-a-comprehensive-evaluation.pdf
2. Toritsemogba TO, Oluwafunmi AE, Opeoluwa A, Evangel CA, Ifeoma PO. A comprehensive review of telemedicine technologies: Past, present, and future prospects. Int Med Sci Res J. 2024;4(2):183–93.

3. Pinnock H, Hanley J, McCloughan L, Todd A, Krishan A, Lewis S, et al. Effectiveness of telemonitoring integrated into existing clinical services on Hospital admission for chronic obstructive pulmonary disease exacerbation: Researcher blind, multicentre, randomised controlled trial. Br Med J. 2013;347(3): f6070–f6070.
4. Akar JG, Bao H, Jones P, Wang Y, Chaudhry SI, Varosy P, et al. Use of remote monitoring of newly implanted cardioverter-defibrillators: Insights from the patient related determinants of ICD remote monitoring (PREDICT RM) study. Circulation. 2013;128(22):2372–83.
5. Davydov G. Revolutionizing healthcare delivery: The transformative impact of telemedicine in the post-pandemic era. J Int Med Grad[Internet]. 2023 [cited 2025 Feb 26];2(1). Available from: http://www.jimgs.com/ojs/index.php/journal/article/view/90
6. Ananthakrishnan M, Chaitra CM. A qualitative study of community health officers' experiences with telemedicine in a rural block of Madhya Pradesh. PARIPEX INDIAN J Res. 2023;12(8):15–7.
7. Singhal A, Riley JP, Cowie MR. Benefits and challenges of telemedicine for heart failure consultations: A qualitative study. BMC Health Serv Res. 2023;23(1):847.
8. Bange EM, Li Y, Kumar P, Doucette A, Gabriel P, Parikh R, et al. The association between telemedicine, advance care planning, and unplanned hospitalizations among high-risk patients with cancer. Cancer. 2024;130(4):636–44.
9. Mamta Kale EA. Telemedicine revolution: Bridging gaps in access to healthcare. Int J Recent Innov Trends Comput Commun. 2023;11(7):394–9.
10. McGrowder DA, Miller FG, Vaz K, Anderson Cross M, Anderson-Jackson L, Bryan S, et al. The utilization and benefits of telehealth services by health care professionals managing breast cancer patients during the COVID-19 pandemic. Healthcare. 2021;9(10):1401.
11. Kissi J, Dai B, Dogbe CS, Banahene J, Ernest O. Predictive factors of physicians' satisfaction with telemedicine services acceptance. Health Informatics J. 2020;26(3):1866–80.
12. Ahila A, Dahan F, Alroobaea R, Alghamdi WY, Mustafa Khaja M, Hajjej F, et al. A smart IoMT based architecture for e-healthcare patient monitoring system using artificial intelligence algorithms. Front Physiol. 2023;14:1125952.
13. Iranpak S, Shahbahrami A, Shakeri H. Remote patient monitoring and classifying using the internet of things platform combined with cloud computing. J Big Data. 2021;8(1):120.
14. Obira JO, Sinde R. Development of a sensor-based heartbeat and body temperature monitoring system for remote chronic patients. Eng Technol Appl Sci Res. 2021;11(4):7375–80.
15. Qureshi MA, Qureshi KN, Jeon G, Piccialli F. Deep learning-based ambient assisted living for self-management of cardiovascular conditions. Neural Comput Appl. 2022;34(13):10449–67.
16. Hwang DK, Hsu CC, Chang KJ, Chao D, Sun CH, Jheng YC, et al. Artificial intelligence-based decision-making for age-related macular degeneration. Theranostics. 2019;9(1):232–45.
17. Sugandi MS, Mulyana D, Elita RFM, Rusmana A. Lessons learned from the pandemic in Indonesia: Enhancing doctor-patient medical communication to improve diagnostic certainty in telemedicine. J Law Sustain Dev. 2023;11(11):e2076.
18. Prasanth LL, Uma E Revolutionizing neurostimulator care: Enhancing remote health monitoring through SDN-cloud networks. Telecommun Sys: Model Analysis Design Manag. 2025;88(1):1–16.
19. Raghavan A, Demircioglu MA, Taeihagh A. Public health innovation through cloud adoption: A comparative analysis of drivers and barriers in Japan, South Korea, and Singapore. Int J Environ Res Public Health. 2021;18(1):334.
20. Shankar GS, Onyema EM, Kavin BP, Gude V, Prasad BS. Breast cancer diagnosis using virtualization and extreme learning algorithm based on deep feed forward networks. Biomed Eng Comput Biol. 2024;15:11795972241278907.
21. Kotzias K, Bukhsh FA, Arachchige JJ, Daneva M, Abhishta A. Industry 4.0 and healthcare: Context, applications, benefits and challenges. IET Softw. 2023;17(3):195–248.
22. Zhao H, Zhang Z, Tang J. Enhancing rural healthcare through internet-based remote collaborative outpatient services: A comprehensive evaluation in Changzhi, Shanxi Province. Medicine (Baltimore). 2024;103(36):e39614.
23. Yang DM, Chang TJ, Hung KF, Wang ML, Cheng YF, Chiang SH, et al. Smart healthcare: A prospective future medical approach for COVID-19. J Chin Med Assoc. 2023;86(2):138–46.
24. Zhang J, Yang H. A privacy-preserving remote heart rate abnormality monitoring system. IEEE Access. 2023;11:97089–98.
25. Wang S, Gao T, Zhang Y. Searchable and revocable multi-data owner attribute-based encryption scheme with hidden policy in cloud storage. Lin F, editor. PLOS One. 2018;13(11):e0206126.
26. Gallardo MO, Dela Torre J, Ebardo R. The role of initial trust in the behavioral intention to use telemedicine among Filipino older adults. Gerontol Geriatr Med. 2024;10:23337214231222110.
27. Hoque SI, Karim AM, Hossen MR, Arjumand D. Evaluation of patients' satisfaction in telemedicine service quality: A case study on Maizbhanderi Foundation, Fatikchari, Bangladesh. Am Econ Social Rev. 2021;8(1):1–10.

28. Vaidya V, Patil V, Oswal J, Narula A, Khare Y, Patil P, et al. Healthcare in the modern era: Launching a telemedicine-based OPD consultation in rural Pune (process, results, and challenges). Cureus [Internet]. 2024 [cited 2025 Feb 26];16(5). https://www.cureus.com/articles/253600-healthcare-in-the-modern-era-launching-a-telemedicine-based-opd-consultation-in-rural-pune-process-results-and-challenges.pdf
29. Hendy A, Farghaly Abdelaliem SM, Zaher A, Sadek BN, Nashwan AJ, Al-Jabri MMA, Ahmeda A, Hendy A, Alabdullah AAS, Sinnokrot SM. Telehealth satisfaction among patients with chronic diseases: a cross-sectional analysis. PeerJ. 2025;13:e19245.
30. Shareef SM, Goud BP, Domalapally B, Ashraf TN, Prakash GB, Bhavika D, et al. Assessing telemedicine competency among doctors in a tertiary care hospital: A questionnaire-based cross-sectional study. Cureus [Internet]. 2024 [cited 2025 Feb 26];16(4). https://www.cureus.com/articles/236892-assessing-telemedicine-competency-among-doctors-in-a-tertiary-care-hospital-a-questionnaire-based-cross-sectional-study.pdf
31. Bhaskar S, Bradley S, Chattu VK, Adisesh A, Nurtazina A, Kyrykbayeva S, et al. Telemedicine as the new outpatient clinic gone digital: Position paper from the pandemic health system REsilience PROGRAM (REPROGRAM) International Consortium (Part 2). Front Public Health. 2020;8:410.
32. Libreros-Peña L, Quintero JA, Gelves A, Alarcón J, Morales S, Escobar MF, et al. Telemedicine consultation for emergency patients' attention: A clinical experience from a high complex university hospital from Latin America. BMC Health Serv Res. 2023;23(1):559.
33. Gupta PP, Uranw S, Gupta S, Das RK, Bhattarai A, Bhatta N, et al. Study of the impact of a telemedicine service in improving pre-hospital care and referrals to a tertiary care university hospital in Nepal. J Fam Med Prim Care. 2021;10(12):4531–5.
34. Prodhan UK, Rahman MZ, Jahan I. Design and implementation of an advanced telemedicine model for the rural people of Bangladesh. Technol Health Care. 2018;26(1):175–80.
35. Gao J, Fan C, Chen B, Fan Z, Li L, Wang L, et al. Telemedicine is becoming an increasingly popular way to resolve the unequal distribution of healthcare resources: Evidence from China. Front Public Health. 2022;10:916303.
36. Omaghomi TT, Arowoogun JO, Akomolafe O, Odilibe IP, Elufioye OA. Telemedicine in rural Africa: A review of accessibility and impact. World J Adv Res Rev. 2024;21(2):421–31.
37. Park JH, Lee MJ, Tsai MH, Shih HJ, Chang J. Rural, regional, racial disparities in telemedicine use during the COVID-19 pandemic among US adults: 2021 National Health Interview Survey (NHIS). Patient Prefer Adherence. 2023; 17:3477–87.
38. Said S, Khan A, Talreja C, Nazir N, Ibrahim HF, Iqbal S, et al. Bridging healthcare horizons: Illuminating Pakistani medical students' perspectives on telemedicine. Int J Health Med Nurs Pract. 2024;6(3):52–61.
39. Yankappa N, Kumar A, Prasad A, Tiwari L, Kumar P. Clinicodemographic profile and clinical outcome of children presenting to Telemedicine Center at Institute of National Importance of India: A prospective observational study. Int J Telemed Appl. 2024;2024:1–9.
40. Samira A, Ehizogie PA, Bisola OA, Adebukola AA, Emem HU. A review of the challenges and opportunities in implementing health informatics in rural healthcare settings. Int Med Sci Res J. 2024;4(5):606–31.
41. Hubach RD, O'Neil AM, Hamrick J, Ernst C, Stowe M, Stone T, et al. Assessing the amenability of rural MSM to using telemedicine for medical and mental healthcare. Ann LGBTQ Public Popul Health. 2021;2(2):125–34.
42. Walker RC, Tong A, Howard K, Palmer SC. Clinicians' experiences with remote patient monitoring in peritoneal dialysis: A semi-structured interview study. Perit Dial Int J Int Soc Perit Dial. 2020;40(2):202–8.
43. Goto T, Mori K, Nakayama T, Yamamoto J, Shintani Y, Nakasuka K, et al. Transmission rate of remote monitoring and mortality in patients with pacemaker. Circ Rep. 2020;2(9):471–8.
44. Hwang Y, Bae H. The first pilot study on remote monitoring of implantable cardiac electronic devices in South Korea: Rationale and study protocol of a REMOTE- CARE [Internet]. 2023 [cited 2025 Feb 26]. Available from: https://www.researchsquare.com/article/rs-2964150/v1
45. El-Rashidy N, El-Sappagh S, Islam SM, El-Bakry H, Abdelrazek S. Mobile health in remote patient monitoring for chronic diseases: Principles, trends, and challenges. Diagnostics. 2021;11(4):607.
46. Seppen B, Wiegel J, Ter Wee MM, van Schaardenburg D, Roorda LD, Nurmohamed MT, et al. Smartphone-assisted patient-initiated care versus usual care in patients with rheumatoid arthritis and low disease activity: A randomized controlled trial. Arthritis Rheumatol. 2022; 74(11):1737–1745.
47. De Larochellière H, Champagne J, Sarrazin J, Steinberg C, Philippon F, Roy K, et al. Findings of remote monitoring of implantable cardioverter defibrillators during the COVID-19 pandemic. Pacing Clin Electrophysiol. 2020;43(11):1366–72.
48. Idris U, Abdulkarim L, Arkwright B. Systematic overview of remote patient monitoring in Nigeria (West Africa). Telehealth Med Today [Internet]. 2022 [cited 2025 Feb 26];7(3). Available from: https://telehealthandmedicinetoday.com/index.php/journal/article/view/359

49. Lindsay R, Cunnington P, Dixon-Woods M. Why is implementing remote monitoring in virtual wards (hospital at home) for people living with frailty so hard? Qualitative Interview Study [Internet]. 2024 [cited 2025 Feb 26]. Available from: http://medrxiv.org/lookup/doi/10.1101/2024.10.02.24314760
50. Lehmann HI, Sharma K, Bhatia R, Mills T, Lang J, Li G, et al. Real-world disparities in remote follow-up of cardiac implantable electronic devices and impact of the COVID-19 pandemic: A single-center experience. J Am Heart Assoc. 2023;12(3):e027500.
51. Singh R, Sharma HK, Choudhury T, Mor A, Mohanty S, Mohanty SN. Blockchain for IoT-enabled healthcare. EAI Endorsed Trans Pervasive Health Technol [Internet]. 2023 [cited 2025 Feb 26];9. Available from: https://publications.eai.eu/index.php/phat/article/view/3348
52. Kaliyadan F, Al Ameer MA, Al Ameer A, Al Alwan Q. Telemedicine practice in Saudi Arabia during the COVID-19 pandemic. Cureus [Internet]. 2020 [cited 2025;12 Feb 26]; 12. Available from: https://www.cureus.com/articles/46208-telemedicine-practice-in-saudi-arabia-during-the-covid-19-pandemic.pdf
53. Iyer S, Shafi K, Lovecchio F, Turner R, Albert TJ, Kim HJ, et al. The spine physical examination using telemedicine: Strategies and Best practices. Glob Spine J. 2022;12(1):8–14.
54. Nanda M, Sharma R. A review of patient satisfaction and experience with telemedicine: A virtual solution during and beyond COVID-19 pandemic. Telemed E-Health. 2021;27(12):1325–31.
55. Cui F, Ma Q, He X, Zhai Y, Zhao J, Chen B, et al. Implementation and application of telemedicine in China: Cross-sectional study. JMIR MHealth UHealth. 2020;8(10):e18426.
56. Mohammadzadeh N, Rezayi S, Saeedi S. Telemedicine for patient management in remote areas and underserved populations. Disaster Med Public Health Prep. 2023;17:e167.
57. Khoong EC, Butler BA, Mesina O, Su G, DeFries TB, Nijagal M, et al. Patient interest in and barriers to telemedicine video visits in a multilingual urban safety-net system. J Am Med Inform Assoc. 2021;28(2):349–53.
58. Cheng C, Humphreys H, Kane B. Transition to telehealth: Engaging medical students in telemedicine healthcare delivery. Ir J Med Sci. 2022;191(5):2405–22.
59. Ftouni R, AlJardali B, Hamdanieh M, Ftouni L, Salem N. Challenges of telemedicine during the COVID-19 pandemic: A systematic review. BMC Med Inform Decis Mak. 2022;22(1):207.
60. Domingo JD, Karla M. Durian G, Andre H. Lucido M, Kyla Maru A. Magcamit Ma, Louise M. Roque K, Anne Beatrice M. Santos C, et al. Factors affecting the applicability of telemedicine in selected areas of Luzon based on the perceptions of medical professionals. Int J Res Publ [Internet]. 2021 [cited 2025 Feb 26];82(1). Available from: https://www.ijrp.org/paper-detail/2164
61. Garcia-Huidobro D, Rivera S, Valderrama Chang S, Bravo P, Capurro D. System- wide accelerated implementation of telemedicine in response to COVID-19: Mixed methods evaluation. J Med Internet Res. 2020;22(10):e22146.
62. Elam AR, Sidhom D, Ugoh P, Andrews CA, De Lott LB, Woodward MA, et al. Disparities in eye care utilization during the COVID-19 pandemic. Am J Ophthalmol. 2022;233:163–70.
63. Fatabhoy MG, Zhu G, Lajaunie A, Schneiderhan JR, Pierce J. Intimate partner violence and telemedicine usage and satisfaction early in the COVID-19 pandemic. J Am Board Fam Med. 2023;36(5):755–65.
64. Vongsachang H, Lagstein O, Boland MV, Repka MX, Kraus CL, Collins ME. Telemedicine utilization by pediatric ophthalmologists during the COVID-19 pandemic. J Am Assoc Pediatr Ophthalmol Strabismus. 2021;25(5):293–5.
65. Sawicki GS, Van Citters AD, Dieni O, Sabadosa KA, Willis A, Benitez D, et al. Financial impacts of the COVID-19 pandemic on cystic fibrosis care: Lessons for the future. J Cyst Fibros. 2021;20:16–20.
66. Coto J, Restrepo A, Cejas I, Prentiss S. The impact of COVID-19 on allied health professions. PLoS One. 2020;15(10):e0241328.
67. Wosik J, Fudim M, Cameron B, Gellad ZF, Cho A, Phinney D, et al. Telehealth transformation: COVID-19 and the rise of virtual care. J Am Med Inform Assoc. 2020;27(6):957–62.
68. DeHart D, Iachini A, King LB, LeCleir E, Reitmeier M, Browne T. Benefits and challenges of telehealth use during COVID-19: Perspectives of patients and providers in the rural South. Adv Soc Work. 2023;22(3):953–75.
69. Isaranuwatchai W, Redwood O, Schauer A, Van Meer T, Vallée J, Clifford P. A remote patient monitoring intervention for patients with chronic obstructive pulmonary disease and chronic heart failure: Pre-post economic analysis of the smart program. JMIR Cardio. 2018;2(2):e10319.
70. Zhang Y, Pena MT, Fletcher LM, Swint JM, Reneker JC. Cost of remote patient monitoring for cardiovascular disease: A systematic review protocol. JBI Evid Synth. 2022;20(6):1585–92.
71. Ward K, Vagholkar S, Sakur F, Khatri NN, Lau AYS. Visit types in primary care with telehealth use during the COVID-19 pandemic: Systematic review. JMIR Med Inform. 2022;10(11):e40469.
72. Acharya M, Ali MM, Hayes CJ, Bogulski CA, Magann EF, Eswaran H. Trends in telehealth visits during pregnancy, 2018 to 2021. JAMA Netw Open. 2023;6(4):e236630.

73. Shah SS, Gvozdanovic A, Knight M, Gagnon J. Mobile App–based remote patient monitoring in acute medical conditions: Prospective feasibility study exploring digital health solutions on clinical workload during the COVID crisis. JMIR Form Res. 2021;5(1):e23190.
74. Gajarawala SN, Pelkowski JN. Telehealth benefits and barriers. J Nurse Pract. 2021;17(2):218–21.
75. Alanazi ESS, Alenezi MSK, Alenezi MFM, Albarrak SIR, Aghithi ASM, Alenzi AF, et al. Exploring health informatics for chronic disease management: A systematic review of tools and outcomes. J Ecohumanism. 2024;3(8):960–7.
76. Oesterle TS, Karpyak VM, Coombes BJ, Athreya AP, Breitinger SA, Correa Da Costa S, et al. Systematic review: Wearable remote monitoring to detect non-alcohol/nonnicotine-related substance use disorder symptoms. Am J Addict. 2022;31(6):535–45.
77. Butzner M, Cuffee Y. Telehealth interventions and outcomes across rural communities in the United States: Narrative review. J Med Internet Res. 2021;23(8):e29575.
78. Kolluri S, Stead TS, Mangal RK, Coffee RL, Jr, Littell J, Ganti L. Telehealth in response to the rural health disparity. Health Psychol Res. 2022;10(3):37445. https://doi.org/10.52965/001c.37445
79. Demeke HB, Merali S, Marks S, Pao LZ, Romero L, Sandhu P, et al. Trends in use of telehealth among health centers during the COVID-19 pandemic - United States, June 26–November 6, 2020. MMWR Morb Mortal Wkly Rep. 2021;70(7):240–4.
80. Monaghesh E, Hajizadeh A. The role of telehealth during COVID-19 outbreak: A systematic review based on current evidence. BMC Public Health. 2020;20(1):1193.
81. Geneva TI, Ejike IN, Ebube VE, Mojeed DA. AI-enhanced remote monitoring for chronic disease management in rural areas. Int J Appl Res Soc Sci. 2024;6(8):1824–47.
82. Exploring Health Informatics for Chronic Disease... - Google Scholar [Internet]. [cited 2025 Feb 26]. Available from: https://scholar.google.com/scholar?hl=en&as_sdt=0%2C5&q=Exploring+Health+Informatics+for+Chronic+Disease+Management%3A+A+Systematic+Review+of+Tools+and+Outcomes&btnG=
83. Shafiq M, Choi JG, Cheikhrouhou O, Hamam H. Advances in IoMT for healthcare systems. Sensors. 2023;24(1):10.

CHAPTER 14

AI in Drug Discovery and Pharmaceutical Research

Ahmed Tharwat Al-Namrouti, Reem M. Elsapagh, Nourhan S. Sultan, Aya M. Sherif, Mirna F. Mustafa, and Amira A. Aboali

14.1 INTRODUCTION

Advancements in drug discovery have profoundly impacted the field of medication, transforming formerly deadly diseases into manageable conditions that can be treated as part of routine therapeutic practices. An influential factor contributing to this progress in therapeutics is improving the techniques used to develop and evaluate upcoming medications. Typically, a novel medicine does not replicate an existing drug's chemical structure, which requires identifying a novel molecule [1].

The public sector, including universities and research institutes, mostly discovers new molecules. However, developing new medicines is typically carried out by industrial laboratories. This is because it involves costly chemicals, pharmacologic and toxicological screening, and comprehensive testing. Consequently, recent progress is attributed to the pharmaceutical industry, a multibillion-dollar corporation dedicated to creating and marketing drugs. Novel drug development is a time-consuming and costly process. It involves many studies, including preclinical, pharmacokinetic, pharmacodynamic, and toxicological investigations [2]. These studies encompass various experiments conducted in silico, in vitro, and in vivo. Typically, this entire process takes around 4 years on average. The proposed drug undergoes a series of experiments and characterization called drug screening. Various biologic assays are employed at different levels, including molecular, cellular, organ, and whole animal, to assess drug action and selectivity [3]. The selection and quantity of the primary screening tests are based on the intended pharmacologic objective [4]. For example, it is necessary to initially test anti-infective medications against various pathogenic organisms and hypoglycemic drugs for their effectiveness in reducing blood sugar levels. However, the molecule is commonly examined for various effects to determine drug specificity. This type of testing highlights detrimental impacts. Furthermore, this may lead to a previously unanticipated therapeutic benefit [5]. Choosing compounds for further investigation typically occurs in animal models that simulate human diseases. These excessively long and costly procedures must be altered and, if feasible, streamlined [6].

Artificial intelligence (AI), also called machine intelligence, is computers' cognitive ability. AI refers to any method that enables computers to imitate the human brain, and it can be applied at various stages of drug discovery and development [7]. AI in information synthesis pertains to applications that analyze clinically relevant data to optimize drug development and identify upcoming potential targets. AI can be utilized in drug design to generate and enhance the molecular composition of novel medications. A crucial aspect of drug design is understanding how proteins' exact configurations influence their roles in maintaining health and contributing to disease [8]. Last, AI assists in organizing, optimizing, operating, and recruiting patients for clinical trials [9]. It is commonly combined with enhanced patient monitoring during these trials and medical devices that access specific patient data and provide information for medical decision-making [10]. Furthermore, it is

DOI: 10.1201/9781003533351-14

currently feasible to implement AI techniques to assist with healthcare research and services [11]. One such application uses deep learning (DL) models to provide risk-based recommendations for detecting preventable hospital readmissions [12].

In the past, rational drug discovery utilized different machine intelligence methods to assist traditional studies, which were time-consuming and costly. In recent decades, computer technologies such as quantitative structure-activity relationship modeling have been created to efficiently and inexpensively find possible biologically active molecules from large pools of candidate compounds. Subsequently, as drug development transitioned to the era of extensive data, machine learning (ML) methods advanced into DL approaches. These approaches are highly effective and efficient at handling the vast data produced by modern drug discovery techniques [13].

As previously mentioned, the process of discovering novel medicines is intricate, costly, time-intensive, and demanding. An average of 12 years of uninterrupted research and an estimated cost of $2.7 billion is required to discover an effective medicine through the standard drug development pipeline [14]. Reducing research costs and expediting the development process of an innovative drug has become imperative for pharmaceutical companies and patients. Therefore, computer-aided drug discovery has gained popularity as a potent and promising technology, aiming to expedite, reduce costs, and enhance medication design efficacy [15]. The rapid advancement of computational approaches in drug discovery has played a crucial role in drug design and has yielded valuable insights into different aspects of pharmacotherapy. By utilizing AI techniques, which are becoming more prevalent in drug discovery and development, promising outcomes are anticipated [16].

This chapter explains AI's role in pharmacology, pharmacovigilance (PV), and drug safety. Moreover, this chapter discusses AI applications in drug discovery and pharmaceutical product life cycles. In the end, it describes AI's role in pharmaceutical markets.

14.2 AI IN PHARMACOLOGY

AI is presently integrated into almost all facets of pharmacological research and clinical applications [17]. ML algorithms are employed to enhance therapeutic drug monitoring (TDM). Additionally, natural language processing (NLP) is increasingly leveraged to extract real-world data from electronic health records (EHRs) [18]. AI and ML prove particularly beneficial in the analysis of data derived from diverse sources, including disease attributes, clinical patient profiles, genomic information, and the chemical structures of drugs. Additionally, the concept of polypharmacology plays a vital role in the creation of pharmaceuticals that can act on various biological targets while reducing unintended side effects. Chaudhari et al. discuss the idea of polypharmacology and its role with AI in the development of safer drug molecules [19].

Rajpurkar et al. examine the prospects of employing rational design through polypharmacology to create therapeutic agents that are both more effective and less toxic [20]. In a related study, investigate the application of AI and data analysis in the context of polypharmacy. The focus of the AI component is on the identification and characterization of polypharmacy through algorithms that can forecast health outcomes resulting from various medication combinations [21].

14.2.1 AI in Pharmaceutical Sciences

In this instance, we looked at how DL models are used for de novo drug design, reaction outcome prediction, and synthetic route optimization. Drug discovery could proceed more quickly and the design of new drug molecules could be made easier with the combination of AI with chemical knowledge and data-driven approaches. The work by Park and colleagues [22]. is centered on the use of DL in medicinal chemistry and drug discovery, particularly the synthesis of novel medicinal compounds. The application of generative models and reinforcement learning to de novo drug design and optimization is examined.

Jiménez-Luna et al. [23] talk about how to integrate chemical knowledge with ML algorithms to create automated and effective synthesis planning. It has many uses, such as creating new chemical structures, forecasting chemical characteristics, and streamlining artificial pathways. The study emphasizes how DL can speed up the drug discovery process. Additionally, it offers ML models for forecasting the yield and regioselectivity of organic reactions. It talks about how molecular descriptors and reaction fingerprints can be used as input features to forecast the results of reactions. The study shows how ML can direct efforts in synthetic chemistry [24].

14.2.2 AI in Drug Treatment Optimization

TDM helps adjust the dosage of drugs that have a narrow therapeutic range. Statistical models are frequently used to analyze TDM data to determine drug levels and the best treatment approach for individuals. However, the application of AI in this field is not as developed as in drug discovery. This is mainly due to the need for extensive clinical data to train the models, which is not always accessible [25, 26].

Among the most popular docking tools are DOCK [27–29], Glide, and AutoDock. In the context of ML, random forests (RFs) and support vector machines (SVMs) have been used to improve docking scoring functions 14. Aladdin, a method that combines ML and docking, and the RF classifier were shown to improve docking accuracy in VEGFR2, p38α MAPK, and GCR. It also resolved docking issues and scoring functions such as solvation and protein flexibility. Furthermore, DL, a kind of ML that improves docking results, has been applied extensively in drug development and design [7, 30].

As input data, protein voxels and ligand fingerprints are processed by a study's DL neural network design, which outputs linear data by DockBench, including RMSDmin, RMSDave, and nRMSD [31]. LBVS is the initial choice in situations where the target compound's three-dimensional (3D) structure is unknown. It is predicated on the theory that similar structures will result in similar cellular effects. The application of AI in QSAR-based long-base visibility systems has proven successful thus [32, 33]. The QSAR-based LBVS builds upon AI algorithms such as ANN, RF, SVM, Bayesian algorithm, deep neural network (DNN), and kNN, which bear similarities to the SBVS techniques we have been discussing [34].

14.2.3 AI in Medical Diagnosis

Medical diagnoses involving skin cancer, neurological conditions, strokes, Alzheimer's disease, acute ischemic stroke, and other conditions can be successfully made with AI. Infectious diseases are treated with NLP, the first reading device to offer clinicians exceptional flexibility in analyzing the descriptors of chest X-rays, and cancer is diagnosed with LSSVM, an AI-based least square SVM. Alzheimer's disease has been identified in its early stages using [35, 36], different SVM combinations. An enhanced probabilistic neural network (EPNN) can also be used to diagnose Parkinson's disease. The study of diabetic patients with referable diabetic retinopathy (RDR) involved the design of convolutional neural networks (CNNs) [37].

14.3 AI IN DRUG REPURPOSING

Finding novel therapeutic applications for medications that are currently on the market and have already received approval for a different condition or indication is known as drug repurposing. Researchers investigate the potential of already-approved medications to treat a variety of ailments by taking advantage of their well-established pharmacokinetics, mechanisms of action, and safety profiles [38]. Doing so can drastically cut down on the time, expense, and risk involved in introducing a new medication to the market. Repurposing may also help find different ways to treat illnesses for which there are unmet medical needs. Because of its capacity to analyze vast volumes of data and find new correlations and patterns, AI has grown in importance in the drug repurposing industry [39]. Table 14.1 lists the three ways AI aids in drug reprofiling: matrix-based, network-based, and feature-based techniques.

14.4 AI IN THE LIFECYCLE OF PHARMACEUTICAL PRODUCTS

AI has a significant impact on the entire pharmaceutical product development process, from basic research to patient care. It can assist in rational drug design, decision-making, personalized medicine, and management of clinical data [46, 47]. Eularis has created E-VAI, an AI platform that utilizes ML algorithms and an intuitive user interface to generate sales predictions for pharmaceutical products based on market competition, key stakeholders, and current market share [48]. This is beneficial for marketing professionals to effectively allocate resources, enhance sales, and make informed investment decisions.

14.4.1 AI in Drug Discovery

The enormous chemical space, which includes over 10^{60} molecules, supports the creation of numerous drug compounds [49]. However, the drug development process is currently limited by the lack of advanced technologies, making it time-consuming and expensive. With the help of AI, this particular issue can be solved [50]. AI can identify potential drug compounds

TABLE 14.1 Approaches of AI in Drug Repurposing

Approaches of AI in Drug Repurposing	Network-Based Approach	Feature-Based Approach	Matrix-Based Approach
Advantage	1. Helps find links between drugs, diseases, and biological pathways.	Allows for discovering repurposing opportunities by analyzing biological markers shared between drugs and diseases.	1. This method uses matrix operations and algorithms to find patterns and connections in the data.
	2. Allows us to discover drug-disease connections that might not be obvious with standard methods.	This method emphasizes gathering useful data from various sources to forecast drug-disease links.	2. It enables the combination and examination of various biological data to reveal possible drug-disease links.
	3. Initially, biological networks are formed, followed by data integration.		3. This process uncovers hidden patterns, helping to identify potential opportunities for drug repurposing.
	The network is then analyzed, leading to drug identification and subsequent verification and clinical assessment.		
Reference	[40, 41]	[42, 43]	[44, 45]

and accelerate the validation of drug targets, as well as the optimization of drug structure design [49, 51].

AI faces data challenges due to the scale, growth, and diversity of data. For drug development, traditional ML struggles with large datasets. While QSAR-based models can predict compounds and physicochemical parameters, they can't predict complex biological properties and encounter issues like small training sets and experimental errors. New AI approaches like DL show promise for drug molecule safety and efficacy evaluations based on big data modeling. In 2012, several studies found that DL outperformed traditional methods in predicting drug candidate properties [52–54].

The virtual chemical space is vast, resembling a map of molecules that shows their distributions and properties. The goal of chemical space representation is to collect location-based molecular information to look for bioactive chemicals. Virtual screening (VS) assists in selecting the right molecules for further testing. Many open-access chemical spaces exist, including ChemBank, PubChem, DrugBank, and ChemDB.

There are several computer-based methods for screening compounds and selecting potential drug molecules. These methods analyze chemical structures and properties more efficiently, which helps to quickly eliminate non-lead compounds and reduce costs. To identify lead compounds, algorithms such as coulomb matrices and molecular fingerprint identification take into consideration physical, chemical, and toxicological features [55]. Various parameters, like predictive models, molecule similarity, generation process, and in silico approaches, can be used to predict a compound's chemical structure [51]. Paul et al introduced DeepVS, which performed well by docking 40 receptors and 2950 ligands, with 95,000 decoys tested [56]. Another approach used a multiobjective automated replacement algorithm to optimize a cyclin-dependent kinase-2 inhibitor's potency profile by evaluating shape similarity, biochemical activity, and physicochemical properties [57].

AI-based QSAR techniques, including decision trees, RFs, SVMs, and linear discriminant analysis (LDA), have been developed from QSAR modeling tools to identify probable drug candidates. QSAR analysis can be accelerated with these methods [58, 59]. Comparing anonymous substances to conventional methods for biological activity rating, King et al. discovered that six AI systems showed a statistically insignificant difference [60].

14.4.2 AI in Designing Drug Molecules

14.4.2.1 Prediction of the Target Protein Structure

When developing a drug, accurately predicting the target protein's structure is essential. AI, such as AlphaFold, can assist in this process by predicting 3D protein structures. It has demonstrated beneficial effects, correctly predicting 25 of 43 structures [61].

In a study by AlQurashi, an RNN called recurrent geometric network (RGN) was used to predict protein structure. Although AlphaFold might be more accurate than AlQurashi's strategy, especially when dealing with sequences that resemble reference structures, AlQurashi's method was predicted to be faster [62]. Another study used MATLAB and a three-layered NN toolbox to predict the 2D structure of a protein. It was reported that the prediction accuracy for the 2D structure was 62.72% [63].

14.4.2.2 Predicting Drug–Protein Interactions

Drug–protein interactions are crucial in therapy success. AI methods accurately predict these interactions, aiding in drug repurposing and preventing polypharmacology [61]. For example, Vora et al. used an SVM approach to discover new compounds and their interactions with crucial targets based on protein sequences and structural characteristics [61]. Ye et al. employed RF models to predict drug-protein interactions, hastening the drug discovery process [64]. Additionally, Xiao et al. developed iDrugTarget, outperforming existing predictors in accuracy and consistency. AI's predictive ability assists in drug repurposing, qualifying existing drugs for Phase II clinical trials, and reducing costs compared to launching new drugs. The "Guilt by Association" approach forecasts drug-disease associations using ML approaches like SVM, NN, logistic regression, and DL [64].

Polypharmacology—a condition in which a medication interacts with several receptors to produce unwanted side effects—can be predicted by drug–protein interactions [65]. AI can design new molecules based on polypharmacology and generate safer drugs. AI platforms like SOM and extensive databases can link compounds to multiple targets and off-targets. Bayesian classifiers and SEA algorithms can establish connections between drug pharmacological profiles and potential targets.

14.4.3 AI in Drug Screening

Drug discovery is a time-consuming and expensive process. Algorithms are used in VS to assess activity. Biopharmaceutical companies collaborate with IT firms to develop platforms for therapy discovery. Predicting physicochemical properties is crucial for understanding drug interactions. AI-based tools like ML play a significant role in drug development by forecasting compound behavior. Effective drug delivery depends on the drug molecules' ability to attach to target proteins or receptors. A significant objective is to predict drug-target binding affinity (DTBA). AI-based methods can measure binding affinity by analyzing drug and target features or similarities. There are web applications like ChemMapper and SEA that can predict drug-target interactions. ML and DL approaches have been used to determine DTBA, evaluate drug-protein similarity, and consider feature-based and similarity-based interactions [49]. The ability to predict drug toxicity is crucial to avoid harmful effects and reduce the cost of drug discovery. For this, there are web-based programs like Toxtree, admetSAR, pkCSM, and LimTox. Advanced AI-based approaches and the ML algorithm DeepTox can efficiently predict toxicity based on input features and chemical descriptors. DeepTox outperformed other methods by identifying various features within the molecules and predefined toxicophore features [33].

14.5 ROLE OF ARTIFICIAL INTELLIGENCE IN PHARMACOVIGILANCE AND DRUG SAFETY

PV, the science encompassing the detection, assessment, understanding, and prevention of adverse drug events, plays a pivotal role in safeguarding public health throughout a drug's lifecycle [66, 67]. With the arduous journey of developing a new drug spanning a decade or more, the integration of AI emerges as a transformative force in expediting and enhancing critical processes within drug discovery and clinical trials. Pharmaceutical enterprises are increasingly turning to AI not only to streamline research and development operations but also to avert costly errors that may impede progress. Through a data-driven paradigm, AI adeptly navigates vast datasets to pinpoint patient cohorts that stand to benefit most from specific clinical trials, thereby optimizing medical interventions and setting the stage for personalized healthcare initiatives [46].

In the realm of drug safety and PV, ML heralds a new era of proactive risk mitigation and adverse event prediction. AI applications extend across diverse domains, from identifying patients primed for treatment success to extracting pivotal insights from Individual Case Safety Reports (ICSRs). By leveraging AI algorithms, stakeholders can predict patient attrition rates in clinical trials, anticipate adverse events, and even forecast drug efficacy with remarkable accuracy. DL algorithms further empower pharmaceutical entities in predicting molecular structures with analogous functionalities, thereby expediting drug discovery processes and enhancing post-market surveillance initiatives [68]. Automated adverse event detection, a cornerstone application of AI in PV, efficiently identifies safety signals within expansive datasets, bolstering regulatory compliance and fortifying patient safety measures [69]. The convergence of AI and PV not only revolutionizes safety monitoring practices but also underscores the imperative of real-time data analysis and early intervention strategies [70]. By harnessing AI to scrutinize vast troves of unstructured data sourced from EHRs and medical literature, stakeholders can swiftly detect adverse drug reactions (ADRs), ascertain risk-benefit profiles, and proactively address emerging safety concerns [71]. The readiness of AI algorithms for PV hinges on a multifaceted evaluation framework encompassing algorithmic performance, transparency, explainability, and robustness in real-world scenarios. As the landscape of PV continues to evolve, the harmonization of AI best practices and standards will be instrumental in fortifying drug safety frameworks, bolstering public health, and instilling confidence in the integrity of healthcare systems worldwide [72, 73].

14.6 ROLE OF AI IN DRUG FORMULATION AND MANUFACTURING

Every stage of the pharmaceutical product life cycle, from drug discovery to post-marketing surveillance, can seamlessly integrate AI to enhance efficacy [22]. AI plays a pivotal role in supporting decision-making processes, facilitating rational drug design, personalizing treatment regimens, leveraging clinical data for future drug development, and optimizing drug structure design [47]. The pharmaceutical industry's adoption of AI is partly attributed to the time and cost constraints associated with developing novel therapeutic compounds [74]. Technologies like artificial neural networks (ANNs), fuzzy logic, and neuro-fuzzy logic are instrumental in swiftly identifying hit materials, validating drug targets, and refining drug structures, thereby potentially reducing costs and timelines in discovering groundbreaking molecules. Despite these advantages, AI faces challenges in navigating complex, diverse, and ambiguous datasets [75].

The evolution from ML to DL has equipped AI with the capability to handle vast amounts of data generated during drug discovery and development [59]. Techniques such as XenoSite, FAME, and SMARTCyp harness AI to identify drug metabolism sites, while virtual chemical space exploration unveils molecular topographies and aids in the selection of bioactive compounds through VS. Open-access chemical databases like PubChem, ChemBank, DrugBank, and ChemDB serve as valuable resources in this pursuit. Various tools and frameworks, including reinforcement learning for optimizing drug combinations, DeepChem for DL in chemistry, and Neural Graph Fingerprints for encoding molecular structures, illustrate the breadth of AI applications in pharmaceutical research and development, spanning from formulation to toxicity prediction and drug optimization [76]. Leveraging NLP tools, cheminformatics resources, and predictive modeling techniques, AI catalyzes innovation across pharmaceutical subfields, revolutionizing drug manufacturing and streamlining formulation techniques for next-generation drug products [77].

14.7 AI IN NATURAL LANGUAGE PROCESSING IN THE FIELD OF PHARMACY

NLP, a subset of AI, utilizes information technologies to process human language, comprehends it to a certain extent, and applies it across various domains. In recent times, NLP has predominantly transitioned to DNNs, notably large language models (LLMs) pre-trained on extensive text datasets to capture diverse linguistic nuances, general knowledge, and domain-specific information. Within the realm of pharmacology, NLP plays a critical role in addressing key tasks and applications through a spectrum of methodologies, encompassing traditional approaches and cutting-edge DL techniques. Named entity recognition (NER) stands out as a prominent NLP method used to categorize named entities within text. In the pharmacological context, these entities can range from cells, genes, and proteins to diseases and drugs. Relation Extraction is another vital NLP process that involves discerning semantic connections between entities, such as ADRs, drug–drug interactions (DDI), and associations between medications and their properties. ADRs identification, crucial in pharmaceutical safety, relies on NLP for the detection of harmful reactions linked to drugs and diseases, often leveraging advanced language models like BERT [78].

Literature-based drug discovery (LBD) serves as a valuable technique for extracting new insights from scientific texts. By employing varied NLP tasks like entity recognition, relation extraction, and reaction extraction, researchers can sift through vast volumes of pharmacological literature to extract pertinent knowledge. Question answering (QA) tasks within NLP facilitate the summarization of pharmaceutical literature and can address inquiries related to emerging health challenges like COVID-19, even across multiple languages, including low-resource languages like Persian [79, 80]. Leveraging NLP to access and analyze unstructured data from EHRs presents a promising avenue for enhancing patient care, evaluating medication efficacy in real time, and supporting informed decision-making within pharmacy and therapeutics committees [81].

14.8 AI IN QUALITY CONTROL AND QUALITY ASSURANCE

Pharmaceutical quality assurance is a critical ingredient of the broader discipline of pharmaceutical quality management. The field of quality assurance is rapidly evolving due to the widespread use of AI and ML technologies, providing new opportunities to enhance pharmaceutical quality assurance practices [82]. These state-of-the-art technologies have the power to completely transform traditional methods by offering enhanced capabilities in data analysis, predictive modeling, and decision support [83].

Pharmaceutical companies stand to benefit significantly from integrating AI and ML into Six Sigma documentation techniques, leading to improved efficiency, streamlined processes, and enhanced quality assurance efforts. These technologies hold the potential to accelerate innovation and improve outcomes across various aspects of pharmaceutical research and development, from identifying potential drug candidates to optimizing industrial processes and predicting adverse reactions [84].

In the realm of pharmaceutical quality assurance, AI and ML find wide-ranging applications across several Quality Management System (QMS) components, offering innovative ways to boost productivity, accuracy, and compliance. Analyzing production processes and product quality represents a primary use of AI and ML in quality assurance, enabling pharmaceutical firms to proactively identify abnormalities and ensure product consistency [85]. By leveraging ML algorithms for automated inspection procedures, quality control processes can be enhanced, leading to increased speed, accuracy, and reduced human error in pharmaceutical production [86].

Furthermore, AI and ML solutions play a significant role in enhancing quality assurance procedures not only in manufacturing but also in clinical trials and regulatory compliance [87]. By employing ML models to analyze clinical trial data and regulatory papers, pharmaceutical companies can identify patient subpopulations, forecast treatment outcomes, and expedite regulatory compliance activities [88]. These technologies facilitate proactive quality control measures by predicting future quality problems through the analysis of historical data, ultimately leading to improved decision-making and product quality.

In conclusion, the integration of AI and ML technologies in pharmaceutical quality control and quality assurance processes presents a wealth of opportunities for enhancing data integrity, automation, predictive analytics, and decision-making, thereby revolutionizing current practices in the pharmaceutical industry.

14.9 PHARMACEUTICAL MARKET OF AI

Pharmaceutical companies are increasingly leveraging AI to reduce financial costs and mitigate failure risks associated with drug development. The AI market is anticipated to grow to $5 billion by 2024, a significant rise from $200 million in 2015 and $700 million in 2018 [89]. With a projected annual growth rate of 40% from 2017 to 2024, AI is poised to revolutionize the pharmaceutical and healthcare industries. Many pharmaceutical firms have invested in AI and continue to do so. They have also formed partnerships with AI companies to create vital healthcare solutions. A notable example is the collaboration between the Royal Free London NHS Foundation Trust and Google subsidiary DeepMind Technologies to address acute kidney injury [89].

14.9.1 AI in Market Positioning

Market positioning involves crafting a distinct identity for a product in the market to attract consumers, making it a critical part of most business strategies for companies aiming to establish a unique identity [90, 91]. This method was employed in marketing the pioneer brand Viagra, which was targeted not only for treating men's erectile dysfunction but also for addressing other quality-of-life issues.

14.9.2 AI in Market Prediction and Analysis

A company's success depends on the ongoing development and expansion of its business. Despite having significant financial resources, the pharmaceutical industry's R&D output is declining due to the failure to adopt new marketing technologies [92]. The progress in digital technologies, known as the "Fourth Industrial Revolution," is facilitating innovative digital marketing through a multicriteria decision-making approach. This approach gathers and analyzes statistical and mathematical data, incorporating human insights to develop AI-based decision-making models that explore new marketing methods [93].

AI-based software can also predict sales and conduct market analysis. It engages consumers and raises awareness among physicians by showing ads that link directly to the product site with a single click [94]. Additionally, these methods utilize NLP tools to analyze keywords entered by customers, linking them to the likelihood of purchasing the product [95, 96]. Market prediction is crucial for pharmaceutical distribution companies, which can utilize AI through tools like "Business Intelligent Smart Sales Prediction Analysis." This approach combines time series forecasting with real-time applications to help pharmaceutical companies forecast product sales in advance, avoiding the costs associated with excess inventory and preventing customer loss due to shortages.

14.9.3 AI in Product Cost

Based on market analysis and the costs incurred in developing a pharmaceutical product, a company sets the final price. The key to using AI for this task lies in its ability to replicate human expert reasoning to evaluate the factors influencing the product's post-manufacturing price [97]. These factors include research and development expenses, the country's stringent price regulations, the length of the exclusivity period, the market share of the new drug a year before patent expiration, the price of the reference product, and price-fixing policies that determine the pricing of both branded and generic drugs [98].

14.9.4 Post-Market Surveillance

Once a drug is approved for the market, AI can be used to monitor it for potential harm. This helps in the early identification of patient safety issues and enhances patient protection [99]. The US Food and Drug Administration recognizes that AI-driven technologies can improve the reliability and efficacy of medical products but emphasizes the importance of safeguarding patients during their development and use. The FDA plans to work with the industry to establish regulations that encourage innovation while ensuring patient safety [100].

14.10 ONGOING CHALLENGES IN ADOPTING AI

The success of AI relies heavily on having ample data for training, which can be costly and must be dependable for accurate predictions. Challenges in AI adoption in pharmaceuticals include a shortage of skilled staff, budget constraints for smaller firms, concerns about job displacement, doubts about AI-generated data, and the opacity of AI decision-making processes [13].

Automation in drug development, manufacturing, supply chains, clinical trials, and sales will eventually occur, but these areas are classified as "narrow AI." This type of AI is trained with a large volume of data, making it specialized for specific tasks. Consequently, human intervention remains crucial for the successful implementation, development, and operation of AI systems. The concern about unemployment may be unfounded, as AI is primarily replacing repetitive tasks, thereby allowing human intelligence to focus on developing more complex insights and fostering creativity [22].

AI has already been adopted by many pharmaceutical companies, and it is anticipated that AI-based solutions in the pharmaceutical sector will generate revenues of $2.199 billion by 2022. Between 2013 and 2018, the industry invested over $7.20 billion in more than 300 deals.

Pharmaceutical organizations need to understand the potential of AI technology for solving problems once implemented and set realistic goals. To fully leverage AI, companies must develop skilled data scientists and software engineers with a strong understanding of AI technology and the company's business targets and R&D goals [22].

14.11 AI-BASED MODELS

AI models are gaining traction in drug discovery and development due to their ability to predict biological or chemical properties. By training on extensive historical data, these models can forecast the behavior of new compounds with speed and accuracy [49, 101].

This provides a notable advantage over traditional methods, which are often time-consuming, costly, and require extensive trial and error. AI models can expedite the drug discovery process and lower the chances of developing ineffective or toxic drugs [49, 102].

The field of AI focuses on creating machines that possess intelligence, especially intelligent computer programs [103].

The pharmaceutical sector is experiencing a transformative shift due to the application of AI models in drug formulation and manufacturing processes. AI's advanced data mining and algorithm analysis tools are revolutionizing various aspects of the industry, enabling more efficient and cost-effective medication development [73].

Supervised learning is an ML approach where an algorithm is trained on a labeled dataset, meaning the correct output is already known. By analyzing patterns and relationships in this labeled data, the algorithm learns to map input data to the appropriate output. This method is widely used in applications like image recognition, NLP, and predictive modeling. Task-driven strategies set specific goals to achieve desired results from given inputs, using labeled data to train algorithms for tasks such as data classification or outcome prediction. The main supervised learning tasks are classification, which predicts labels, and regression, which predicts quantities. Various techniques are available for these tasks, depending on the data and problem domain, including Naïve Bayes, K-nearest neighbors, SVMs, ensemble learning, RF, linear regression, and support vector regression, among others [104].

Supervised learning algorithms play a crucial role in various aspects of the pharmaceutical industry [105]. In drug discovery and design, these algorithms predict the activity or properties of new drug candidates by learning from known compounds and their activities [106]. In pharmaceutical manufacturing, supervised learning aids in predictive maintenance and quality control by analyzing manufacturing processes and sensor data to foresee equipment failures and product quality issues [107]. For drug target identification, these algorithms analyze biological data to find potential targets by learning patterns related to drug response or disease progression [108]. In disease diagnosis and prognosis, they classify diseases and predict patient outcomes by training on medical data. Additionally, supervised learning helps detect adverse events in PV by identifying patterns in adverse event reports [109]. Last, predictive modeling in clinical trials uses historical data to forecast patient responses and treatment outcomes, guiding trial design and optimizing patient selection [110].

Unsupervised learning is a type of ML where the algorithm is not given as labeled data and must independently identify patterns and relationships within the dataset. This method is commonly used in exploratory data analysis to uncover hidden structures or clusters. Known as a "data-driven methodology," it aims to extract patterns, structures, or insights from unannotated data. Common unsupervised tasks include clustering, dimensionality reduction, visualization, finding association rules, and anomaly detection. These tasks can be addressed using techniques like clustering algorithms (e.g., hierarchical clustering, K-means, K-medoids, single linkage, complete linkage, BOTS), association learning algorithms, and feature selection and extraction methods (e.g., Pearson correlation, principal component analysis) based on the characteristics of the data [111].

Unsupervised learning techniques in AI are valuable for pharmaceutical applications, especially in exploratory analysis, pattern recognition, and data visualization. Clustering algorithms group data points by similarities, aiding in target identification, patient stratification, and recognizing distinct classes of compounds or diseases [112]. Dimensionality reduction methods like PCA and t-SNE reduce the complexity of high-dimensional data while preserving key information, facilitating the visualization and exploration of datasets [113]. Anomaly detection algorithms identify rare or unusual data points, useful for spotting adverse events, potential safety concerns, and data quality issues [114]. Association rule mining, such as the Apriori algorithm, uncovers relationships between items in datasets, aiding in understanding drug–drug interactions and co-occurrence patterns between medical conditions and medications [115]. Topic modeling algorithms, like LDA, extract themes from large text datasets, helping analyze scientific literature, clinical trial reports, or social media data to identify research themes, emerging trends, or patient sentiments [116, 117].

Unsupervised learning techniques provide valuable insights and support exploratory analysis in pharmaceutical applications. However, interpreting the results from these methods often requires domain expertise and further validation to derive actionable knowledge and ensure the reliability of the findings [111].

In pharmaceutical product development, various AI and ML models are employed to optimize and innovate processes.

- ANNs assist in predicting the release kinetics of drugs from different dosage forms, facilitating optimal formulations under varied conditions [118].
- Genetic algorithms use natural selection and genetics principles to refine formulation compositions and dosage characteristics [119].
- SVMs help predict relationships between formulation variables, optimizing dosage form design [120].
- Particle swarm optimization (PSO) is applied to enhance particle size distribution and dissolution profiles [121].
- Hybrid models, combining ANNs with genetic algorithms, efficiently search the formulation space for optimal solutions.
- Computational fluid dynamics (CFD) focuses on understanding fluid flow and mixing within dosage manufacturing [122, 123].
- Monte Carlo simulations are utilized to optimize drug products by assessing uncertainties and variabilities [123].
- Response surface methodology (RSM) statistically optimizes dosage forms by analyzing multiple variables [124–126].
- Expert systems leverage rule-based and fuzzy logic to simulate human decision-making in formulation processes [22].
- Last, multivariate analysis techniques, such as PCA and PLS, identify critical formulation variables, aiding in reducing dimensionality and enhancing performance [127–129].

14.12 AI APPROACHES IN PHARMACEUTICAL ANALYSIS

Pharmaceutical analysis encompasses the processes of identifying, determining, quantifying, and purifying pharmaceutical raw materials, which are crucial in drug discovery. The two primary experimental methods in this field are qualitative and quantitative analyses. Despite their high accuracy, these techniques are costly when screening new drug candidates from a large number of natural products. In contrast, computational methods incur minimal costs. Therefore, AI techniques have been integrated into pharmaceutical analysis to complement traditional experimental methods [103].

14.12.1 Prediction of Drug Toxicity

Toxicity prediction in drug development measures the adverse effects of chemicals [130]. Evaluating toxicity is a critical step in drug discovery, aimed at identifying harmful substances for humans [131].

Traditional in vivo testing involves animal experiments, which increase costs. Computational methods offer a low-cost, efficient alternative for predicting chemical toxicity. Consequently, AI-based methods have been developed for this purpose [132]. To compare the effectiveness of various computational toxicity prediction methods, the scientific community introduced the "Toxicology in the 21st Century (Tox21)" challenge [132]. DeepTox, an ensemble model for toxicity prediction, utilizes a three-layer DNN [133]. After data cleaning and quality control, chemicals are encoded with 0D to 3D molecular descriptors as DNN inputs. The DeepTox pipeline is optimized by adjusting hyperparameters like the number of hidden units, learning rate, and dropout rate. Comparative analysis using the Tox21 dataset shows that DeepTox excels in toxicity prediction [133].

14.12.2 Prediction of Drug Bioactivity

In practice, many drugs derived from natural products lack bioactivity, rendering them ineffective. Therefore, evaluating drug bioactivity has become a crucial area in drug discovery. While in vitro and in vivo experiments can simulate the functions of molecules in the human body, they are still costly and time-consuming. Consequently, AI techniques have become a popular alternative for predicting drug bioactivities, such as anticancer, antiviral, and antibacterial activities, due to their cost-effectiveness and time efficiency [103, 134, 135]. For instance, Smith et al. developed a directed message passing neural network capable of predicting antibacterial activity [136]. They first created a molecular graph for each molecule based on its SMILES representation and derived feature vectors from atomic features (like the number of bonds and atomic number) and bond features (such as bond type and stereochemistry) [136]. By applying the message passing operation multiple times, the optimized feature vector was then input into a feedforward neural network, which predicted the antibacterial probability of a molecule [136].

14.12.3 Prediction of Physicochemical Properties of Drugs

Physicochemical properties are fundamental characteristics of drugs, essential for comprehending and modeling their actions. Solubility is particularly important among these properties, as it influences the pharmacokinetic properties and formulations of drugs. Despite its significance, experimental techniques for predicting solubility are laborious and expensive, hindering rapid prediction [137, 138]. Consequently, significant efforts have been directed toward developing AI-based models for solubility prediction [103]. Olayan et al. evaluated various DL techniques, including fully connected neural networks, RNNs, graph neural networks, and SchNet, along with different molecular representation methods such as molecular descriptors, SMILES, molecular graphs, and 3D atomic coordinates, for predicting solubility [137].

14.13 SUMMARY

AI revolutionized drug and pharmaceutical research by providing a unique solution that develops and enhances the drug process. The AI paved the way for the new direction of deep discovery to be faster, more accurate, low-cost with a low side effect, and more effective. All of the advancements and current technologies in AI, resulting in the development of better medicines for patients, progress to evolve, and its influence on the pharmaceutical sector will expand, driving further developments and creating new vistas for advancement in research in the future.

ACKNOWLEDGMENTS

I would like to thank my colleagues for their wonderful collaboration. They supported me greatly and were always willing to help me.

REFERENCES

1. Sirois, C., Khoury, R., Durand, A., Deziel, P.-L., Bukhtiyarova, O., Chiu, Y., ... Simard, M. (2021). Exploring polypharmacy with artificial intelligence: Data analysis protocol. *BMC Medical Informatics and Decision Making*, *21*(1), 219.
2. Workgroup, E. M., Marshall, S. F., Burghaus, R., Cosson, V., Cheung, S. Y. A., Chenel, M., ... Visser, S. A. G. (2016). Good practices in model-informed drug discovery and development: Practice, application, and documentation. *CPT: Pharmacometrics & Systems Pharmacology*, *5*(3), 93–122.
3. Lage, O. M., Ramos, M. C., Calisto, R., Almeida, E., Vasconcelos, V., & Vicente, F. (2018). Current screening methodologies in drug discovery for selected human diseases. *Marine Drugs*, *16*(8), 279.
4. Waring, M. J., Arrowsmith, J., Leach, A. R., Leeson, P. D., Mandrell, S., Owen, R. M., ... Weir, A. (2015). An analysis of the attrition of drug candidates from four major pharmaceutical companies. *Nature Reviews Drug Discovery*, *14*(7), 475–486.
5. Kort, E., & Jovinge, S. (2021). Drug repurposing: Claiming the full benefit from drug development. *Current Cardiology Reports*, *23*(6), 62.
6. Griesinger, C., Desprez, B., Coecke, S., Casey, W., & Zuang, V. (2016). Validation of alternative in vitro methods to animal testing: Concepts, challenges, processes and tools. In C. Eskes & M. Whelan (Eds.), *Validation of Alternative Methods for Toxicity Testing* (pp. 65–132). Cham, Switzerland: Springer International Publishing.
7. Gupta, R., Srivastava, D., Sahu, M., Tiwari, S., Ambasta, R. K., & Kumar, P. (2021). Artificial intelligence to deep learning: Machine intelligence approach for drug discovery. *Molecular Diversity*, *25*(3), 1315–1360.
8. Singh, S. (2024). Leveraging AI and machine learning in Six-Sigma documentation for pharmaceutical quality assurance. *Chinese Journal of Applied Physiology*, e20240005.
9. Askin, S., Burkhalter, D., Calado, G., & El Dakrouni, S. (2023). Artificial intelligence applied to clinical trials: Opportunities and challenges. *Health and Technology*, *13*(2), 203–213.
10. Lysaght, T., Lim, H. Y., Xafis, V., & Ngiam, K. Y. (2019). AI-assisted decision-making in healthcare. *Asian Bioethics Review*, *11*(3), 299–314.
11. Raies, A. B., & Bajic, V. B. (2016). In silico toxicology: Computational methods for the prediction of chemical toxicity. *WIREs Computational Molecular Science*, *6*(2), 147–172.

12. Zhao, W., Chen, J. J., Perkins, R., Liu, Z., Ge, W., Ding, Y., & Zou, W. (2015). A heuristic approach to determine an appropriate number of topics in topic modeling. *BMC Bioinformatics*, *16*(13), S8.
13. Lamberti, M. J., Wilkinson, M., Donzanti, B. A., Wohlhieter, G. E., Parikh, S., Wilkins, R. G., & Getz, K. (2019). A study on the application and use of artificial intelligence to support drug development. *Clinical Therapeutics*, *41*(8), 1414–1426.
14. Cui, W., Aouidate, A., Wang, S., Yu, Q., Li, Y., & Yuan, S. (2020). Discovering anti-cancer drugs via computational methods. *Frontiers in Pharmacology*, *11*, 733.
15. Macalino, S. J. Y., Gosu, V., Hong, S., & Choi, S. (2015). Role of computer-aided drug design in modern drug discovery. *Archives of Pharmacal Research*, *38*(9), 1686–1701.
16. Sarker, I. H. (2022). AI-based modeling: Techniques, applications and research issues towards automation, intelligent and smart systems. *SN Computer Science*, *3*(2), 158.
17. Harrer, S., Shah, P., Antony, B., & Hu, J. (2019). Artificial intelligence for clinical trial design. *Trends in Pharmacological Sciences*, *40*(8), 577–591.
18. Vamathevan, J., Clark, D., Czodrowski, P., Dunham, I., Ferran, E., Lee, G., … Zhao, S. (2019). Applications of machine learning in drug discovery and development. *Nature Reviews Drug Discovery*, *18*(6), 463–477.
19. Chaudhari, R., Fong, L. W., Tan, Z., Huang, B., & Zhang, S. (2020). An up-to-date overview of computational polypharmacology in modern drug discovery. *Expert Opinion on Drug Discovery*, *15*(9), 1025–1044.
20. Rajpurkar, P., Chen, E., Banerjee, O., & Topol, E. J. (2022). AI in health and medicine. *Nature Medicine*, *28*(1), 31–38.
21. Singh, S., Kaur, N., & Gehlot, A. (2024). Application of artificial intelligence in drug design: A review. *Computers in Biology and Medicine*, *179*, 108810.
22. Park, Y., Goto, D., Yang, K., Downton, K., Lecomte, P., Olson, M., & Mullins, C. D. (2016). A literature review of factors affecting price and competition in the global pharmaceutical market. *Value in Health*, *19*, A265.
23. Jiménez-Luna, J., Grisoni, F., & Schneider, G. (2020). Drug discovery with explainable artificial intelligence. *Nature Machine Intelligence*, *2*, 573–584.
24. Verma, H., Mlynar, J., Schaer, R., Reichenbach, J., Jreige, M., Prior, J., … Depeursinge, A. (2023). Rethinking the role of AI with physicians in oncology: Revealing perspectives from clinical and research workflows. In Proceedings of the 2023 CHI Conference on Human Factors in Computing Systems (pp. 1–19). ACM.
25. Gelder, T., & Vinks, A. A. (2021). Machine learning as a novel method to support therapeutic drug management and precision dosing. *Clinical Pharmacology & Therapeutics*, *110*(2), 273–276.
26. Sisodiya, S. M. (2021). Precision medicine and therapies of the future. *Epilepsia*, *62*(S2), S90–S105.
27. Ewing, T. J. A., Makino, S., Skillman, A. G., & Kuntz, I. D. (2001). DOCK 4.0: Search strategies for automated molecular docking of flexible molecule databases. *Journal of Computer-Aided Molecular Design*, *15*(5), 411–428.
28. Friesner, R. A., Banks, J. L., Murphy, R. B., Halgren, T. A., Klicic, J. J., Mainz, D. T., … Shenkin, P. S. (2004). Glide: A new approach for rapid, accurate docking and scoring. 1. Method and assessment of docking accuracy. *Journal of Medicinal Chemistry*, *47*(7), 1739–1749.
29. Trajanov, D., Trajkovski, V., Dimitrieva, M., Dobreva, J., Jovanovik, M., Klemen, M., … Robnik-Šikonja, M. (2023). Review of natural language processing in pharmacology. *Pharmacological Reviews*, *75*(4), 714.
30. Pereira, J. C., Caffarena, E. R., & dos Santos, C. N. (2016). Boosting docking-based virtual screening with deep learning. *Journal of Chemical Information and Modeling*, *56*(12), 2495–2506.
31. Jiménez-Luna, J., Cuzzolin, A., Bolcato, G., Sturlese, M., & Moro, S. (2020). A deep-learning approach toward rational molecular docking protocol selection. *Molecules*, *25*(11), 2487.
32. Dai, W., & Guo, D. (2019). A ligand-based virtual screening method using direct quantification of generalization ability. *Molecules*, *24*(13) 2414.
33. Yadav, S., Singh, A., Singhal, R., & Yadav, J. P. (2024). Revolutionizing drug discovery: The impact of artificial intelligence on advancements in pharmacology and the pharmaceutical industry. *Intelligent Pharmacy*, *2*(3), 367–380.
34. Lima, A. N., Philot, E. A., Trossini, G. H. G., Scott, L. P. B., Maltarollo, V. G., & Honorio, K. M. (2016). Use of machine learning approaches for novel drug discovery. *Expert Opinion on Drug Discovery*, *11*(3), 225–239.
35. Sun, Y., Peng, Y., Chen, Y., & Shukla, A. J. (2003). Application of artificial neural networks in the design of controlled release drug delivery systems. *Advanced Drug Delivery Reviews*, *55*(9), 1201–1215.
36. van der Lee, M., & Swen, J. J. (2023). Artificial intelligence in pharmacology research and practice. *Clinical and Translational Science*, *16*(1), 31–36.
37. Gulshan, V., Peng, L., Coram, M., Stumpe, M. C., Wu, D., Narayanaswamy, A., … Webster, D. R. (2016). Development and validation of a deep learning algorithm for detection of diabetic retinopathy in retinal fundus photographs. *JAMA*, *316*(22), 2402–2410.
38. Cheng, F., Liu, C., Jiang, J., Lu, W., Li, W., Liu, G., & Tang, Y. (2012). Prediction of drug-target interactions and drug repositioning via network-based inference. *PLOS Computational Biology*, *8*(5), e1002503.

39. Venkatapurapu, S. P., Iwakiri, R., Udagawa, E., Patidar, N., Qi, Z., Takayama, R., … Fernandez, J. (2022). A computational platform integrating a mechanistic model of Crohn's disease for predicting temporal progression of mucosal damage and healing. *Advances in Therapy*, *39*(7), 3225–3247.
40. Bi, W. L., Hosny, A., Schabath, M. B., Giger, M. L., Birkbak, N. J., Mehrtash, A., … Aerts, H. J. W. L. (2019). Artificial intelligence in cancer imaging: Clinical challenges and applications. *CA: A Cancer Journal for Clinicians*, *69*(2), 127–157.
41. Jarada, T. N., Rokne, J. G., & Alhajj, R. (2020). A review of computational drug repositioning: Strategies, approaches, opportunities, challenges, and directions. *Journal of Cheminformatics*, *12*(1), 46.
42. Meng, Y., Wang, Y., Xu, J., Lu, C., Tang, X., Peng, T., … Yang, J. (2024). Drug repositioning based on weighted local information augmented graph neural network. *Briefings in Bioinformatics*, *25*(1), bbad431.
43. Noguchi, Y., Ueno, A., Otsubo, M., Katsuno, H., Sugita, I., Kanematsu, Y., … Teramachi, H. (2018). A new search method using association rule mining for drug-drug interaction based on spontaneous report system [original research]. *Frontiers in Pharmacology*, *9*, 197.
44. Wang, K., & Li, M. (2023). Fusion-based deep learning architecture for detecting drug-target binding affinity using target and drug sequence and structure. *IEEE Journal of Biomedical and Health Informatics*, *27*(12), 6112–6120.
45. Wu, F., Zhou, Y., Li, L., Shen, X., Chen, G., Wang, X., … Huang, Z. (2020). Computational approaches in preclinical studies on drug discovery and development. *Frontiers in Chemistry*, *8*, 726.
46. Blasiak, A., Khong, J., & Kee, T. (2019). CURATE. AI: Optimizing personalized medicine with artificial intelligence. *SLAS Technology: Translating Life Sciences Innovation*, *25*(2), 95–105.
47. Duch, W., Swaminathan, K., & Meller, J. (2007). Artificial intelligence approaches for rational drug design and discovery. *Current Pharmaceutical Design*, *13*(14), 1497–1508.
48. Baronzio, G., Parmar, G., & Baronzio, M. (2015). Overview of methods for overcoming hindrance to drug delivery to tumors, with special attention to tumor interstitial fluid. *Frontiers in Oncology*, *5*, 165.
49. Mak, K.-K., & Pichika, M. R. (2019). Artificial intelligence in drug development: Present status and future prospects. *Drug Discovery Today*, *24*(3), 773–780.
50. Milgrom, P., & Tadelis, S. (2019). 23. How artificial intelligence and machine learning can impact market design. In A. Agrawal, J. Gans, & A. Goldfarb (Eds.), *The Economics of Artificial Intelligence: An Agenda* (pp. 567–586). Chicago: University of Chicago.
51. Schneider, P., Walters, W. P., Plowright, A. T., Sieroka, N., Listgarten, J., Goodnow, R. A., … Schneider, G. (2020). Rethinking drug design in the artificial intelligence era. *Nature Reviews Drug Discovery*, *19*(5), 353–364.
52. Ciallella, H. L., & Zhu, H. (2019). Advancing computational toxicology in the big data era by artificial intelligence: Data-driven and mechanism-driven modeling for chemical toxicity. *Chemical Research in Toxicology*, *32*(4), 536–547.
53. Zhavoronkov, A., Vanhaelen, Q., & Oprea, T. I. (2020). Will artificial intelligence for drug discovery impact clinical pharmacology? *clinical pharmacology. & Therapeutics*, *107*(4), 780–785.
54. Zhu, H. (2020). Big data and artificial intelligence modeling for drug discovery. *Annual Review of Pharmacology and Toxicology*, *60*(1), 573–589.
55. Chan, H. C. S., Shan, H., Dahoun, T., Vogel, H., & Yuan, S. (2019). Advancing drug discovery via artificial intelligence. *Trends in Pharmacological Sciences*, *40*(8), 592–604.
56. Paul, D., Sanap, G., Shenoy, S., Kalyane, D., Kalia, K., & Tekade, R. K. (2021). Artificial intelligence in drug discovery and development. *Drug Discovery Today*, *26*(1), 80–93.
57. Firth, N. C., Atrash, B., Brown, N., & Blagg, J. (2015). MOARF, an integrated workflow for multiobjective optimization: Implementation, synthesis, and biological evaluation. *Journal of Chemical Information and Modeling*, *55*(6), 1169–1180.
58. Jain, N., Gupta, S., Sapre, N., & Sapre, N. S. (2015). In silico de novo design of novel NNRTIs: A bio-molecular modelling approach [10.1039/C4RA15478A]. *RSC Advances*, *5*(19), 14814–14827.
59. Yu, H., Chen, J., Xu, X., Li, Y., Zhao, H., Fang, Y., … Wang, Y. (2012). A systematic prediction of multiple drug-target interactions from chemical, genomic, and pharmacological data. *PLoS One*, *7*(5), e37608.
60. King, R. D., Hirst, J. D., & Sternberg, M. J. E. (1995). Comparison of artificial intelligence methods for modeling pharmaceutical QSARS. *Applied Artificial Intelligence*, *9*(2), 213–233.
61. Vora, L. K., Gholap, A. D., Jetha, Thakur, R. R., Solanki, H. K., & Chavda, V. P. (2023). Artificial intelligence in pharmaceutical technology and drug delivery design. *Pharmaceutics*, *15*(7), 1916.
62. AlQuraishi, M. (2019). End-to-end differentiable learning of protein structure. *Cell Systems*, *8*(4), 292–301.
63. Avdagic, Z., Purisevic, E., Omanovic, S., & Coralic, Z. (2009). Artificial intelligence in prediction of secondary protein structure using CB513 database. *Summit on Translational Bioinformatics*, *2009*, 1.
64. Ye, Z., & Ouyang, D. (2021). Prediction of small-molecule compound solubility in organic solvents by machine learning algorithms. *Journal of Cheminformatics*, *13*(1), 98.
65. Li, X., Xu, Y., Cui, H., Huang, T., Wang, D., Lian, B., … Xie, L. (2017). Prediction of synergistic anti-cancer drug combinations based on drug target network and drug induced gene expression profiles. *Artificial Intelligence in Medicine*, *83*, 35–43.

66. Mukhamediev, R. I., Popova, Y., Kuchin, Y., Zaitseva, E., Kalimoldayev, A., Symagulov, A., … Yelis, M. (2022). Review of artificial intelligence and machine learning technologies: Classification, restrictions, opportunities and challenges. *Mathematics, 10*(15), 2552.
67. Stokes, J. M., Yang, K., Swanson, K., Jin, W., Cubillos-Ruiz, A., Donghia, N. M., … Bloom-Ackermann, Z. (2020). A deep learning approach to antibiotic discovery. *Cell, 180*(4), 688–702.
68. Aronson, J. K. (2022). Artificial intelligence in pharmacovigilance: An introduction to terms, concepts, applications, and limitations. *Drug Safety, 45*(5), 407–418.
69. Meyboom, R. H. B., Egberts, A. C. G., Gribnau, F. W. J., & Hekster, Y. A. (1999). Pharmacovigilance in perspective. *Drug Safety, 21*(6), 429–447.
70. Deye, N., Vincent, F., Michel, P., Ehrmann, S., da Silva, D., Piagnerelli, M., … For the SRLF Trial Group (2016). Changes in cardiac arrest patients' temperature management after the 2013 "TTM" trial: Results from an international survey. *Annals of Intensive Care, 6*(1), 4.
71. Reddy, A. S., & Zhang, S. (2013). Polypharmacology: Drug discovery for the future. *Expert Review of Clinical Pharmacology, 6*(1), 41–47.
72. Hamid, A. A. A., Rahim, R., & Teo, S. P. (2022). Pharmacovigilance and its importance for primary health care professionals. *Korean Journal of Family Medicine, 43*(5), 290–295.
73. Xuan, P., Song, Y., Zhang, T., & Jia, L. (2019). Prediction of potential drug- disease associations through deep integration of diversity and projections of various drug features. *International Journal of Molecular Sciences, 20*(17), 4102.
74. Wang, X., Li, Q., Liu, Y., Du, Z., & Jin, R. (2022). Drug repositioning of COVID-19 based on mixed graph network and ion channel. *Mathematical Biosciences and Engineering, 19*(4), 3269–3284.
75. Martin, L., Hutchens, M., Hawkins, C., & Radnov, A. (2017). How much do clinical trials cost? *Nature Reviews Drug Discovery, 16*(6), 381–382.
76. Al Kuwaiti, A., Nazer, K., Al-Reedy, A., Al-Shehri, S., Al-Muhanna, A., Subbarayalu, A. V., … Al-Muhanna, F. A. (2023). A review of the role of artificial intelligence in healthcare. *Journal of Personalized Medicine, 13*(6), 951.
77. Landin, M., & Rowe, R. (2013). Artificial neural networks technology to model, understand, and optimize drug formulations. *Formulation Tools for Pharmaceutical Development* (pp. 7–37). Elsevier, Oxford, UK. ISBN: 9781907568992.
78. Toker, D., Tozan, H., & Vayvay, Ö (2013). A decision model for pharmaceutical marketing and a case study in Turkey. *Ekonomska Istrazivanja, 26*, 101–114.
79. Harper, M. (2006). Introducing speech and language processing. *Computational Linguistics, 32*, 137–142.
80. Streefland, M. B. (2018). Why are we still creating individual case safety reports? *Clinical Therapeutics, 40*(12), 1973–1980.
81. Del Rio-Bermudez, C., Medrano, I. H., Yebes, L., & Poveda, J. L. (2020). Towards a symbiotic relationship between big data, artificial intelligence, and hospital pharmacy. *Journal of Pharmaceutical Policy and Practice, 13*(1), 75.
82. Antony, J., & Coronado, R. B. (2002). Design for six sigma. *Manufacturing Engineer, 81*(1), 24–26.
83. Chopra, S., & Meindl, P. (2007). Supply chain management. Strategy, planning & operation. In C. Boersch & R. Elschen (Eds.), *Das Summa Summarum Des Management: Die 25 Wichtigsten Werke Für Strategie, Führung Und Veränderung* (pp. 265–275). Wiesbaden, Germany: Gabler.
84. Dashpute, S., Pansare, J., Deore, Y., Pansare, M., Sonawane, P., & Jadhav, S. (2023). Artificial intelligence and machine learning in the pharmaceutical industry. *International Journal of Pharmacy and Pharmaceutical Research, 28*, 111–131.
85. Singh, J., Flaherty, K., Sohi, R. S., Deeter-Schmelz, D., Habel, J., Le Meunier- FitzHugh, K., … Onyemah, V. (2019). Sales profession and professionals in the age of digitization and artificial intelligence technologies: Concepts, priorities, and questions. *Journal of Personal Selling & Sales Management, 39*(1), 2–22.
86. Beam, A. L., & Kohane, I. S. (2018). Big data and machine learning in health care. *JAMA, 319*(13), 1317–1318.
87. Jiang, F., Jiang, Y., Zhi, H., Dong, Y., Li, H., Ma, S., … Wang, Y. (2017). Artificial intelligence in healthcare: Past, present and future. *Stroke and Vascular Neurology, 2*(4), 230.
88. Finlayson Samuel, G., Subbaswamy, A., Singh, K., Bowers, J., Kupke, A., Zittrain, J., … Saria, S. (2021). The clinician and dataset shift in artificial intelligence. *New England Journal of Medicine, 385*(3), 283–286.
89. Ali, K. A., Mohin, S. K., Mondal, P., Goswami, S., Ghosh, S., & Choudhuri, S. (2024). Influence of artificial intelligence in modern pharmaceutical formulation and drug development. *Future Journal of Pharmaceutical Sciences, 10*(1), 53.
90. Jalkala, A., & Keränen, J. (2014). Brand positioning strategies for industrial firms providing customer solutions. *The Journal of Business and Industrial Marketing, 29.*
91. Kalafatis, S., Tsogas, M., & Blankson, C. (2000). Positioning strategies in business markets. *Journal of Business & Industrial Marketing, 15*, 416–437.
92. Syam, N., & Sharma, A. (2018). Waiting for a sales renaissance in the fourth industrial revolution: Machine learning and artificial intelligence in sales research and practice. *Industrial Marketing Management, 69*, 135–146.

93. Shiammala, P. N., Duraimutharasan, N. K. B., Vaseeharan, B., Alothaim, A. S., Al- Malki, E. S., Snekaa, B., … Selvaraj, C. (2023). Exploring the artificial intelligence and machine learning models in the context of drug design difficulties and future potential for the pharmaceutical sectors. *Methods, 219*, 82–94.
94. Milgrom, P. R., & Tadelis, S. (2019). *The economics of artificial intelligence: An Agenda* (pp. 567–585). Chicago, IL: University of Chicago Press; for the National Bureau of Economic Research.
95. Davenport, T., Guha, A., Grewal, D., & Bressgott, T. (2020). How artificial intelligence will change the future of marketing. *Journal of the Academy of Marketing Science, 48*(1), 24–42.
96. Sweilam, N. H., Tharwat, A. A., & Abdel Moniem, N. K. (2010). Support vector machine for diagnosis cancer disease: A comparative study. *Egyptian Informatics Journal, 11*(2), 81–92.
97. Duran, O., Rodriguez, N., & Consalter, L. A. (2009). Neural networks for cost estimation of shell and tube heat exchangers. *Expert Systems With Applications, 36*(4), 7435–7440.
98. Panapitiya, G., Girard, M., Hollas, A., Sepulveda, J., Murugesan, V., Wang, W., & Saldanha, E. (2022). Evaluation of deep learning architectures for aqueous solubility prediction. *ACS Omega, 7*(18), 15695–15710.
99. Murali, K., Kaur, S., Prakash, A., & Medhi, B. (2019). Artificial intelligence in pharmacovigilance: Practical utility. *Indian Journal of Pharmacology, 51*(6), 373–376.
100. Shi, G., Lin, L., Liu, Y., Chen, G., Luo, Y., Wu, Y., & Li, H. (2021). Pharmaceutical application of multivariate modelling techniques: A review on the manufacturing of tablets [10.1039/D0RA08030F]. *RSC Advances, 11*(14), 8323–8345.
101. Bao, H., Zhao, J., Zhao, X., Zhao, C., Lu, X., & Xu, G. (2023). Prediction of plant secondary metabolic pathways using deep transfer learning. *BMC Bioinformatics, 24*(1), 348.
102. Wan, F., & Zeng, J. (2016). Deep learning with feature embedding for compound- protein interaction prediction. *bioRxiv*, 086033.
103. Chen, W., Liu, X., Zhang, S., & Chen, S. (2023). Artificial intelligence for drug discovery: Resources, methods, and applications. *Molecular Therapy: Nucleic Acids, 31*, 691–702.
104. Salas, M., Petracek, J., Yalamanchili, P., Aimer, O., Kasthuril, D., Dhingra, S., … Bostic, T. (2022). The use of artificial intelligence in pharmacovigilance: A systematic review of the literature. *Pharmaceutical Medicine, 36*(5), 295–306.
105. Dara, S., Dhamercherla, S., Jadav, S. S., Babu, C. H. M., & Ahsan, M. J. (2022). Machine learning in drug discovery: A review. *Artificial Intelligence Review, 55*(3), 1947–1999.
106. Kavasidis, I., Lallas, E. N., Gerogiannis, V., Charitou, T., & Anthony, K. (2023). Predictive maintenance in pharmaceutical manufacturing lines using deep transformers. *Procedia Computer Science, 220*, 576–583.
107. Bagherian, M., Sabeti, E., Wang, K., Sartor, M. A., Nikolovska-Coleska, Z., & Najarian, K. (2021). Machine learning approaches and databases for prediction of drug–target interaction: A survey paper. *Briefings in Bioinformatics, 22*(1), 247–269.
108. Kumar, Y., Koul, A., Singla, R., & Ijaz, M. F. (2023). Artificial intelligence in disease diagnosis: A systematic literature review, synthesizing framework and future research agenda. *Journal of Ambient Intelligence and Humanized Computing, 14*(7), 8459–8486.
109. Chapman, A. B., Peterson, K. S., Alba, P. R., DuVall, S. L., & Patterson, O. V. (2019). Detecting adverse drug events with rapidly trained classification models. *Drug Safety, 42*(1), 147–156.
110. Elkin, M. E., & Zhu, X. (2021). Predictive modeling of clinical trial terminations using feature engineering and embedding learning. *Scientific Reports, 11*(1), 3446.
111. Visan, A. I., & Negut, I. (2024). Integrating artificial intelligence for drug discovery in the context of revolutionizing drug delivery. *Life, 14*(2), 233.
112. Karim, M. R., Beyan, O., Zappa, A., Costa, I. G., Rebholz-Schuhmann, D., Cochez, M., & Decker, S. (2021). Deep learning-based clustering approaches for bioinformatics. *Briefings in Bioinformatics, 22*(1), 393–415.
113. Trott, O., & Olson, A. J. (2010). AutoDock Vina: Improving the speed and accuracy of docking with a new scoring function, efficient optimization, and multithreading. *Journal of Computational Chemistry, 31*(2), 455–461.
114. Goldstein, M., & Uchida, S. (2016). A comparative evaluation of unsupervised anomaly detection algorithms for multivariate data. *PLoS One, 11*(4), e0152173.
115. Nagarajan, N., Yapp, E. K. Y., Le, N. Q. K., Kamaraj, B., Al-Subaie, A. M., & Yeh, H.-Y. (2019). Application of computational biology and artificial intelligence technologies in cancer precision drug discovery. *BioMed Research International, 2019*(1), 8427042.
116. Liu, L., Tang, L., Dong, W., Yao, S., & Zhou, W. (2016). An overview of topic modeling and its current applications in bioinformatics. *SpringerPlus, 5*(1), 1608.
117. Zhang, L., Tan, J., Han, D., & Zhu, H. (2017). From machine learning to deep learning: Progress in machine intelligence for rational drug discovery. *Drug Discovery Today, 22*(11), 1680–1685.
118. Su, D., Xu, Y., Yu, T., Siddique, F. B., Barezi, E. J., & Fung, P. (2020). CAiRE-COVID: A question answering and query-focused multi-document summarization system for COVID-19 scholarly information management. *arXiv Preprint Arxiv:2005.03975.*

119. Bannigan, P., Aldeghi, M., Bao, Z., Häse, F., Aspuru-Guzik, A., & Allen, C. (2021). Machine learning directed drug formulation development. *Advanced Drug Delivery Reviews*, *175*, 113806.
120. Mishra, D. V. (2018). Artificial intelligence: The beginning of a new era in pharmacy profession. *Asian Journal of Pharmaceutics (AJP)*, *12*(02), 72–76.
121. Sellwood, M. A., Ahmed, M., Segler, M. H. S., & Brown, N. (2018). Artificial intelligence in drug discovery. *Future Medicinal Chemistry*, *10*(17), 2025–2028.
122. Böhling, P., Khinast, J. G., Jajcevic, D., Davies, C., Carmody, A., Doshi, P., ... Sarkar, A. (2019). Computational fluid dynamics-discrete element method modeling of an Industrial-Scale Wurster Coater. *Journal of Pharmaceutical Sciences*, *108*(1), 538–550.
123. Eberle, L., Sugiyama, H., & Schmidt, R. (2014). Improving lead time of pharmaceutical production processes using Monte Carlo simulation. *Computers & Chemical Engineering*, *68*, 255–263.
124. Madgulkar, A. R., Bhalekar, M. R., Kolhe, V. J., & Kenjale, Y. D. (2009). Formulation and optimization of sustained release tablets of venlafaxine resinates using response surface methodology. *Indian Journal of Pharmaceutical Sciences*, *71*(4), 387.
125. Mahapatra, A. P. K., Saraswat, R., Botre, M., Paul, B., & Prasad, N. (2020). Application of response surface methodology (RSM) in statistical optimization and pharmaceutical characterization of a patient compliance effervescent tablet formulation of an antiepileptic drug levetiracetam. *Future Journal of Pharmaceutical Sciences*, *6*(1), 82.
126. Malakar, J., Nayak, A. K., & Goswami, S. (2012). Use of response surface methodology in the formulation and optimization of bisoprolol fumarate matrix tablets for sustained drug release. *International Scholarly Research Notices*, *2012*(1), 730624.
127. Ferreira, A., & Tobyn, P. (2015). Multivariate analysis in the pharmaceutical industry: Enabling process understanding and improvement in the PAT and QbD era. *Pharmaceutical Development and Technology*, *20*(5), 513–527.
128. Kundu, P., Paul, V., Kumar, V., & Mishra, I. M. (2015). Formulation development, modeling and optimization of emulsification process using evolving RSM coupled hybrid ANN-GA framework. *Chemical Engineering Research and Design*, *104*, 773–790.
129. Sengupta, S., Basak, S., & Peters, R. A. (2019). Particle swarm optimization: A survey of historical and recent developments with hybridization perspectives. *Machine Learning and Knowledge Extraction*, *1*(1), 157–191.
130. Guengerich, F. P. (2011). Mechanisms of drug toxicity and relevance to pharmaceutical development. *Drug Metabolism and Pharmacokinetics*, *26*(1), 3–14.
131. Basile, A. O., Yahi, A., & Tatonetti, N. P. (2019). Artificial intelligence for drug toxicity and safety. *Trends in Pharmacological Sciences*, *40*(9), 624–635.
132. Pinzi, L., & Rastelli, G. (2019). Molecular docking: Shifting paradigms in drug discovery. *International Journal of Molecular Sciences*, *20*(18), 4331.
133. Mayr, A., Klambauer, G., Unterthiner, T., & Hochreiter, S. (2016). DeepTox: Toxicity prediction using deep learning [original research]. *Frontiers in Environmental Science*, *3*, 80.
134. Agüero-Chapin, G., Galpert-Cañizares, D., Domínguez-Pérez, D., Marrero-Ponce, Y., Pérez-Machado, G., Teijeira, M., & Antunes, A. (2022). Emerging computational approaches for antimicrobial peptide discovery. *Antibiotics*, *11*(7), 936.
135. Huang, R., Xu, M., Zhu, H., Chen, C. Z., Zhu, W., Lee, E. M., ... Zheng, W. (2021). Biological activity-based modeling identifies antiviral leads against SARS-CoV-2. *Nature Biotechnology*, *39*(6), 747–753.
136. Smith, G. F. (2022). Artificial intelligence in drug safety and metabolism. In A. Heifetz (Ed.), *Artificial Intelligence in Drug Design* (pp. 483–501). New York, NY: Springer.
137. Olayan, R. S., Ashoor, H., & Bajic, V. B. (2018). DDR: Efficient computational method to predict drug–target interactions using graph mining and machine learning approaches. *Bioinformatics*, *34*(7), 1164–1173.
138. Yang, X., Wang, Y., Byrne, R., Schneider, G., & Yang, S. (2019). Concepts of artificial intelligence for computer-assisted drug discovery. *Chemical Reviews*, *119*(18), 10520–10594.

CHAPTER 15

Healthcare Data Security and Privacy in the AI Era

Indresh Kumar, Bhavna Dhingra, and Shikha Malik

15.1 INTRODUCTION

The healthcare industry has benefited greatly from the revolutionary changes brought about by the start of the artificial intelligence (AI) era. Because of its ability to process massive amounts of data and generate insights, AI has the potential to revolutionise patient care, increase diagnostic accuracy, and enhance operational efficiency [1]. Protecting sensitive health data is crucial because breaches can have serious effects, such as identity theft, a decline in patient trust, and legal issues.

There are serious worries about possible prejudice when health data is integrated with other systems. For example, financial organisations may use health insurance data to determine loan eligibility, interpreting low health indicators as a default risk [2]. This transfer of health information to non-healthcare entities could result in discriminatory practices in employment or other social and welfare benefits.

Equally critical are the issues surrounding data security. The Aadhaar system in India, for example, has experienced several security breaches, with minimal efforts made to enhance its security framework. Similarly, concerns have been raised about the security of the DigiLocker, which is proposed to be the model for Health Locker, especially regarding the biometric information it holds [3]. The initiation of the eSign electronic signature framework by the Indian government, which involves third parties, also poses a risk to sensitive health data. Although the National Digital Health Blueprint (NDHB) mentions establishing a security policy as well as a Security Operations Centre (SOC), it lacks clear protocols for responding to privacy breaches [4]. Furthermore, cyberattacks on healthcare institutions pose a threat to data integrity, with attackers capable of locking systems and demanding ransoms to restore access. Factors contributing to such vulnerabilities include outdated technology and insufficient cybersecurity training among healthcare staff. A notable incident occurred in June 2018 when Mumbai's Mahatma Gandhi Memorial Hospital suffered a ransomware attack, resulting in the loss of important data, despite the fact that no monetary loss was documented [5].

This chapter examines the complex issues of healthcare data privacy and security in the AI era. We go over the legal frameworks controlling data protection, the special difficulties presented by AI technology, and the best practices for protecting medical records [6]. By doing a thorough investigation, we hope to offer a clear understanding of the steps required to protect healthcare data in a world that is becoming more and more AI-driven [7].

15.2 THE EVOLUTION OF AI IN HEALTHCARE WITH DATA SECURITY AND PRIVACY CHALLENGES

In healthcare, AI has advanced remarkably, helping with clinical decision-making, forecasting patient outcomes, and customising treatment regimens. Numerous aspects of healthcare have integrated AI technology, such as machine learning algorithms, computer vision, and natural language processing. From crude computational models to complex systems that are changing medical procedures, AI has quickly revolutionised the healthcare industry. This evolution has been driven by advancements in data science, machine learning, and computing power, resulting in

DOI: 10.1201/9781003533351-15

cutting-edge applications that improve patient care, diagnosis, and treatment [3]. AI systems examine patient data to offer physicians evidence-based suggestions that support diagnosis and treatment choices. These technologies are able to identify correlations and patterns that human practitioners might not notice right away. The interpretation of medical images, such as computed tomography (CT) scans, magnetic resonance imaging (MRI), and X-rays, is enhanced by AI systems. They frequently surpass human radiologists in identifying specific disorders due to their excellent precision in identifying anomalies [8]. AI can forecast illness outbreaks, readmissions, and the probability of complications by examining past patient data. This makes resource allocation and proactive actions possible. This strategy reduces side effects and increases therapeutic effectiveness [6].

15.2.1 Early Beginnings

The introduction of rule-based systems in the 1950s and 1960s marked the beginning of AI's journey in healthcare. These early systems, including MYCIN and Dendral, were made to help with diagnostic procedures. MYCIN, for instance, was a pioneer in expert systems, using predefined rules to diagnose bacterial infections and suggest antibiotics. However, these systems were limited by their dependence on manually inputted rules and could not learn from new data.

15.2.2 Machine Learning and Big Data

The capabilities of AI underwent a dramatic change in the 1990s with the advent of machine learning. Machine learning algorithms, as opposed to rule-based systems, could learn from data and use patterns to predict outcomes and make decisions. The explosion of healthcare data, driven by the digitisation of medical records and advances in imaging technologies, provided the necessary fuel for these algorithms. AI systems began to excel in areas such as medical imaging, where they could help radiologists and identify anomalies by analysing large volumes of data.

15.2.3 Deep Learning and Advanced Applications

Deep learning is a branch of machine learning that emerged in the 2010s and models intricate patterns in data using neural networks. Because deep learning has made it possible to create AI systems that can interpret medical images with almost human precision, it has had a particularly significant impact on the healthcare industry. AI systems are currently utilised, for example, to identify eye conditions from retinal scans, detect tumours in mammograms, and even forecast patient outcomes using electronic medical records. Additionally, AI's uses have surpassed diagnostics. AI models in drug discovery examine biological data to find possible therapeutic targets, hastening the creation of novel medications. AI-driven personalised medicine also creates more targeted and efficient therapies by customising medicines for each patient based on their DNA profiles and other information.

15.2.4 AI in Clinical Practice

As the technology has advanced, the application of AI in clinical practice has increased. Clinicians can now concentrate more on patient care by using AI-powered virtual assistants to aid them with administrative duties. Real-time recommendations based on patient data are provided by AI-powered decision support systems, which help medical personnel make informed decisions. Additionally, the use of AI-powered predictive analytics to identify patients at risk of conditions like sepsis or readmission enables early interventions.

In conclusion, the continuous innovation in healthcare brought about by the growth of AI has altered how physicians diagnose, treat, and care for their patients. AI has a huge and mostly unrealised potential to improve healthcare outcomes and increase the effectiveness of medical practice as it develops.

15.3 REGULATORY FRAMEWORKS, PRIVACY AND SECURITY

Nowadays, with so much sensitive patient data being processed, transferred, and kept digitally, healthcare data security has become a major problem. The increasing reliance on electronic health records (EHRs) and connected medical devices has brought attention to the need for robust regulatory frameworks to protect healthcare data and the rise in cyber threats. Safe data sharing throughout the healthcare system is made possible by these frameworks, which also ensure patient data availability, confidentiality, and integrity.

15.3.1 Health Insurance Portability and Accountability Act

One of the most important legislative frameworks controlling the security of healthcare data in the

United States is HIPAA, which was passed into law in 1996. National guidelines for the protection of health information are the first step. Covered entities (such as clearinghouses, health plans, and healthcare providers) are required to implement administrative, technological, and physical safeguards to ensure the confidentiality, availability, and integrity of electronic protected health information (ePHI). The HIPAA Security Rule mandates this. Access restrictions, encryption, audit controls, and secure transmission techniques are some of these protections. The HIPAA Privacy Rule establishes guidelines for protecting patient privacy and disclosing protected health information. Patient health information may only be used and shared for permitted purposes, such as treatment, payment, and healthcare operations.

15.3.2 GDPR (General Data Protection Regulation)

Regardless of the association's location, all organisations handling the personal data of EU persons are subject to the GDPR, a comprehensive data protection law that went into effect in 2018. This covers insurance companies, healthcare providers, and any other organisations that handle health data. Health information is considered "special category data," which means it needs more protections under the GDPR. Individuals' express consent must be obtained before organisations can process their health data, unless doing so is required for public health or medical reasons. In order to secure health data, GDPR also mandates that organisations make use of suitable organisational and technical safeguards, like pseudonymisation and encryption. According to GDPR, companies have 72 hours from the time they learn of a data breach involving personal information to notify the appropriate supervisory body. Notifying the impacted individuals is also necessary if the breach presents a significant risk to their rights and liberties.

15.3.3 Act for the Protection of Electronic Documents and Personal Information (PIPEDA)

Healthcare organisations and other private sector companies that collect, use, or disclose personal data in the course of their operations are subject to Canada's federal privacy law, sometimes known as PIPEDA. Healthcare organisations must put in place the proper protections in accordance with PIPEDA to prevent unauthorised access, disclosure, alteration, and destruction of personal health information. These protections, which could include administrative, technical, and physical precautions, must be commensurate with the sensitivity of the data. PIPEDA emphasises how important it is to obtain individuals' informed consent before collecting, utilising, or disclosing their personal information. It also allows users to check their health information and request modifications if necessary.

15.3.4 National Health Services Data Security Standards

The UK's National Health Service (NHS) has established data security standards to protect patient information and ensure that it is used securely across all of its services. Such standards are part of a broader legal framework, which includes the Data Protection Act 2018 (the UK's implementation of GDPR) and other laws.

The NHS handles patient-identifiable information in accordance with the Caldicott Principles. These guidelines stress the necessity of using only the bare minimum of information, justifying the use of patient data, and making sure that only those who need to know may access it. Every year, healthcare organisations are required by the NHS to complete the Data Security and Protection Toolkit. Organisations can use this self-assessment tool to determine areas for improvement and assess their compliance with data protection rules.

15.3.5 Other Relevant Regulations

California consumers have rights over their personal information under the 2020 California Consumer Privacy Act (CCPA), which includes the ability to opt out of data sales, erase personal data, and know what data is being collected. To protect personal data within their borders, certain nations, such as Singapore and Thailand, have passed their own data protection legislation based on the GDPR, known as the Personal Data Protection Act (PDPA) [8].

India's Ministry of Health and Family Welfare released a draft of the proposed Digital Information Security in Healthcare Act (DISHA 2018) for public review in March 2018. In order to facilitate the digital exchange of personal health records between hospitals and clinics, this legislation provides a rights-based framework for medical privacy [5]. It grants patients the right to confidentiality, security, and privacy. Patients have the option to decline consent, and the draft law requires the owner's specific prior authorisation before any digital health data is transmitted [9].

Additionally, the draft legislation does not address the status of the data when consent is withdrawn and does not include measures for situations in which digital health data is gathered without the owner's consent [10]. Since recent research shows that triangulating many data points can identify individuals, data anonymisation is insufficient in an AI-equipped society [11]. Furthermore, it can be difficult to assess data usage or revoke consent for its use when using machine learning models because it is difficult to understand how certain data is interpreted and used to create algorithms [12].

The present draft of the Personal Data Protection bill, which the Indian parliament has not yet adopted, likewise suggests a consent-based paradigm. Health information is classified as "sensitive" and "critical," and its use requires the owner's consent [13]. The measure ignores permission and rights at various points in the data life cycle, which is problematic given that three parties are involved in the co-creation of health data: The patient, the hospital or physician, and the payer. Additionally, it is unclear from consent-based frameworks how much patient data is shared with private organisations including hospitals, pharmacies, and insurers.

15.4 ENSURING DATA SECURITY IN THE AI ERA

The security of sensitive health data is becoming more difficult to maintain as AI continues to pervade the healthcare industry. Strong security measures must be put in place to guard against illegal access, cyberattacks, and data breaches.

15.4.1 Encryption of Data

Data is protected by encryption, a basic security feature that transforms data into an unintelligible format that can only be decoded with a decryption key. Data must be encrypted while in transit and at rest in order to prevent unwanted access [14]. Data kept on servers, databases, and other storage devices must be encrypted. Methods like database encryption and full-disk encryption are frequently employed. To do this, data sent via networks must be encrypted to guard against eavesdropping and interception. Commonly used secure communication protocols include Virtual Private Networks (VPNs) and Transport Layer Security (TLS) [15].

15.4.2 Access Controls

For sensitive health information to be accessible to only those who are authorised, access restrictions are essential. Strict access control implementation involves three key components: (i) audit trails, (ii) authorization, and (iii) authentication. Authentication ensures that operators' identities are verified through methods such as multi-factor authentication (MFA), biometrics, and PINs. Authorization is applied by granting users access to specific data and systems according to their job roles and responsibilities, with role-based access control being one of the most commonly used approaches [16]. Audit trails are maintained by preserving user activity logs, which help track access and detect any unauthorized activity.and investigation of possible security events are aided by audit trails [17].

15.4.3 Data De-identification and Anonymisation

Techniques such as anonymisation and de-identification are used to disguise or eliminate Personally Identifiable Information (PII) from datasets, making it more challenging to link the data to specific individuals. In order to prevent re-identification, this procedure entails permanently deleting any identifying information. While effective, it may reduce the utility of the data for certain analytical purposes [18]. This process involves removing or masking identifying information while retaining enough data to allow for meaningful analysis [19].

15.4.4 Design Security

Security features must be integrated into AI system design and development from the start in order to minimise vulnerabilities and guarantee data protection. Secure Coding Techniques: Following safe coding guidelines and standards can help avoid common vulnerabilities like cross-site scripting (XSS) and SQL injection [14]. Frequent evaluation of security: To find and reduce any threats, perform routine security assessments using techniques like vulnerability scanning, penetration testing, and code reviews. Continuous monitoring: Installing tools for ongoing observation in order to promptly detect and resolve security events. Examples of these include intrusion detection systems (IDS) and security information and event management (SIEM) systems [13].

15.5 PRIVACY CONSIDERATIONS IN AI-DRIVEN HEALTHCARE

AI's reliance on enormous amounts of data presents serious privacy issues. The advantages of AI must be balanced with the requirement to preserve patient privacy, which calls for a careful strategy. Due to the extreme sensitivity of healthcare data, breaches may affect a person's work prospects, autonomy, and dignity [20]. Over 35,000 patients' medical details, including their HIV status, were sensationalised in 2016 after the records of an investigative laboratory in Mumbai were compromised [3]. It's possible that many impacted people are still ignorant of this incident because the database was breached several times in the past, sometimes up to three times per week, and the lab did nothing to secure the data [21]. This framework does, however, come with a number of difficulties. First and foremost, the ability of the entire population to make knowledgeable decisions regarding the collection and use of their data is necessary in order to gain meaningful permission. Due to complicated and time-consuming terms and conditions, some patients may develop consent fatigue, whereas many patients have low literacy and educational skills [10].

15.5.1 AI-Driven Healthcare with Informed Consent

Informed consent, which ensures that patients are aware of and provide their approval to the use of their personal information and AI in their treatment, is a critical ethical requirement in the healthcare industry. Addressing the particular difficulties and factors pertaining to informed consent is crucial as AI is used into healthcare more and more. The use of AI in patient care must be explained in detail to patients, including the data being gathered, the goal of its application, and any possible hazards and advantages. Patients should receive information in a manner and format that they can easily grasp, avoiding technical jargon and guaranteeing understanding.

Consent should be obtained for specific uses of AI, such as diagnostic support, treatment recommendations, or data analysis. In some cases, broader consent might be needed for research purposes, where the exact use of AI and data might not be fully defined at the time of consent. Patients must have the freedom to choose whether or not to participate in AI-driven healthcare, without any coercion or undue influence. It is important to let patients know that they can change their minds at any moment and that their information won't be utilised again. Patients require guarantees that the security of their data will be maintained against breaches and illegal access. Data should be de-identified or anonymised whenever feasible to safeguard patient identity.

Consent should be a continual process, with patients being re-informed and re-consented as needed, as AI systems develop and new applications for data are found. Patients should have the ability to voice their opinions about AI-driven healthcare through systems so that ethical issues and future developments can be informed. Giving patients the tools they need to understand more about AI and how it affects healthcare may foster trust and enable well-informed decision-making.

Implementing informed consent for AI presents difficulties: It can be challenging to describe how AI algorithms operate and their possible effects on patient care due to their complexity and opacity. AI systems continuously learn and evolve, which may change their behaviour and impact over time, complicating the consent process. Providing too much technical detail can overwhelm patients, while too little may leave them uninformed. Use analogies, visual aids, and simple language to explain AI concepts to patients. Implement consent forms that offer information in layers, allowing patients to choose the level of detail they wish to explore. Keep patients informed about any significant changes in the use of AI in their care, and re-consent them as needed.

In AI-driven healthcare, informed consent is crucial to upholding ethical norms, patient autonomy, and trust. Healthcare providers can successfully incorporate AI while upholding patient rights and expectations by guaranteeing openness, choice, and ongoing interaction.

15.5.2 Minimisation of Data Collection

In AI-driven healthcare, minimising data gathering is a fundamental tenet to safeguard patient privacy and lower the risk of data breaches [22]. To do this, just the information required for a particular purpose must be gathered, and any information gathered must be managed appropriately [23]. Here is a summary of the main ideas and difficulties surrounding minimising data gathering in the context of healthcare powered by AI.

Only gather information for precise, well-defined goals. For instance, only gather the data necessary to achieve the goal of improving diagnostic accuracy. Make sure that only the bare minimum of information is gathered in order to accomplish the intended result. Steer clear of collecting information that may not be immediately applicable.

To preserve patient identification, remove PII from datasets whenever feasible. Integrate information from several sources into a broad format that retains important insights while avoiding the exposure of specific data pieces. From the beginning, take privacy into account when designing and developing AI systems. This includes creating systems that run on very little data. Make use of default settings that encourage the acquisition of as little data as possible and need users to consent to further data collection rather than to opt out.

Establish guidelines for the maximum amount of time data can be kept, making sure that when it is no longer required for the designated purpose, it is anonymised or deleted. To verify adherence to data minimisation guidelines and find ways to further cut back on data acquisition, conduct routine audits of data. Make sure users understand what information is being gathered, why, and how it will be utilised. Prior to collecting handlers' data, obtain their express consent. Make it possible for people to view, edit, and remove their data. Permit handlers to withdraw their consent to collect data at any moment. Encrypt data both in transit and at rest to prevent breaches and unauthorised access. To ensure that only allowed persons can access sensitive data, strict access restrictions are put in place.

Data is necessary for AI systems to operate efficiently, and striking the correct balance between data collecting and privacy protection can be difficult. AI systems' data needs may change as they develop, requiring constant modifications to data collection procedures. Minimising data gathering and maintaining compliance with many data protection laws, such GDPR or HIPAA, can be challenging. Construct AI systems that can adjust to shifting data needs without necessitating an excessive amount of data collection. Make use of strategies such as differential privacy, which let data be analysed while lowering the possibility of identifying specific people. In AI-driven healthcare, minimising data gathering is essential to preserving patient privacy and confidence. Healthcare practitioners can use AI efficiently while protecting patient rights and lowering risks by following the principles of purpose limitation, employing data reduction strategies, and putting strong privacy and security safeguards in place.

15.5.3 Algorithmic Bias Mitigation and Fairness

In order to avoid biased outcomes and safeguard patient rights, it is imperative that AI algorithms be fair and mitigate bias. Checking AI algorithms for bias on a regular basis and taking corrective action to fix any problems found. One aspect of this is assessing how AI affects various demographic groups [24]. By employing representative and diverse training datasets, bias can be reduced and AI systems can be ensured to function equitably across various populations [25]. Enhancing the explainability of AI systems to allow for scrutiny and accountability. This involves providing clear explanations for AI-driven decisions and enabling users to challenge and appeal those decisions [26].

All people and groups should be treated equally by AI systems, free from prejudice based on protected traits such as age, gender, race, or socioeconomic status. AI systems should strive for equitable results in addition to equal treatment, guaranteeing that every patient receives the same standard of care and health advantages.

Arises from the use of biased or unrepresentative datasets, which can reflect historical inequalities and stereotypes. Occurs when the design or training of an AI model introduces bias, either intentionally or unintentionally. Consequences from how AI arrangements are implemented and used in actual settings, which whitethorn differ from controlled trying environments.

15.5.4 Challenges in Ensuring Fairness and Mitigating Bias

Healthcare data often contains biases that reflect existing social and healthcare disparities. For example, minority groups may be underrepresented in clinical datasets. Predictions and results can be skewed by imbalanced datasets, where some groups are over-represented or under-represented. It can be challenging to comprehend how judgements are produced and to spot bias because many AI models, such as deep neural networks, are intricate and challenging to understand. Even well-intentioned AI systems can have unintended consequences, perpetuating or amplifying existing biases. AI systems may not account for cultural differences and variations in health beliefs, behaviours, and practices. Health conditions and treatment responses can vary across different populations and geographic regions, necessitating context-aware AI models.

Collect data from diverse populations to ensure representation of all groups, including historically marginalised and underserved communities. Use techniques such as oversampling and undersampling to create balanced datasets, reducing the risk of bias due to data imbalance. To identify and quantify bias, audit and assess AI models on a regular basis. To evaluate the effect on various groups, use fairness metrics. Make AI models understandable and interpretable to increase transparency. Utilise resources such as model explainability methodologies to gain insight into the decision-making process. Design and train AI models with fairness limitations in mind. These constraints can ensure that the model's predictions do not disproportionately affect certain groups. Use adversarial training techniques to minimise bias. This involves training models to perform well across different subgroups and to be robust against biased predictions. Engage with patients and communities to gather feedback on AI systems, ensuring that their perspectives and needs are considered.

Ensuring algorithmic fairness and mitigating bias in AI-driven healthcare are essential for achieving equitable healthcare outcomes and maintaining trust in AI technologies. By focusing on diverse data collection, transparency, ethical oversight, and continuous improvement, healthcare providers, and AI developers can work together to create fair, unbiased, and effective AI systems.

15.6 CASE STUDIES

Innovations and solutions in the field of AI have been spurred by the difficulties related to data security and privacy. During the deployment of AI technology, a number of important problems have surfaced, necessitating the revision of procedures and regulations. These difficulties and the ensuing developments to overcome them are demonstrated by the significant cases listed below. Here are a few notable case studies.

15.6.1 Case Study First: Google Health and the NHS

An AI approach for identifying acute kidney injury was developed in 2016 by Google DeepMind in partnership with the NHS in the United Kingdom. The project raised serious privacy issues even as it showed off AI's promise in healthcare [21].

The partnership involved the transfer of millions of patient records without explicit patient consent, raising questions about data privacy and transparency. The data-sharing arrangement was deemed to be in violation of data protection rules by the United Kingdom's Information Commissioner's Office. Consequently, the NHS and Google DeepMind had to enhance their data handling procedures and guarantee adherence to privacy laws.

15.6.2 Case Study Second: IBM Watson for Oncology

The goal of IBM Watson for Oncology was to analyse enormous volumes of patient data and medical literature in order to deliver personalised cancer treatment recommendations. Notwithstanding its promise, Watson for Oncology was criticised for its opaqueness and challenges incorporating into therapeutic procedures. Furthermore, the diversity and quality of the training data were questioned [2]. The example demonstrated the value of establishing transparency, incorporating physicians in the creation and validation of AI systems, and regularly assessing the efficacy and equity of AI algorithms.

15.6.3 Case Study: HealthMap's Epidemic Tracking

HealthMap, created by Boston Children's Hospital, analyses data from several sources, such as news articles, social media posts, and public health reports, using AI to detect and monitor disease outbreaks [1].

Privacy issues and the possibility of false information are brought up by the analysis of data from news sources and social media. It's crucial to strike a balance between using data to advance public health and attending to privacy issues [27].

Because AI may improve diagnosis, treatment, and operational efficiency, it has the potential to completely transform the healthcare sector. It does, however, also present serious privacy and data security issues. The future of healthcare data security and privacy in the AI era is as follows.

15.7 ROAD AHEAD FOR HEALTHCARE DATA SECURITY AND PRIVACY

AI has the ability to completely transform the healthcare sector by enhancing diagnosis, care, and operational effectiveness. However, it also brings significant challenges regarding data security and privacy. Here's a look at the road ahead for healthcare data security and privacy in the AI era:

i. Improved Data Encryption and Protection: To prevent unwanted access to patient data, more robust encryption mechanisms are being implemented for data in transit and at rest, reducing the possibility of data breaches by making sure that data is encrypted from the time of collection until it is stored and analysed [28].

ii. Strict Access Controls: To reduce exposure, restrict access to sensitive data according to the duties and responsibilities of medical personnel. By using MFA, healthcare systems and data can be more securely accessed, limiting access to only those who are authorised.

iii. Data De-Identification and Anonymisation: To preserve patient privacy while enabling insightful analysis, PII should be eliminated from datasets used for AI training, putting in place de-identification techniques that adjust to changing privacy laws and standards in order to maintain compliance.

iv. Regulation Compliance: Making sure AI systems and data management procedures abide by laws pertaining to data protection, such as the GDPR and HIPAA, carrying out routine evaluations and audits to confirm adherence to data protection regulations and pinpoint areas in need of development.

v. Transparency and Explainability: Creating AI models that offer concise justifications for their judgements and forecasts promotes trust and guarantees that both patients and clinicians can comprehend how they work. To foster confidence and guarantee informed consent, patients and stakeholders should be made fully aware of how patient data is used, stored, and safeguarded.

vi. Ethical Issues and Bias Reduction: Handling AI bias putting methods in place to identify and lessen biases in AI algorithms in order to stop discriminatory actions and guarantee that every patient is treated fairly, establishing moral standards for the creation and application of AI to allay worries about data ownership, consent, and privacy.

vii. Management and Incident Response: Establishing and upholding strong incident response protocols to promptly handle security issues or data breaches and minimise possible harm, giving healthcare workers continual instruction on data security best practices and how to spot and handle possible attacks.

viii. Blockchain Technology Integration: Examining the role of blockchain technology, which provides an immutable and transparent record of data transactions, in enhancing data security and integrity. Smart contracts on blockchain platforms can be used to automate and secure data access and usage agreements, ensuring reliability and reducing the risk of unauthorized manipulation [29].

ix. Patient Empowerment and Control: Implementing secure patient portals allows individuals to view, control, and manage their own health information. These portals ensure that patients' decisions are respected by providing explicit options to consent or decline the use of their data in AI systems [30].

x. Cross-Sector Collaboration: Developing and implementing cross-sector standards for data security and privacy requires active engagement with stakeholders across different industries. To stay ahead of emerging threats and challenges, healthcare providers, technology developers, and regulatory agencies should collaborate closely and share expertise, fostering a unified approach to safeguarding sensitive information [29, 31].

Healthcare data security and privacy in the AI era require a multifaceted strategy that strikes a balance between innovation and stringent security safeguards. The healthcare sector can leverage AI's potential while protecting patient data by strengthening encryption, putting strong access restrictions in place, abiding by legal requirements, and encouraging openness and moral behaviour. Maintaining the integrity and reliability of healthcare data systems will require constant adaptation to changing threats, legal requirements, and technological breakthroughs [32].

15.8 SUMMARY

AI's potential to improve patient outcomes and revolutionise medical practice is enormous. But it also calls for a watchful approach to privacy and data security. The healthcare sector may fully utilise AI while protecting private patient data by following strict legal requirements, putting in place thorough security measures, and attending to ethical issues. Building a robust and reliable healthcare ecosystem in the AI era will require a dedication to data protection and privacy.

REFERENCES

1. Hirani, R., Noruzi, K., Khuram, H., Hussaini, A. S., Aifuwa, E. I., Ely, K. E., Lewis, J. M., Gabr, A. E., Smiley, A., Tiwari, R. K., & Etienne, M. (2024). Artificial intelligence and healthcare: A journey through history, present innovations, and future possibilities. *Life (Basel, Switzerland)*, *14*(5), 557. https://doi.org/10.3390/life14050557
2. IBM Watson Health. (2012). *Case study: IBM Watson for Oncology and personalized cancer treatment.* https://www.ibm.com/industries/healthcare
3. Bak, M., Madai, V. I., Fritzsche, M. C., Mayrhofer, M. T., & McLennan, S. (2022). You can't have AI both ways: Balancing health data privacy and access fairly. *Frontiers in Genetics*, *13*, 929453. https://doi.org/10.3389/fgene.2022.929453
4. Mehrtak, M., SeyedAlinaghi, S., MohsseniPour, M., Noori, T., Karimi, A., Shamsabadi, A., Heydari, M., Barzegary, A., Mirzapour, P., Soleymanzadeh, M., Vahedi, F., Mehraeen, E., & Dadras, O. (2021). Security challenges and solutions using healthcare cloud computing. *Journal of Medicine and Life*, *14*(4), 448–461. https://doi.org/10.25122/jml-2021-0100
5. Ueda, D., Kakinuma, T., Fujita, S., Kamagata, K., Fushimi, Y., Ito, R., Matsui, Y., Nozaki, T., Nakaura, T., Fujima, N., Tatsugami, F., Yanagawa, M., Hirata, K., Yamada, A., Tsuboyama, T., Kawamura, M., Fujioka, T., & Naganawa, S. (2024). Fairness of artificial intelligence in healthcare: Review and recommendations. *Japanese Journal of Radiology*, *42*(1), 3–15. https://doi.org/10.1007/s11604-023-01474-3
6. Al Kuwaiti, A., Nazer, K., Al-Reedy, A., Al-Shehri, S., Al-Muhanna, A., Subbarayalu, A. V., Al Muhanna, D., & Al-Muhanna, F. A. (2023). A review of the role of artificial intelligence in healthcare. *Journal of Personalized Medicine*, *13*(6), 951. https://doi.org/10.3390/jpm13060951
7. Saccà, R., Turrini, R., Ausania, F., Turrina, S., & De Leo, D. (2024). The ménage à trois of healthcare: The actors in after-AI era under patient consent. *Frontiers in Medicine*, *10*, 1329087. https://doi.org/10.3389/fmed.2023.1329087
8. Mocydlarz-Adamcewicz, M., Bajsztok, B., Filip, S., Petera, J., Mestan, M., & Malicki, J. (2023). Management of onsite and remote communication in oncology hospitals: Data protection in an era of rapid technological advances. *Journal of Personalized Medicine*, *13*(5), 761. https://doi.org/10.3390/jpm13050761
9. Basu, K., Sinha, R., Ong, A., & Basu, T. (2020). Artificial intelligence: How is it changing medical sciences and its future? *Indian Journal of Dermatology*, *65*(5), 365–370. https://doi.org/10.4103/ijd.IJD_421_20
10. Lotan, E., Tschider, C., Sodickson, D. K., Caplan, A. L., Bruno, M., Zhang, B., & Lui, Y. W. (2020). Medical imaging and privacy in the era of artificial intelligence: Myth, fallacy, and the future. *Journal of the American College of Radiology*, *17*(9), 1159–1162. https://doi.org/10.1016/j.jacr.2020.04.007
11. Google DeepMind. (2016). *Case study: Google DeepMind and the NHS collaboration on acute kidney injury detection.* https://deepmind.google/discover/blog/deepminds-work-in-2016-a-round-up/
12. Jaime, F. J., Muñoz, A., Rodríguez-Gómez, F., & Jerez-Calero, A. (2023). Strengthening privacy and data security in biomedical microelectromechanical systems by IoT communication security and protection in smart healthcare. *Sensors (Basel, Switzerland)*, *23*(21), 8944. https://doi.org/10.3390/s23218944
13. Bekbolatova, M., Mayer, J., Ong, C. W., & Toma, M. (2024). Transformative potential of AI in healthcare: Definitions, applications, and navigating the ethical landscape and public perspectives. *Healthcare (Basel, Switzerland)*, *12*(2), 125. https://doi.org/10.3390/healthcare12020125
14. Almalawi, A., Khan, A. I., Alsolami, F., Abushark, Y. B., & Alfakeeh, A. S. (2023). Managing security of healthcare data for a modern healthcare system. *Sensors (Basel, Switzerland)*, *23*(7), 3612. https://doi.org/10.3390/s23073612
15. Bhatia, S., Lassmann, B., & Cohn, E. (2021). Using digital surveillance tools for near real- time mapping of the risk of infectious disease spread. *npj Digital Medicine*, **4**, 73. https://doi.org/10.1038/s41746-021-00442-3

16. Tom, E., Keane, P. A., Blazes, M., Pasquale, L. R., Chiang, M. F., Lee, A. Y., Lee, C. S., & AAO Artificial Intelligence Task Force. (2020). Protecting data privacy in the age of AI- enabled ophthalmology. *Translational Vision Science & Technology, 9*(2), 36. https://doi.org/10.1167/tvst.9.2.36
17. Rahman, M. A., Victoros, E., Ernest, J., Davis, R., Shanjana, Y., & Islam, M. R. (2024). Impact of artificial intelligence (AI) technology in healthcare sector: A critical evaluation of both sides of the coin. *Clinical Pathology (Thousand Oaks, Ventura County, Calif.), 17*, 2632010X241226887. https://doi.org/10.1177/2632010X241226887
18. Xiong, H., Chen, H., Xu, L., Liu, H., Fan, L., Tang, Q., & Cho, H. (2022). A survey of data element perspective: Application of artificial intelligence in health big data. *Frontiers in Neuroscience, 16*, 1031732. https://doi.org/10.3389/fnins.2022.1031732
19. Wang, C., Zhang, J., Lassi, N., & Zhang, X. (2022). Privacy protection in using artificial intelligence for healthcare: Chinese Regulation in comparative perspective. *Healthcare (Basel, Switzerland), 10*(10), 1878. https://doi.org/10.3390/healthcare10101878
20. Bhagat, S. V., & Kanyal, D. (2024). Navigating the future: The transformative impact of artificial intelligence on Hospital management- A comprehensive review. *Cureus, 16*(2), e54518. https://doi.org/10.7759/cureus.54518
21. Farhud, D. D., & Zokaei, S. (2021). Ethical issues of artificial intelligence in medicine and healthcare. *Iranian Journal of Public Health, 50*(11), i–v. https://doi.org/10.18502/ijph.v50i11.7600
22. Korkmaz, S. (2024). Artificial intelligence in healthcare: A revolutionary ally or an ethical dilemma. *Balkan Medical Journal, 41*(2), 87–88. https://doi.org/10.4274/balkanmedj.galenos.2024.2024-250124
23. Mennella, C., Maniscalco, U., De Pietro, G., & Esposito, M. (2024). Ethical and regulatory challenges of AI technologies in healthcare: A narrative review. *Heliyon, 10*(4), e26297. https://doi.org/10.1016/j.heliyon.2024.e26297
24. Yelne, S., Chaudhary, M., Dod, K., Sayyad, A., & Sharma, R. (2023). Harnessing the power of AI: A comprehensive review of its impact and challenges in nursing science and healthcare. *Cureus, 15*(11), e49252. https://doi.org/10.7759/cureus.49252
25. Maleki Varnosfaderani, S., & Forouzanfar, M. (2024). The role of AI in hospitals and clinics: Transforming healthcare in the 21st century. *Bioengineering (Basel, Switzerland), 11*(4), 337. https://doi.org/10.3390/bioengineering11040337
26. Yadav, N., Pandey, S., Gupta, A., Dudani, P., Gupta, S., & Rangarajan, K. (2023). Data privacy in healthcare: In the era of artificial intelligence. *Indian Dermatology Online Journal, 14*(6), 788–792. https://doi.org/10.4103/idoj.idoj_543_23
27. Alanazi, A. (2023). Clinicians' views on using artificial intelligence in healthcare: Opportunities, challenges, and beyond. *Cureus, 15*(9), e45255. https://doi.org/10.7759/cureus.45255
28. Bajwa, J., Munir, U., Nori, A., & Williams, B. (2021). Artificial intelligence in healthcare: Transforming the practice of medicine. *Future Healthcare Journal, 8*(2), e188–e194. https://doi.org/10.7861/fhj.2021-0095
29. Davenport, T., & Kalakota, R. (2019). The potential for artificial intelligence in healthcare. *Future Healthcare Journal, 6*(2), 94–98. https://doi.org/10.7861/futurehosp.6-2-94
30. Elendu, C., Amaechi, D. C., Elendu, T. C., Jingwa, K. A., Okoye, O. K., John Okah, M., Ladele, J. A., Farah, A. H., & Alimi, H. A. (2023). Ethical implications of AI and robotics in healthcare: A review. *Medicine, 102*(50), e36671. https://doi.org/10.1097/MD.0000000000036671
31. Zonoobi, M., Elahi, S., Khansari, M., Hassanzadeh, A., & Saheb, T. (2023). Health data sharing with the goal of value creation: Trying to develop a framework using qualitative content analysis. *Archives of Academic Emergency Medicine, 12*(1), e6. https://doi.org/10.22037/aaem.v12i1.2091
32. Zuhair, V., Babar, A., Ali, R., Oduoye, M. O., Noor, Z., Chris, K., Okon, I. I., & Rehman, L. U. (2024). Exploring the impact of artificial intelligence on global health and enhancing healthcare in developing nations. *Journal of Primary Care & Community Health, 15*, 21501319241245847. https://doi.org/10.1177/21501319241245847

CHAPTER 16

AI-Driven Virtual Health Assistants

Enhancing Patient Communication

Parashuram Kallem, Aya Kamal Al-Zabidi, and Hirah Nadeem

16.1 INTRODUCTION

16.1.1 Definition

Artificial intelligence (AI)-powered virtual health assistants (VHAs) represent sophisticated software systems crafted to assist patients and healthcare providers through various functions. These virtual assistants harness AI technologies, consists of machine learning (ML), natural language processing (NLP), and data analytics, to furnish tailored health information, monitor patient health, and refine communication between patients and healthcare professionals. Accessible via diverse platforms such as mobile applications, websites, and standalone devices, these assistants deliver continual support and direction. By emulating human-like interactions, they effectively bridge the divide between patients and healthcare services, thereby fostering ongoing and proactive health management [1–5].

16.1.1.1 Background

As healthcare systems worldwide evolve, adopting digital platforms has become crucial in meeting the growing demands of a more connected and informed patient population. The shift from traditional face-to-face consultations to digital communication, such as telemedicine and real-time data sharing, is driven by the need for efficiency, accuracy, and accessibility. This transition is significant in enhancing healthcare access for patients in remote or underserved areas. Virtual health technology began with essential telemedicine services, where healthcare consultations and advice were delivered over the phone or via simple online platforms. These early systems primarily focused on facilitating remote consultations and providing medical guidance but were often constrained by technological limitations and connectivity issues. However, advances in digital communication, cloud computing, and mobile technology paved the way for more sophisticated virtual health solutions, establishing a foundation for integrating advanced AI technologies into these platforms. So, the integration of AI marked an important advancement in the capabilities of VHAs. Additionally, advancements in data analytics allow VHAs to provide detailed health insights by analyzing extensive datasets from electronic health records (EHRs), wearable devices, and other sources. This progression has dramatically improved their efficiency and effectiveness in delivering high-quality patient care, making AI an integral part of the digital transformation in healthcare [1–4].

16.1.2 Importance of Patient Communication

Patient communication goes beyond simply sharing information; it's about building a relationship grounded in trust, empathy, and understanding. In today's fast-paced healthcare environment, where time with patients is often limited, high-quality communication is more crucial than ever. It ensures patients understand their health conditions and feel supported throughout their care. This is especially important for managing chronic illnesses, where continuous dialogue is needed to monitor progress, adjust treatments, and provide emotional reassurance. Moreover, good communication helps alleviate patient anxiety by making complex medical information more precise and less intimidating through compassionate conversations. As healthcare increasingly adopts digital tools, maintaining empathy and clarity

DOI: 10.1201/9781003533351-16

in communication remains essential. These technologies help bridge gaps in healthcare access for remote or underserved populations by providing 24/7 support, enabling real-time data collection, and facilitating timely, personalized care [3, 4].

16.1.3 Overview of AI-Driven Virtual Health Assistants

AI-driven VHAs are becoming essential in modern healthcare as they provide advanced support beyond fundamental interactions. These assistants use sophisticated algorithms to learn from user interactions, improving their ability to offer personalized care over time. VHAs are crucial in the patient journey, from initial symptom assessment and triage to continuous health monitoring and follow-up after treatment. For example, they assist in managing chronic conditions like diabetes by tracking glucose levels and delivering real-time feedback to patients and healthcare providers. They also help alleviate the workload of healthcare experts by managing regular duties, enabling doctors and nurses to aim on complex cases. VHAs are particularly valuable for those who meet challenges get into traditional healthcare services, for instance those with mobility issues or living in isolated places. Since these tools progress, their effect on patient communication and care will likely grow, solidifying their role in digital health strategies [2–4].

16.2 THE ROLE OF AI IN ENHANCING PATIENT COMMUNICATION

16.2.1 Personalized, Timely, and Cost-Effective Health Advice

In the past decade, the healthcare corporation has go through a major revolution with the help of artificial intelligence (AI), revolutionizing the delivery of personalized, timely, and cost-effective advice and support to patients. At the forefront of this transformation are AI-driven VHAs that use vast patient data, including medical history and treatment records, to provide highly individualized recommendations. This tailored approach ensures that advice is specific to each patient's needs, ultimately improving the quality of care [6]. For example, advanced VHAs can analyze symptoms, treatment history, and current health conditions to offer real-time guidance [7]. This immediate access to information empowers patients to make informed decisions and reduces healthcare costs by minimizing the need for regular in-person meetings. Furthermore, the integration of AI in healthcare streamlines processes and alleviates administrative burdens, allowing healthcare providers to focus on direct patient need [5].

16.2.2 Interactive and Responsive Communication

AI technologies, particularly those incorporating NLP, have greatly improved patient communication through enhanced interactivity and responsiveness. VHAs in modern healthcare are developed to effectively engage patients by customizing responses to individual preferences and health literacy levels [3–5]. This adaptability improves patient engagement and adherence to treatment plans. For example, AI systems can recognize emotional cues in patient interactions, offering tailored support based on their emotional state. This proactive approach is critical for managing chronic conditions and addressing specific patient concerns. By personalizing interactions, AI not only helps patients adhere to their treatment regimens but also strengthens the patient-provider relationship, enhancing the bond healthcare professionals share with their patients [1]. This improved communication can direct to safer conditions and improved patient happiness.

16.2.3 Bridging Communication Gaps

AI's use in healthcare has enabled overcoming communication barriers between patients and healthcare providers. Traditional obstacles, such as language differences and varying levels of health literacy, can impede effective care [8]. AI-powered solutions can address these challenges by offering real-time translation services and tailoring information according to cultural and literacy needs. For example, AI systems can provide information in multiple languages and adjust the complexity of explanations based on the patient's understanding [7]. This inclusivity confirms that all patients, whatever of their background, have the necessary information to be updated results about their health. Additionally, AI-driven virtual assistants and chatbots can manage frequent inquiries, enabling healthcare professionals to dedicate their attention to more complex care aspects and improve productivity [6]. This enhances healthcare delivery efficiency and improves patient experience by providing clear and easily accessible information.

16.3 TECHNOLOGY BEHIND AI-DRIVEN VIRTUAL HEALTH ASSISTANTS

16.3.1 Natural Language Processing Virtual Health Assistants

Integrating NLP into VHAs represents a significant advancement in healthcare technology. This innovation aims to enhance patient engagement and simplify access to healthcare information. With NLP algorithms, these assistants can recognize and report to natural communication queries, making patient interactions with healthcare systems additional intuitive and user-friendly. By providing instant answers to questions about medical conditions, treatment options, and general health information, NLP-powered virtual assistants empower patients by promoting education and self-management. This technology facilitates quick and convenient access to healthcare resources, supports better patient adherence to treatment plans, and allows for a more personalized and engaging healthcare experience, ultimately improving patient satisfaction and outcomes. For example, advanced NLP models like BERT and GPT can handle complex text data from EHRs, enabling efficient clinical coding, entity recognition, and sentiment analysis, improving the quality of patient interactions and care coordination [8, 9]. Figure 16.1 illustrates an ethical framework for integrating AI and NLP into healthcare communication [8, 9].

16.3.2 Benefits of NLP in Healthcare

16.3.2.1 Improved Efficiency

NLP enhances operational efficiency in healthcare by automating processes, particularly in clinical documentation. Traditional record-keeping and data-entry methods are often tedious, slow, and likely to faults. NLP can automate these processes, and meaningful information can be extracted and summarized from clinical notes, reports, and other unstructured data sources. This streamlines documentation and ensures that patient data is maintained accurately and consistently. As a result, healthcare experts can concentrate more on patient care, minimizing administrative tasks and optimize the use of resources [8–10].

16.3.2.2 Enhanced Decision-Making

NLP is a powerful tool that aids healthcare providers in making well-informed decisions by analyzing large datasets and extracting relevant insights. Given the volume and complexity of data in healthcare, it can be challenging for physicians to recognize patterns or glean actionable insights. NLP algorithms extract valuable information from unstructured data, such as medical literature and patient records. This capability allows healthcare professionals to understand patient histories, treatment outcomes, and the latest medical research better. By providing timely, evidence-based

FIGURE 16.1 Ethical framework for integrating AI and NLP into healthcare communication [8, 9].

insights, NLP contributes to perfect diagnoses, specified care tasks, and increased patient care [8–10].

16.3.2.3 Patient-Centric Care

NLP has significantly impacted personalized medicine, fostering a more patient-centered approach to care. By analyzing personal patient records—including clinical notes, genetic data, and treatment histories—NLP enables healthcare experts to alter analysis to each patient's characteristics and needs. Unlike a one-size-fits-all type, this specific method considers the specific nuances of a patient's health profile. NLP helps identify relevant biomarkers, genetic factors, and treatment responses, empowering medical professionals to develop more effective and minimally invasive therapies. This contributes to a patient-centric healthcare model characterized by customized treatments, optimized outcomes, and greatly enhanced patient satisfaction [9].

16.3.3 Challenges and Considerations in Using NLP in Healthcare

NLP in healthcare shows several challenges. Data privacy and security are critical concerns owing to the confidential nature of patient data. Caring against unauthorized access, data breaches, or misuse requires robust quantities for instance encryption, anonymization, secure access controls, and regular security audits. Integration with existing systems is also facing due to the diverse data formats, architectures, and interoperability standards within healthcare infrastructure. Effective integration requires collaboration with healthcare IT professionals, standardization of data formats, and phased implementation approaches to minimize operational disruptions. Additionally, ethical considerations must be addressed, including mitigating biases in NLP algorithms, ensuring algorithmic fairness and transparency, and obtaining informed patient approval for data utilization. Developing transparent norms and governance frameworks is essential to maintain ethical standards and patient trust in NLP applications [8, 9].

16.3.4 Machine Learning

ML, a branch of AI, is centered on developing algorithms that allow machines to independently learn and adapt from data, without explicit programming for each task. The core principle behind ML is that machines should be able to access, analyze, and interpret data independently to detect patterns and relationships. By leveraging various algorithms, techniques, and methods, ML detects underlying correlations within datasets, creating descriptive and predictive models to process and utilize these insights effectively. ML encompasses several related areas, including data mining, pattern recognition, and deep learning. While these fields often overlap, deep learning is considered a newer branch of ML. It uses complex computational algorithms and large datasets to model intricate patterns and relationships that are often too complex for traditional ML techniques to handle. This capacity allows deep learning to define and understand more sophisticated and nuanced data relationships, making it especially powerful for image recognition, NLP, and autonomous driving tasks. As a result, ML and its subfields are increasingly critical in developing advanced AI applications that require sophisticated data analysis and decision-making capabilities [3].

16.3.5 Telemedicine and Chronic Disease Management

VHAs are increasingly crucial in telemedicine and chronic disease management by providing remote patient monitoring and personalized care. These AI-driven tools use advanced algorithms to monitor patients' health in real-time, offering continuous support and timely interventions. For example, VHAs can remind patients to take their medications, suggest lifestyle changes, and provide dietary recommendations tailored to individual health needs. They also facilitate virtual consultations, allowing healthcare providers to assess and adjust treatment plans remotely, thus minimizing the requirement for regular in-person visits. This personalized communication fosters greater patient engagement and adherence to treatment regimens, which are crucial in managing chronic conditions such as diabetes, hypertension, and cardiovascular diseases. By enabling remote monitoring, VHAs help sense early signs of health deterioration and alert patients and healthcare providers, allowing for prompt action and reducing the risk of complications. This continuous monitoring is especially valuable for patients with chronic diseases, who require regular tracking of vital signs and symptoms to prevent hospitalizations and manage their conditions effectively. Moreover, the use of VHAs in telemedicine offers a more accessible and cost-effective solution, ensuring that patients receive high-quality care while minimizing the burden on healthcare systems [11].

16.4 APPLICATIONS AND CASE STUDIES

VHAs are creating a remarkable effect on the healthcare platform by changing how patients interact with medical services. These advanced tools, powered by AI and ML, not only signify technological progress but also play an important part in closing gaps, streamlining methods, and improving the accessibility and customization of healthcare to cater to each person's individual needs.

For instance, Babylon Health has developed an AI-driven symptom checker that functions as a digital-first responder. This user-friendly tool allows individuals to seek medical advice via their smartphones conveniently. By inquiring about symptoms, providing initial guidance, and suggesting next steps, this VHA acts as a portable medical companion, empowering individuals to address their health concerns without being constrained by appointment waiting times. This immediate access to medical advice offers a sense of reassurance and confidence in managing one's health [2]. Likewise, Kaiser Permanente offers "Kp.org," a comprehensive virtual assistant that goes beyond typical chatbot functions. This digital health partner provides personalized care, assisting patients in tasks ranging from scheduling appointments to managing medical records and offering personalized health recommendations. Tailored advice and timely reminders significantly facilitate the management of chronic conditions, providing a similar experience to having a dedicated health coach available around the clock. This personalized care and attention make the healthcare journey more manageable and less daunting, making patients feel valued and understood [12].

Northwell Health demonstrates the invaluable role of VHAs in managing chronic diseases by employing a chatbot to support patients before and after surgery. Acting as a supportive ally, this virtual assistant delivers crucial information and aids in the patient's recovery process, keeping individuals well-informed and at ease. This reassurance and comfort contribute to a smoother and less stressful recovery experience, making the healthcare journey more manageable and less overwhelming [13]. Moreover, virtual assistants and chatbots show a significant impact in addressing patient queries about dental treatments and care. They explain the roles, developments, profits, and probable risks of dental procedures. Additionally, they provide personalized guidance and reminders on oral hygiene exercises, for instance brushing techniques, flossing, and food practices, effectively assisting patients in managing their dental health [10].

16.5 CHALLENGES AND ETHICAL AND REGULATORY CONSIDERATIONS

Despite its numerous advantages, adopting VHAs is accompanied by significant issues. Data security is a primary concern, given that a large portion of the generated data contains sensitive information. It is crucial to implement proper measures to safeguard this data, as it is vital for the functionality of VHAs. Vulnerability to potential attacks such as data breaches, ransomware, or insider threats poses a serious risk to patient data, and ensuring its protection is not only essential to privacy standards but also crucial for maintaining patient trust, a cornerstone of effective healthcare. Leakage of sensitive information could lead to adverse consequences for patients and healthcare institutions alike [14].

Another challenge is the presence of algorithmic bias within VHAs, which mirrors the preferences show in the records used to instruct them. This issue is of utmost value, as it exactly effects the quality of care given by VHAs. If the training data exhibits bias, the VHA's patient interactions may also reflect bias. This can result in suboptimal care for patient demographics not adequately represented in the training data. Deal with this concern needs meticulous selection of training data and, importantly, ongoing supervision of AI systems to confirm fairness and accuracy in patient interactions. This ongoing supervision should provide reassurance about the quality of care, making the audience feel more secure about the use of AI in healthcare [15].

The integration of AI in VHA presents many ethical and legislative issues that must be carefully considered. When AI is given full decision-making authority over a patient's treatment plan, ethical problems arise. While AI has the capability to study large volumes of data immediately, it cannot fully understand the moral dilemmas or individual circumstances that

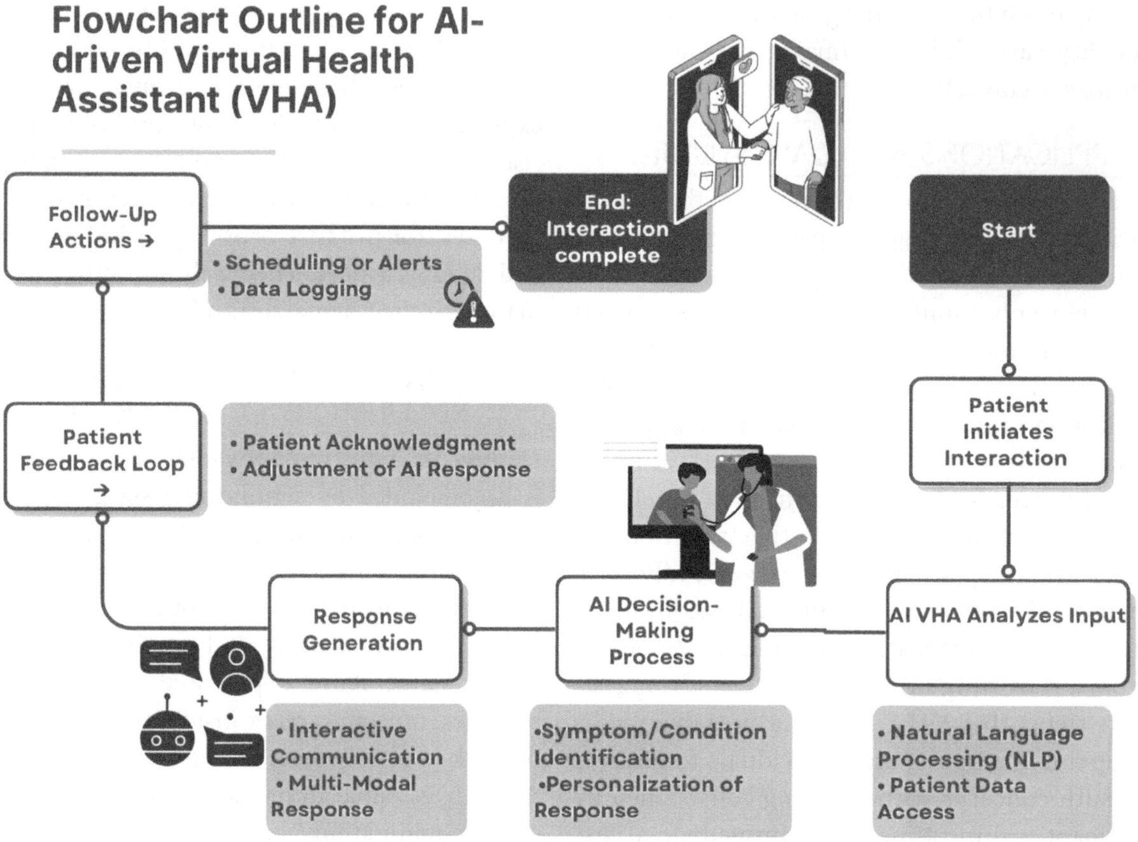

FIGURE 16.2 Flowchart outline for a basic application of an AI integration to deliver virtual health assistants to patients.

often play a significant task in healthcare assessments. For instance, AI might recommend a treatment based on data-driven outcomes but fail to consider a patient's beliefs or community conditions. Therefore, it's not just important, but essential to establish ethical guidelines governing the use of AI in healthcare. These strategies are crucial to focus patient welfare and consent when AI systems are used, making the audience feel the gravity of the situation.

In addition to ethical concerns, there are legislative challenges in implementing AI in the healthcare industry. AI technologies must operate within a global environment that is slow to adapt to new technology. Before AI systems can be used in clinical practice, they must be tested for safety, effectiveness, and reliability. Government regulators need to reconsider their roles and concepts to account for the unique features of AI, such as learning elements. The dynamic nature of AI poses challenges in assigning responsibility, particularly when issues arise with the output delivered. Legislation addressing these concerns should consider both the opportunities AI brings to healthcare and the need to maintain ethical standards and patient safety [16]. AI systems in healthcare must be developed with transparent formulas, secure data, and fair treatment of potential biases. Patient rights advocates and legislators need to focus these problems to confirm that the true potential of AI in the healthcare industry can be realized without compromising people's trust [16] (Figure 16.2).

16.6 FUTURE DIRECTIONS

16.6.1 Personalized Virtual Health Coaches

Future VHAs could serve as personalized health coaches, guiding patients through lifestyle changes and wellness plans while monitoring their health metrics. These assistants could offer customized advice on nutrition, exercise, mental health, and preventive care by integrating data from EHRs, wearable devices, and even genetic information. Leveraging AI's ability to learn and adapt, it can provide real-time coaching to motivate patients, promote healthy behaviors, and help reduce the risk of chronic diseases.

16.6.2 AI-Enhanced Remote Diagnostics and Treatment Planning

Future VHAs could use advanced AI to move beyond simple symptom checking and provide remote diagnostic capabilities. They could offer initial diagnoses and personalized treatment plans by analyzing data from sources like medical imaging, lab results, and genetic information, especially useful in areas lacking access to specialists. VHAs could work with remote medical teams to adjust treatments dynamically, ensuring ongoing, adaptive care based on patient data.

16.6.3 Emotionally Intelligent VHAs for Mental Health Support

Next-generation VHAs could incorporate emotional intelligence to support mental health better. These assistants could assess a patient's emotional state and provide appropriate responses or interventions by analyzing voice tone, language patterns, and facial expressions. They could offer immediate support or connect patients with human therapists if distress is detected, creating a more responsive and empathetic mental health support system.

16.6.4 Predictive Health Crisis Management and Early Warning Systems

VHAs could serve as early warning systems for health crises, detecting emerging threats such as disease outbreaks or adverse drug reactions by analyzing patient data trends. Individually, they could predict potential flare-ups in chronic conditions like asthma or diabetes, enabling timely intervention by patients and caregivers before situations worsen.

16.7 SUMMARY

The healthcare communication landscape is evolving with the integration of AI in the form of VHAs, offering personalized and timely health advice to patients. This chapter demonstrates the role of AI-driven VHAs in improving patient outcomes through constant access to healthcare data and personalized communication. Available 24/7, VHAs provide tailored recommendations, helping patients follow treatment plans and manage their health effectively. Additionally, they play a vital role in telemedicine by assisting with remote patient examining and chronic disease controlling, thus improving healthcare delivery. However, tasks including data security, algorithm bias, and regulatory concerns still pose hurdles. Looking ahead, AI-powered virtual assistants are expected to dominate outpatient, preventive, and long-term care, making healthcare more personalized and accessible. VHAs are set to become personalized health coaches, providing tailored advice on lifestyle and wellness by integrating data from wearable devices and health records. They could also offer remote diagnostic capabilities by analyzing medical data like imaging and lab results, especially benefiting areas lacking specialists. Emotionally intelligent VHAs may support mental health by assessing emotional states and offering timely interventions. Additionally, VHAs could act as early warning systems for health crises or chronic condition flare-ups, enabling timely responses from patients and caregivers to prevent worsening conditions.

REFERENCES

1. Adikari, A., de Silva, D., Moraliyage, H., Alahakoon, D., Wong, J., Gancarz, M., Chackochan, S., Park, B., Heo, R., & Leung, Y. (2022). Empathic conversational agents for real-time monitoring and co-facilitation of patient-centered healthcare. *Future Generation Computer Systems*, *126*, 318–329.
2. Azevedo, D., Legay, A., & Kieffer, S. (2022). User reception of Babylon Health's chatbot. In *Proceedings of the 14th International Conference on Health Informatics* (pp. 134–141). https://doi.org/10.5220/0010803000003124
3. Badawy, M., Ramadan, N., & Hefny, H. A. (2023). Healthcare predictive analytics using machine learning and deep learning techniques: A survey. *Journal of Electrical Systems and Information Technology*, *10*, 40. https://doi.org/10.1186/s43067-023-00108-y
4. Chavali, D., Dhiman, V. K., & Katari, S. C. (2024). AI-powered virtual health assistants: Transforming patient engagement through virtual nursing. *International Journal of Pharmaceutical Sciences*, *2*(2), 613–624. https://www.ijpsjournal.com
5. Hägglund, M., McMillan, B., Whittaker, R., & Blease, C. (2022). Patient empowerment through online access to health records. *BMJ*, *378*, e071531. https://doi.org/10.1136/bmj-2022-071531
6. Mittal, M., Battineni, G., Singh, D., Nagarwal, T., & Yadav, P. (2021). Web-based chatbot for frequently asked queries (FAQ) in hospitals. *Journal of Taibah University Medical Sciences*, *16*(5), 740–746. https://doi.org/10.1016/j.jtumed.2021.06.002
7. Thorat, V., Rao, P., Joshi, N., Talreja, P., & Shetty, A. R. (2024). Role of artificial intelligence (AI) in patient education and communication in dentistry. *Cureus*, *16*(5), e59799. https://doi.org/10.7759/cureus.59799

8. Sarella, P., & Kumar, A. (2024). AI-driven natural language processing in healthcare: Transforming patient-provider communication. *Indian Journal of Pharmacy Practice* [Internet]. *17*(1), 21–26. http://dx.doi.org/10.5530/ijopp.17.1.4
9. Takale, D. G. (2024). A study of natural language processing in healthcare. *Journal of Web Applications and Cyber Security, 2*(2), 1–6. https://qtanalytics.in/publications/index.php/JoWACS/article/view/134
10. Pithpornchaiyakul, S., Naorungroj, S., Pupong, K., & Hunsrisakhun, J. (2022). Using a chatbot as an alternative approach for in-person toothbrushing training during the COVID-19 pandemic: Comparative study. *Journal of Medical Internet Research, 24*(10), e39218. https://doi.org/10.2196/39218
11. Ma, Y., Zhao, C., Zhao, Y., Lu, J., Jiang, H., Cao, Y., & Xu, Y. (2022). Telemedicine application in patients with chronic disease: A systematic review and meta-analysis. *BMC Medical Informatics and Decision Making.* https://doi.org/10.1186/s12911-022-01845-2
12. Lieu, T. A., Warton, E. M., Levan, C., et al. (2022). Association of medical assistant– supported virtual rooming with successful video visit connections. *JAMA Internal Medicine, 182*(6), 680–682. https://doi.org/10.1001/jamainternmed.2022.1032
13. Ma, D., Orner, D., Ghaly, M. M., Parashar, B., Ames, J. W., Chen, W. C., Potters, L., & Teckie, S. (2021). Automated health chats for symptom management of head and neck cancer patients undergoing radiation therapy. *Oral Oncology, 122,* 105551. https://doi.org/10.1016/j.oraloncology.2021.105551
14. Veluru, C. (2024). Impact of artificial intelligence and generative AI on healthcare: Security, privacy concerns, and mitigations. *Journal of Artificial Intelligence & Cloud Computing, 3,* 1–6. https://doi.org/10.47363/JAICC/2024(3)347
15. Mittermaier, M., Raza, M. M., & Kvedar, J. C. (2023). Bias in AI-based models for medical applications: Challenges and mitigation strategies. *npj Digital Medicine, 6*(113). https://doi.org/10.1038/s41746-023-00858-z
16. Naik, N., Hameed, B. M. Z., Shetty, D. K., Swain, D., Shah, M., Paul, R., Aggarwal, K., Ibrahim, S., Patil, V., Smriti, K., Shetty, S., Rai, B. P., Chlosta, P., & Somani, B. K. (2022). Legal and ethical considerations in artificial intelligence in healthcare: Who takes responsibility? *Front Surg, 14*(9), 862322. https://doi.org/10.3389/fsurg.2022.862322.

CHAPTER 17

The Role of AI in Mental Health Diagnosis and Treatment

Ahmed Tharwat Al-Namrouti, Tasneem Mahmoud, Asmaa Sh. Saadallah, Ahmed Samy, Reem M. Elsapagh, Tiba J. K. Al-Imari, Aya yasser elsayed Mohamed, and Mona Muhe Eldeen Eshag

17.1 INTRODUCTION

AI is becoming recognized as a transformational tool in mental health treatment offering fresh prospects for diagnosis and treatment of mental health issues due to its unequaled accuracy and adaptation [1]. AI can search enormous amounts of data to identify trends suggestive of mental health issues by means of modern algorithms and machine learning (ML) techniques. To detect early symptoms of disorders including depression and anxiety, for example, ML algorithms can analyze social media activity, electronic health records (EHRs), and other digital footprints [2]. AI models have shown in studies to obtain diagnostic accuracy either equal to or even greater than human doctors [1].

Furthermore, demonstrating potential in detecting mental health problems is AI's use in natural language processing (NLP) [3]. NLP techniques can look at speech and text to identify linguistic signals connected to mental health problems. For instance, specific word choice, speech pattern, and even frequency of social media posts can expose a person's emotional condition [3]. Early intervention is the foundation of effective therapy; these AI-driven diagnostic tools can assist to provide this early intervention.

In terms of treatment, AI-powered applications are revolutionizing the supply of mental health care. Chatbots including Woebot and Wysa provide users real-time assistance and instruction using cognitive-behavioral approaches [4]. Especially in disadvantaged areas, these AI applications offer a type of therapy accessible anywhere and at any time, therefore helping to close the gap in mental health care access [5]. AI may thus maximize treatment plans by means of patient data analysis to predict which therapies are most likely to be successful, so modifying care and improving outcomes.

AI can also support tracking of therapy progress and patient adherence to treatment plans. Wearable technology and smartphone apps can compile continuously data on a patient's physiological reactions, degree of exercise, and mood [6]. Examining this data will help AI to identify patient reaction to treatment and guide required improvements [7]. Such continuous monitoring helps with early recurrence diagnosis as well as with prompt changes in therapeutic plans.

However, integration of AI into mental health care does not come without challenges. Among the many important ethical questions are ensuring data privacy and addressing probable AI system bias. For instance, if the training data for AI models is not representative of diverse populations, the resulting algorithms could be biassed, therefore affecting therapy. Working collaboratively, technologists, clinicians, and ethicists will help to ensure that AI capabilities are developed and applied correctly [1].

AI in mental health treatment seems bright given ongoing research and development designed to increase the accuracy and accessibility of these technologies. Growing complexity of AI promises to make

DOI: 10.1201/9781003533351-17

mental health treatment more tailored and effective, hence improving the quality of life for those with mental health issues [8].

This chapter explore AI solutions that can be introduce early diagnoses hence improving the treatment outcomes as well. Other AI-enabled applications such as Woebot and Wysa assist in real-time and help in the treatment of underserved areas. AI can also monitor therapy efforts and compliance, thus informing changes in focus of therapy. However, ethical issues include data stewardship and bias.

17.2 OVERVIEW OF ARTIFICIAL INTELLIGENCE USE IN MENTAL HEALTH

AI has rapidly permeated various sectors, with mental health being a prominent field of application. AI's capabilities in predictive analytics, pattern recognition, and NLP present significant opportunities for diagnosing, treating, and managing mental health conditions. AI is progressively assuming a central role in mental health, providing novel interventions and predictive analytics for earlier identification of mental health issues and conditions. AI's incorporation into this field services seeks to transform care by improving accessibility, personalization, and treatment efficacy. Nevertheless, this technological progress also introduces challenges and potential adverse effects on mental well-being [9, 10].

The integration of AI in mental health services has been a subject of significant interest, with systematic reviews and meta-analyses evaluating its effectiveness. AI-enabled chatbots, which is a promising tool, help in psychological interventions for various disorders and employ ML models to simulate therapeutic conversations [11].

17.3 ARTIFICIAL INTELLIGENCE'S APPLICATIONS IN HUMAN MENTAL HEALTH

The applications range from computational psychiatry, which employs quantitative techniques to improve our understanding, detection, and management of mental illnesses, up to the utilization of EHRs, data of brain imaging, scales of mood rating, novel monitoring systems (e.g., video, mobile phone), and social networking websites for classification, predictions, or mental health illnesses subgrouping including schizophrenia and depression [12, 13].

A Turkish study on AI mental health uses signifies the advantages and disadvantages of AI applications at various stages of mental health services. It emphasizes the importance of understanding the future roles and outcomes of AI through the opinions of mental health professionals engaged with AI [14].

ML, a subset of AI, has been instrumental in detecting, diagnosing, and treating mental health problems. A systematic review of the human-computer interaction literature on ML in mental health reveals an increasing exploration of how ML can assist in effective mental health care [15]. Furthermore, AI-based conversational agents have been shown to alleviate psychological distress and improve well-being [16].

The role of AI in computational psychiatry is another area of focus. Quantitative techniques leveraged by AI are disrupting traditional approaches to understanding, detecting, and treating mental illnesses. This becomes notably apparent in the use of EHRs, brain imaging, scales of mood rating, and innovative monitoring systems for predictive analytics [17, 18].

17.3.1 Artificial Intelligence Diagnosis and Screening

AI technologies have demonstrated significant potential in the diagnosis of mental health disorders. ML algorithms, especially those utilizing supervised learning, have proven effective in recognizing patterns that indicate mental health conditions across diverse sources of data for example: clinical records, wearable devices, and social media interactions. For example, Das and Naskar employed deep learning methods to analyze speech patterns for early detection of depression, achieving high levels of accuracy and reliability in clinical settings [19].

17.3.2 Predictive Analytics

Predictive analytics is another critical application of AI in mental health. By analyzing historical and real-time data, AI systems can predict the onset, progression, and outcomes of mental health conditions. In a significant study, Koutsouleris et al. demonstrated that ML models could predict the transition from high-risk mental states to psychosis with a considerable degree of accuracy [20]. This capability is invaluable for early intervention strategies, potentially mitigating the severity of mental health issues.

17.3.3 Natural Language Processing

NLP proves highly advantageous for mental health applications due to the crucial role of language in expressing and diagnosing psychological conditions. AI-driven NLP systems can analyze text derived from therapy sessions, social media updates, and patient records to detect indications of mental health disorders. Inkster et al. demonstrated the potential of NLP in identifying suicidal thoughts expressed in social media posts, showcasing a proactive approach to crisis intervention [21].

17.3.4 Therapeutic Interventions

AI is revolutionizing therapeutic interventions by fostering the creation of digital mental health tools like chatbots and virtual therapists, which provide accessible and scalable mental health support. An illustrative instance involves the AI-driven chatbot called 'Woebot.' Woebot employs cognitive-behavioral therapy (CBT) principles to promptly aid users. Research by Fitzpatrick et al. revealed that individuals who used Woebot experienced significant reductions in anxiety and depression symptoms [22].

17.3.5 Personalization of Treatment

Customized treatment plans play a vital role in delivering effective mental health care. AI algorithms have the capability to analyze individual patient data, enabling the recommendation of personalized treatment strategies. For example, Chekroud et al. devised a predictive model to pinpoint patients who would derive the greatest benefit from particular antidepressants, thereby improving treatment effectiveness and minimizing the need for trial-and-error in medication [23].

17.3.6 Ethical Regard

AI incorporation in mental health presents substantial ethical challenges. Concerns regarding the algorithmic bias risk, security of data, and the privacy need to be carefully managed to ensure AI is deployed responsibly. Fiske et al. highlighted the importance of establishing strong ethical frameworks and regulatory measures to safeguard patient rights and promote equitable use of AI systems [23, 24].

17.3.7 Challenges

Although AI holds promise in mental health, it encounters significant challenges. Issues such as data quality and representation are paramount, given that biased or incomplete data may result in inaccurate predictions and advice. Moreover, the interpretability of AI models is a persistent concern, as many algorithms operate opaquely, making their decision-making processes difficult to discern. Topol underscores the importance of improving transparency and explainability in AI systems to foster trust and acceptance within clinical settings [1].

17.3.8 Future Directions

The future prospects of AI in mental health appear bright, propelled by continuous technological and methodological advancements. Effective collaboration among experts in technology, clinical settings, and ethical perspectives is crucial for refining AI applications and tackling current challenges. The integration of AI into everyday mental health practices is expected to broaden, promising enhanced precision, efficiency, and personalized treatment options [25].

17.4 ARTIFICIAL INTELLIGENCE USE IN DIAGNOSIS AND TREATMENT OF SUBSTANCE ABUSE DISORDERS

Substance use disorders (SUDs) are complex conditions marked by the inability to control substance use, such as drugs and alcohol, despite adverse consequences. These disorders frequently occur alongside mental health issues, posing significant diagnostic and treatment challenges for healthcare providers [26]. Recent advancements in AI are transforming these challenges by improving diagnostic accuracy, analysis of behavior, predicting treatment outcomes, and personalizing treatment plans. The integration of AI into healthcare is bringing a revolutionary shift, especially in the diagnosis and treatment of SUDs. AI's ability to process large datasets and recognize intricate patterns holds great promise for enhancing diagnosis precision and treatment effectiveness [27, 28].

17.4.1 Artificial Intelligence in Diagnosis of Substance Use Disorders

AI has demonstrated substantial efficacy in diagnosing SUDs through advanced algorithms of ML that analyze diverse datasets as, genetic information, EHRs, and behavioral data. For example, Kashyap et al. developed the first opioid prescribing behavior, utilized a deep learning approach to analyze EHRs for early detection of SUDs focusing on opioids, achieving higher accuracy compared to traditional diagnostic methods [29, 28].

This method involves training AI models on large datasets to recognize patterns associated with substance use, thus enabling earlier and more precise diagnosis.

Dong et al. [27], also focused on opioids prescription EHRs, developed a temporal deep learning model to predict overdose risk predictive feature as medications and vital signs. achieve promising results on identifying risk of opioid overdose of patients. This provides an alternative approach to improving the decision in health settings for early detection and management to reduce risk of overdose of opioids [30].

Emerging AI technologies, such as ML and NLP, exhibit significant potential in the diagnosis of SUDs. These advanced tools analyze extensive datasets to detect patterns and symptoms that human clinicians might overlook. For instance, AI algorithms can evaluate EHRs, clinical notes, and even social media posts and comments activity to detect early signs of SUDs [31, 29].

Garg et al. [41] developed a ML risk model for fentanyl drug use, the model designed to detect Reddit posts or comments with help if clinicians who labeled these activities risky. Both the accuracy and sensitivity were 76%. Allow for identification of risk to inform outreach and establish intervention strategies to those patients [31].

An interesting diagnostic use of AI is its ability to handle unstructured data, such as patient interviews and notes from therapy sessions. NLP tools can parse this data to identify linguistic markers associated with substance use, improving early detection and intervention strategies [29]. Additionally, AI systems can continuously update and refine their models with new data, enhancing their diagnostic accuracy over time [32].

17.4.2 Predictive Analytics in SUBs

Predictive analytics is a crucial application of AI in the context of SUDs. AI models can identify individuals at high risk of developing SUDs by analyzing historical data and recognizing risk factors. Leucht et al. [74] demonstrated AI use of AI in prediction the likelihood of relapse in patients undergoing treatment for substance abuse. Their model, which based on intensive longitudinal data patient histories, treatment adherence, and social determinants of health, significantly improved relapse prediction compared to traditional statistical methods [33, 34].

In their study, Ahmed et al. [35] examine different techniques for developing prediction models, emphasizing the difficulties researchers encounter in choosing the most effective method. It presents super learning (SL), a strategy that integrates several prediction algorithms to produce a final model that performs at least as well as any single algorithm. The research compares SL with logistic regression, penalized regression, random forests, and deep learning neural networks to predict successful treatment outcomes for SUDs using a nationwide dataset of almost 100,000 patients. The findings indicate that SL generally surpasses most models, highlighting its strengths in targeted learning and addiction studies, while also offering a comprehensive overview of SL's methodology and advantages [35].

17.4.3 Natural Language Processing for Behavioral Analysis

NLP is a branch of AI, concentrates on the interaction between computers and human language. It proves especially valuable for analyzing unstructured data from clinical notes, patient interviews, and social media. NLP could effectively analyze text data to identify linguistic markers of SUDs, providing a non-invasive and scalable tool for early diagnosis and ongoing monitoring [29]. Goodman-Meza et al. performed a study that assume NLP and ML are better diagnostic metrics to identify patients who abuse drugs by injection and they found that NLP/ML is superior to International Classification of Diseases (ICD)-based coding algorithms at recognition of patients in EHRs. This suggests NLP/ML use to get more improvement in making of clinical decision, healthcare-related research [29].

17.4.4 Artificial Intelligence in Monitoring and Relapse Prevention

AI technologies are also used to monitor patient progress and prevent relapse. ML models can analyze data from wearable devices, mobile apps, and other digital health tools to provide real-time feedback to clinicians. This continuous monitoring allows for proactive adjustments to treatment plans, ensuring that patients receive the most effective care throughout their recovery [34, 35].

Furthermore, predictive analytics can help to identify patients at high risk of relapse by analyzing patterns in their behavior, such as changes in physical activity or social interactions. Early intervention strategies can then be implemented to prevent relapse and support sustained recovery [35].

17.4.5 Ethical Considerations

The integration of AI in mental health and SUDs raises several ethical issues. Ensuring patient privacy and data security is paramount, given the sensitive nature of SUD data. Additionally, addressing potential biases in AI algorithms is crucial to prevent disparities in care. Highlighting the importance of ethical guidelines and robust regulatory frameworks to ensure the fairness and equity of AI applications [32].

17.4.6 Future Directions

The future of AI integration in SUDs appears promising, thanks to continuous technological and methodological advancements. Effective AI application requires multidisciplinary collaborations among experts in technology, clinicians, ethical regard and to tackle existing challenges. Ongoing research aims to enhance these technologies by improving data integration, ensuring algorithm transparency, and developing ethical guidelines for responsible AI use. Innovations, such as incorporating genetic data and advanced modeling techniques, are anticipated to further enhance the precision and personalization of treatment plans [33].

17.5 DIAGNOSIS AND TREATMENT OF ANXIETY DISORDERS

As wearable devices can continuously watch and col lect physiological indicators and data including heart rate, blood oxygen, blood pressure, level, and temperature, they have attracted a lot of interest in the medical community. Wristbands, smartwatches, and eyewear are a few of the wearable devices that are most well known. Wearable technology has made it easier to gather patient data in real time, which when combined with AI technologies can aid in the early diagnosis of many mental illnesses, including anxiety disorder [36]. Wearable technology, however, has been shown to produce a lot of false positives and misinterpret heart rate or other variables as an anxiety symptom [37]. Additionally, ML has been effectively used in neuroimaging analysis, and several abnormal connection patterns between specific brain regions were identified particularly in anxiety disorders [38].

17.5.1 Artificial Intelligence and Diagnosis of Generalized Anxiety Disorders (GAD)

GAD is characterized by intractable continuous distress accompanied by somatic anxiety symptoms [39]. GAD arises due to more than one object or situation, and it produces worry and worries in the brains of suffering patients, and they get concerned about regular events and seemingly unimportant things in life. This type of mental condition is characterized by worry associated with fatigue, agitation, tense muscles, and trouble sleeping [40, 41]. In recent years, ML has advanced significantly to help doctors and psychiatrists diagnose a wider range of anxiety disorders and to classify and contrast them to one another. GAD is one class of anxiety disorders, among others, including social mental disturbance (SAD) and panic disorders [5]. Sau and Bhakta have successfully classified anxiety disorders using five different ML techniques, with the Catboost model achieving the highest accuracy of 89.3% on the testing set [42].

A binary support vector machine (SVM) was used in a study by Hilbert et al. to differentiate between individuals with mental disorders, such as major depressive disorder (MDD) and GAD (as cases), and healthy subjects (non-cases), based on the following information: the cortisol levels, clinical ratings, and gray and white matter from structural magnetic resonance imaging (MRI), the overall accuracy for case classification was 90%, and for disorder classification, it was 67.4% [43]. In another study, Jacobson and Feng gathered movement time series data over the course of a week and utilized it to create ensemble learning models that might potentially predict the severity of GAD symptoms [44].

17.5.2 Artificial Intelligence and Diagnosis of Panic Disorders and Phobias

Social anxiety phobia, which is a subtype of anxiety disorder, affects people who fear social embarrassment when doing any task in front of another group of people, such as giving a lecture [45, 46]. The adaptive neuro-fuzzy inference system model was introduced by Fathi et al. [33] for the diagnosis of SAD. Using five-fold cross-validation, it was discovered that the model's sensitivity, specificity, and accuracy were 97.14%, 100%, and 98.67%, respectively [47].

In a study by Lueken et al. [48] the aim was to distinguish between depressive comorbidity and panic disorder employing a random underestimating tree ensemble in a cross-validation leave-one-out framework. The ML model was trained on functional MRI (fMRI) from 59 panic disorder/agoraphobia patients, including 26 patients with depression comorbidity. During fear conditioning, the model has accurately predicted depression comorbidity in 79% of patients, with sensitivity and specificity of 73% and 85%, respectively [48].

17.6 TREATMENT OF ANXIETY DISORDER USING ARTIFICIAL INTELLIGENCE

Mental health stigma is a barrier to obtaining mental health care, thus there are more individuals in need for mental health therapy but are not receiving it [49]. To overcome this stigmatization, China conducted a multicenter randomized control trial to test the XIAO AN AI-assisted psychotherapy robot. It was intended to use natural interaction technology, mimic human counseling, and provide CBT-based psychotherapy. Nevertheless, it wasn't designed to replace therapists [50]. Another chatbot called Youper was investigated for effectiveness in offering self-guided intervention in emotional problems. The authors essentially tested the cumulative regulation hypothesis, which states that providing just-in-time digital intervention for emotional dysregulation is effective. The users demonstrated a consistent effect and a decrease in anxiety symptoms within the first two weeks, indicating the success of the proposed cumulative regulatory hypothesis [51]. Fitzpatrick et al. [21] used an early chatbot named Woebot. It is a chatbot that was created to automatically send brief messages to users of social media platforms like WhatsApp [21]. It was trained to recognize emotions and acquired skills to reduce the symptoms of anxiety. For phobias, exposure therapy is one of the available therapies. This strategy was proven to be more effective with virtual reality (VR) [52]. Similarly, Parsons and Rizzo used VR exposure treatment to show a significant improvement in anxiety symptoms [53].

17.7 ARTIFICIAL INTELLIGENCE AND MOOD DISORDERS

17.7.1 Major Depressive Disorder

MDD patients usually complain of feeling down, having no energy or desire, having trouble sleeping, having trouble eating, and having suicide thoughts. It is the world's primary cause of disability.

The patient health questionnaire-9 is typically used in primary care settings to screen for and diagnose MDD. It is a self-reporting standardized depression measure [54]. The Montgomery-Asberg Depression Rating Scale, the Rankin Depression Rating Scale, and the Beck-Depression Inventory are examples of widely used diagnostic tools. A lot of work has gone into diagnosing, classifying, and distinguishing depressive disorders from other mental illnesses including anxiety and bipolar disorder (BD) using various AI techniques. According to literature, many data types including voice, genetics, EHRs, physiological variables, environmental, and lifestyle data, were used to train ML models [55].

Nemesure et al.'s study used biological non-psychiatric variables from the EHR, such as blood pressure, heart rate, and housing status, to predict the diagnosis of MDD. High diastolic pressure and satisfaction with living circumstances have been found to be predictive indicators of MDD [56]. According to Nemesure et al., using an EHR can help diagnose MDD patients earlier since doctors will recommend early screening for individuals who show signs of the disorder, which will reduce symptoms and enhance quality of life.

Similar to anxiety illnesses, wearable technologies, such as smartwatches, implanted inside the body, for example, implantable watches or electronic textiles, can, as a result, record someone's physiological parameters, physical activity, and sleep pattern, and therefore, when associated with AI technology, they can be used to diagnose MDD [57]. However, it cannot detect additional data such as social contact, living style, weight loss, and medical history [58]. Albert Haque et al. have employed NLP technologies coupled with voice recognition and computer vision to assess individual language, tone of voice, and facial expression and determine whether individuals suffer from depression.

17.7.2 Treatment of Depression Using Artificial Intelligence

AI chatbots are used in treatment as they are a simulation of psychotherapists, and they are interactive. AI chatbots are capable of handling enormous databases, diagnosing ailments, and offering customized treatment plans [59]. They can also be by NLP technology that can identify human emotion from the responses of the users in the form of text or tone of voice and appropriately provide direction and empathy.

In patients with challenging mental health conditions, psychiatrists use CBT to address such difficult conditions. Woebot was found to be effective in giving tailored CBT using mobile and social media applications [60]. Furthermore, a meta-analysis study revealed that MRI can be accurately analyzed using ML with an appropriate use criterion of 0.84, 79% specificity, and 77% sensitivity [61], compared to prediction using electroencephalogram (EEG) data. This may help to lead the therapy in resistant depression cases toward more invasive techniques such as electroconvulsive therapy with little hesitancy [62].

17.7.3 Bipolar Disorder

Bipolar illness (BP) is described as recurring bouts of depression and mania (highly irritated or euphoric mood) or hypomania (less severe than mania) [62]. BP has been shown to be one of the first 10 disorders that contribute to loss of function among young adults [63].

Misdiagnosis of BP disease is common, with over 60% of BP patients incorrectly seeking therapy for MDD [64]. This could adversely affect the patient's quality of life and prognosis. A review was done to document the usage of ML in bipolar disease diagnosis. SVM methods were identified as the most often utilized ML models in 9 (27%) of the 33 studies for the diagnosis of BD. These were followed by the Gaussian process model (2.6%), artificial neural networks (ANNs), linear regression (3.9%), ensemble models (3.9%), and ANNs (5.15%). Overall, the model had a mean accuracy value of 82.06%, with the highest accuracy being 98% and the lowest being 64%.

In terms of the data used to train the models, it was found that EEG performed with a greater accuracy of 98% than MRI data with SVM [65]. As noted earlier, the BP issue is one of the causes of the loss of functionality. Cognitive capabilities have played a role in discriminating between healthy individuals and BP sufferers. In a study that employed Cambridge Neuropsychological Test Automated Battery (CANTAB) test, data was extracted from 17 individuals with BD (cases) and 19 healthy counterparts (controls). Other data like executive functions and visual memory were identified as factors in the CANTAB test to identify BD from healthy controls [66].

17.7.4 Treatment of Bipolar Disorder Using Artificial Intelligence

Lithium's reacting condition was predicted using ML techniques. In Eugene et al. study, 120 bipolar patients had their peripheral blood samples collected. These samples were then used to create expression profiles and probes, which provided baseline data to distinguish between lithium responders and non-responders. According to the study, male and female BP patients have different transcriptome-level gene signatures, which will predict how well each will respond to lithium before a treatment plan is recommended. The research findings indicate that the RBPMS2 and LILRA5 genes have a sensitivity of 95.83% for predicting male lithium responders, while FHL3, ABRACL, and the NBPF14 genes can predict lithium responders in female with a sensitivity of 91.67% [67].

17.7.5 The Role of Artificial Intelligence in Distinguishing between Depression and Anxiety

Anxiety and depression are commonly associated psychiatric conditions with their symptoms usually overlap. Nonetheless, it's critical to differentiate them clinically because each of them requires a unique approach to treatment. A ML algorithm created by Wang et al. produced generation classification models. These models had accuracy rates of 95.14% and 94.12%, respectively, for predicting anxiety and depression. The Symptom Checklist 90 (SCL-90) scale [68] was used to train the ML algorithm, using only the items that define unique responses to depression and anxiety. SCL-90 is basically a questionnaire used to identify symptoms of different psychological disorders and distinguish patients from healthy individuals. According to a recent analysis, ML models can recognize certain brain areas that are reported to be affected in patients with Anxiety and depression [69].

17.7.6 Artificial Intelligence Use in Diagnosis and Treatment of Mental Disorders

Different fields, including healthcare, are fast being transformed by AI technologies. By using big volumes of data, spotting intricate trends, and offering individualized insights, AI can transform medical diagnosis and treatment [70]. Various AI tools offer significant benefits in the fields of drug discovery and cancer diagnosis [71]. AI is assisting with treatment planning and clinical diagnosis. Conventional diagnosis techniques rely on subjective assessments and experience of the doctors, which might be erroneous. AI improves clinical judgement by means of objectivity and data. AI models benefit from large sets comprising EHRs, medical images, genomic data, other digital health sources [72]. ML methods can find trends and connections in multidimensional data humans would ignore [72].

For many diseases, AI is raising diagnostic accuracy. Deep learning models enable AI to detect diseases from medical imaging [71], extract meaningful information from unstructured clinical notes, and combine multi-omics data for precision diagnosis [72].

17.7.7 Schizophrenia

Affecting about 1% of the world's population, schizophrenia is a serious and persistent mental illness posing major difficulties for people, families, and healthcare systems (schizophrenia, n.d.). Characterized by symptoms including hallucinations, delusions, disordered thinking, and cognitive deficits, schizophrenia needs exact diagnosis and efficient treatment plans to control its incapacitating consequences [73]. Many times, based on subjective clinical judgments and symptom checklists, traditional diagnostic techniques result in possible misdiagnoses and variable treatment outcomes [25]. Moreover, the therapy of schizophrenia usually consists of a trial-and-error approach to medicine, which results in inadequate management of the condition and protracted suffering of the patients [74].

The remarkable progress achieved by artificial intelligence in psychiatry offers novel possibilities for improving the diagnosis and treatment of schizophrenia. AI technologies, particularly those grounded in ML and deep learning, have demonstrated significant potential in swiftly and accurately processing and analyzing extensive volumes of data [70]. Employing their leverage, AI can identify trends and connections across several data sources—including genetic information, neuroimaging data, and EHRs—that might be missed by human doctors. Early identification of schizophrenia is made possible by this capacity, which also enhances diagnosis accuracy and supports quick intervention and better patient outcomes [75].

Genetic factors clearly affect the likelihood of schizophrenia. AI can handle vast amounts of genomic data in search of disease biomarkers. Deep learning techniques have been applied, for example, to investigate single nucleotide polymorphisms and other genetic variations, so providing knowledge of disease susceptibility. MRI and fMRI are among neuroimaging methods that help to further enhance AI's importance in diagnosis. Studies demonstrating high accuracy suggest that AI models can differentiate between people with schizophrenia and healthy controls allowing ML algorithms to examine these photos to identify anomalies linked with schizophrenia [76]. AI may also mine EHRs looking for early symptoms of schizophrenia. For example, NLP systems may examine clinical notes to identify minute linguistic patterns suggestive of disordered thinking or other cognitive problems [77].

Beyond diagnosis, AI presents interesting options for tailored therapy. Managing schizophrenia depends on tailored therapy as different strategies affect different people. By forecasting reactions to drugs and optimizing psychological treatments, AI may customize therapy regimens to fit particular requirements. By analyzing genetic markers associated to drug metabolism and effectiveness, this method—known as pharmacogenomics—helps doctors choose drugs more likely to be helpful for a given patient, hence lowering the trial-and-error time and increasing outcomes [78]. Moreover, AI-powered solutions may track drug adherence and provide reminders, therefore addressing a typical difficulty in controlling.

Moreover, improving psychosocial therapies, which are essential for the inclusive treatment of schizophrenia, are AI-driven platforms. By customizing treatment sessions depending on patient progress and feedback, digital CBT systems may provide a unique therapeutic experience [79]. Integrated with AI, VR treatments provide immersive settings for social skills training and exposure therapy, thereby helping patients control symptoms like social anxiety and hallucinations.

Despite the positive advancements, integrating AI into the clinical practice of diagnosing and treating schizophrenia poses several challenges. Ensuring ethical and successful application of AI involves meticulous management of issues such as data security, algorithmic dysfunction, and the interpretability of AI models [80]. Still, the promising potential of AI in managing schizophrenia offers a hope for more precise, timely, and individualized treatment for this complex illness.

17.8 PERSONALITY DISORDERS

Personality disorders are a spectrum of mental health illnesses distinguished from the expectations of an individual's culture by persistent cognitive and behavioral patterns. These problems seriously interfere with social, vocational, and other vital spheres of life. Conventional approaches for personality disorders can depend on subjective clinical evaluations, organized interviews, and self-report questionnaires, which could result in variance in management approaches [81]. Recent advancements in AI promise to revolutionize the diagnosis and treatment of personality disorders with more precise, efficient, and personalized approaches [32].

AI can analyze huge amounts of data and identify patterns that physicians may not notice [70]. To improve diagnostics accuracy and enable early intervention, these technologies may interpret data from many sources—including EHRs, psychometric tests, genetic information, and neuroimaging data [71].

An important use of AI in personality disorder diagnosis is EHR and clinical note analysis. From unstructured text, NLP systems may recognize symptoms and behavioral patterns suggestive of certain personality disorder [82]. NLP technologies can detect minor linguistic signals and communication patterns that are indicative of conditions such as borderline personality disorder or antisocial personality disorder. AI may provide a complete diagnostic profile by combining data from many clinical sites, hence lowering the dependence on subjective judgments and enhancing diagnosis consistency [83].

With AI improving its diagnostic value, neuroimaging has also grown to be a useful technique in studying personality disorders. Brain scans allow ML algorithms to detect structural and functional problems linked to different personality disorders. By spotting certain brain patterns and connection problems, studies have shown that AI systems can discriminate between people with various personality disorders [42]. This capacity not only improves diagnosis accuracy but also provides an understanding of the fundamental neurobiological causes of many illnesses, hence driving more focused treatments [84].

The therapy of personality problems by AI has similarly great potential. Since people with personality disorders typically react differently to therapeutic approaches, tailored treatment programs are very vital. By forecasting treatment results depending on a person's individual clinical and genetic profile, AI may help to customize these plans. Systems driven by AI can analyze data from previous therapy sessions and patient progress reports to propose modifications in treatment plans such as dialectical behavior therapy, CBT, or psychopharmacological treatments [85].

Furthermore, becoming useful tools for addressing personality disorders are AI-powered digital health treatments. Equipped with AI algorithms, mobile apps, and web platforms may provide real-time monitoring and assistance, hence improving patient involvement and adherence to treatment guidelines. These instruments may provide customized therapeutic activities, mood monitoring, and symptom management techniques, therefore enabling therapy to be more sensitive to particular needs [86]. Furthermore, under investigation are virtual reality (VR) treatments combined with AI for its ability to build immersive therapeutic settings allowing patients to practice social skills and reduce anxiety in a regulated environment.

While AI holds significant potential for treating and diagnosing personality conditions, specific challenges must be addressed to ensure its effective utilization. The development of confidence and acceptance in clinical practice is heavily dependent upon some ethical concerns such as algorithmic bias, data privacy, and interpretability of AI models [87]. The successful integration of AI systems into mental health treatment relies on the assurance that professionals and patients possess knowledge about the decision-making process and that the systems are transparent.

AI has significant potential to revolutionize the diagnosis and treatment of personality disorders, considering all relevant factors. Using cutting-edge algorithms and varied datasets, AI can improve diagnosis accuracy, assist early intervention, and customize treatment strategies. Realizing the ethical and pragmatic possibilities of AI in revolutionizing mental health treatment and enhancing results for people with personality disorders depends on addressing the related issues.

17.9 AI USE IN THE DIAGNOSIS AND TREATMENT

17.9.1 Eating Disorders

Eating disorders (EDs) are intricate metabolic-psychiatric illnesses that are on the rise and are influenced by societal changes. Treatment for these disorders remains challenging, with a lack of strong evidence supporting current care practices [88]. As such, there is a pressing need to explore new interventions. EDs primarily impact the young adults, who are particularly susceptible to the influence of rapidly advancing technologies. Currently, we are entering a new era that holds promise for transforming ED treatment, but it is crucial to consider the associated risks and ethical implications moving forward.

The utilization of AI in the realm of EDs can be classified into two main areas: the first one is the utilization of AI in analyzing large datasets collected from either patient without their direct participation, thereby expanding the scope of epidemiological, and psychosocial studies, and the second one is the active integration of AI-based tools in treatment procedures and daily life. Both aspects have the potential in order to drive therapeutic advancements, yet they may also pose challenges and ethical dilemmas. Therefore, a detailed examination of key aspects of AI implementation in the field of EDs is deemed necessary. As a preventive measure, the utilization of AI by individuals who have not yet sought treatment should be taken into consideration. AI has the capacity to expedite the recognition of symptoms characteristic of EDs or of individuals at high risk for developing EDs.

Early detection plays a pivotal role in effectively managing EDs and can significantly impact the likelihood of successful treatment outcomes. AI algorithms can sift through extensive datasets, such as social media content, online search histories, and smartphone AI apps, to identify potential indicators. Through discerning behavioral trends and linguistic cues, AI has the potential to alert users and possibly notify healthcare professionals, enabling timely intervention at the first stages of the illness. Such early detection can markedly enhance the prospects of successful prognosis. Moreover, AI could offer relevant and precise outputs on treatment options and the efficacy of interventions for EDs in a more rapid and comprehensive manner compared to conventional internet searches [89].

Conversely, a patient who has recently exhibited symptoms of EDs but lacks motivation for treatment may exacerbate the condition by accessing information on purging behaviors, excessive exercise routines, unhealthy diets, and advice directly from AI. Despite the implementation of safety measures by developers to prevent access to potentially harmful content, individuals can easily bypass these safeguards by asking indirect questions. This possibility is especially critical when considering the impact on vulnerable individuals. Misinformation regarding weight, food, and calorie intake can have devastating consequences for those with EDs. It is widely recognized that effective treatment for EDs begins with addressing and correcting misconceptions about metabolism and the body, thereby challenging false beliefs and misinformation. The use of AI in treatment offers several potential benefits. Firstly, AI can serve as a valuable resource for providing support between therapy sessions when triggers emerge (e.g., cravings, unexpected weight changes, pre- and post-meal anxiety, etc.), and for offering problem-solving strategies in the individual's daily environment as they work toward treatment objectives (e.g., meeting weekly weight gain targets) [90].

EDs can impact people living in remote places where access to specialized care is restricted by geographical or financial limitations. AI, on the other hand, can offer accessible and cost-effective support. AI-powered mobile apps and online platforms can provide evidence-based alerts, interventions, psychoeducation for individuals. Nonetheless, challenges such as privacy and data security need to be addressed to safeguard individuals' rights. It is crucial to prioritize the protection of personal information and ensure transparency in AI algorithms to inform patients about data usage. Establishing laws and regulations regarding data privacy in AI applications is imperative. If AI fails to deliver the expected solutions to a severely distressed individual, it could pose risks. Despite the widespread availability of AI, our mission is to thoroughly examine the phenomenon with honesty and empathy, involving individuals affected by EDs to minimize unknown risks. It is our responsibility to educate patients on the responsible use of AI opportunities while learning from their experiences. It is better to adapt to the changes brought by AI rather than be overwhelmed by them [89].

17.9.2 Eating Disorders and Machine Learning

In recent years, generative AI (GAI) has been developed. GAI systems can recognize data and create new content like texts, images, and music. Various GAI systems like Bard, ChatGPT, DeepMind, and Midjourney exist. The editorial and recent articles suggest potential uses and risks of AI in studying, preventing, and treating Eds [91]. These applications extend to other medical fields. GAI will play a significant role in medical research, drug discovery, clinical reports, diagnoses, disease monitoring, patient communication, and robotic surgery. For instance, GPT-4 has shown high accuracy in diagnosing complex cases compared to many physicians. In the realm of eating and weight disorders, the combination of AI and fMRI shows great potential for researching Prader-Willi syndrome, a genetic disorder associated with obesity. AI exhibits promise in the prevention and management of EDs and concerns related to body image [92].

In conclusion, AI has the potential to revolutionize ED treatment. It can help analyze data, detect symptoms early, provide support, create personalized treatment plans, and make care more accessible. However, it is important to use AI ethically and protect patient privacy. We need to conduct research and establish guidelines for its use. It is also crucial to address the ethical and legal concerns surrounding AI in therapy. All parties should collaborate to assess how AI affects the relationship between clinicians and patients. This is a major obstacle that we need to address in the upcoming years [89].

17.9.3 Employing Machine Learning Methods in Disorders Associated with Trauma

ML methods are being more commonly used in the research of trauma-linked conditions like post-traumatic stress disorder (PTSD) and acute stress disorder (ASD). The Diagnostic and Statistical Manual, fifth edition (DMS-5) lists 20 features for PTSD, categorized into four groups: reliving trauma, avoiding reminders, negative feelings, and heightened reactions. According to research, a history of previous trauma or multiple traumatic experiences may raise the likelihood of developing PTSD [93]. Diagnosing PTSD and ASD has been challenging in clinical practice and academic research due to the intricate and multifaceted nature of trauma-related illnesses. The criteria for diagnosis, like those found in the DSM-5, have mostly been created through research with long-term patient groups in specialized medical facilities, limiting our understanding of the disorder's characteristics [94]. Extended exposure and cognitive processing therapy are proven to be the most successful evidence-based treatments for trauma-related disorders, particularly when compared to other therapies.

ML technologies offer the potential to improve disease categorization, risk factor prediction, and treatment results, as well as personalized therapy choices [95]. Due to the diverse nature of PTSD and ASD in clinical and biological terms, it can be challenging to pinpoint the root causes and develop successful treatments and diagnostic methods. ML offers a practical solution for resolving these problems. This study intends to thoroughly examine data regarding the use of ML methods in evaluating PTSD and ASD for categorization, prediction, and therapy selection. A key application of ML in the field of psychiatry involves developing predictive models for mental health disorders using risk factors. Out of the 49 studies that were examined, 33 employed ML for this purpose. Various ML methodologies may be necessary due to the diverse weights of predictors in different populations and screening settings [96]. Our analysis uncovered various studies that focused on demographics, such as combat veterans.

The current research investigated how ML techniques are being applied in PTSD studies. Several factors can cause these tactics to surpass traditional parametric methods. For instance, while customary regression analyses focus solely on the direct impact of variables on an outcome, certain algorithms can reveal hidden interactions and nonlinear relationships between features. Moreover, while a conventional regression utilizing highly correlated independent variables may be precise within one sample, it can show poor performance when utilized on different samples due to overfitting of the model. Contrarily, ML algorithms could lessen the risk of overestimating prediction accuracy. ML methods enable researchers to handle larger and more intricate datasets, as well as integrate information from various sources such as audio/video recordings and biological samples. Nevertheless, it is necessary to tackle various obstacles such as computational capacity, multimodality, validation of models, diversity in phenotypic and etiological factors, rare event assessment, expenses, non-uniform data distribution, diagnosis based on observation, absence of a standardized ML research process, lack of funding, and limited interpretability [97]. Moreover, changing data poses a challenge for ML in this situation. Future studies should assess the comparison of researchers and clinicians to ML models to enhance PTSD diagnosis. Furthermore, the creation of a quality evaluation tool for. It is emphasized that adherence to best practices in the development of ML studies is crucial, encompassing theoretical advancement, familiarity with various techniques, methodological precision, careful selection and management of appropriate data, and effective presentation of results. Computational psychiatry approaches integrate diverse computational levels and data types to enhance comprehension, prediction, and treatment of mental disorders [97]. ML methods enable the identification of relationships within data that may not be discernible through traditional statistical approaches; the presence of etiologic heterogeneity in complex conditions such as PTSD underscores the suitability of these techniques [98]. A key challenge moving forward is the translation of ML models into clinical practice to improve patient outcomes.

AI language predictors (ALPs) have demonstrated potential in forecasting trauma-related results for survivors, as conventional approaches have been difficult to use. Factors like personality traits and life stress have been identified as influencing both mental and physical health results. Nevertheless, the evaluation of these factors usually requires filling out lengthy questionnaires [31]. Another method involves analyzing the daily speech of survivors using ALPs. These models can now identify risk factors in verbal or written messages. ALPs offer fair and reliable evaluations that are automated, cost-effective, and can be easily adjusted to large healthcare systems [99].

17.9.4 Challenges

Mental ill health significantly contributes to the global health burden, potentially causing lifelong disability and/or premature mortality [31]. Research reveal that the prevalence of mental illnesses notably increased during and after the COVID-19 pandemic. Since then, telepsychiatry services have been rapidly growing to meet the increasing demand for mental health care. Although AI has achieved remarkable success in various clinical fields [100], its application in mental health care remains limited due to several unresolved challenges.

17.10 TECHNOLOGICAL CHALLENGES

17.10.1 Data Privacy

Data privacy remains the main concern while using AI in any domain. Some studies reported that most AI apps used for mental healthcare did not comply with privacy laws. Even AI applications that claimed to have a privacy policy admitted to sharing personal information with third parties without users' consent [101]. Sandoval et al. noted that users of mental health apps, particularly those with cognitive impairments, were often unaware of the privacy policies they should review before using the apps. Several measures are recommended to address data privacy concerns in AI applications. The European Data Protection Board has issued guidelines to prevent data breaches [102], highlighting the importance of establishing a top supervisory authority to continuously review, control, and secure personal data used in AI-based apps. Moreover, maintaining sustained technical traffic analysis is advised to ensure the technology complies with privacy regulations. Parties that commit criminal data breach should face strict penalties. Furthermore, users should be educated about their privacy rights when using digital apps.

17.10.2 Algorithm Performance

Accurate performance of AI model relies on clear and extensive datasets. However, in the current clinical reality, mental health data are insufficient, heterogeneous, and complex, limiting the full potential of AI models in psychiatric care [71]. Moreover, the fluctuating nature of mental health symptoms poses another challenge, as deep learning models are often restricted by fixed inputs and do not account for temporality of data. Consequently, ML algorithms are constrained by the time-limited training data, making them unsuitable for capturing the dynamic nature of mental disorders [71]. Besides, training AI algorithms on non-representative populations or uncommon conditions with non-generalizable data lead to failures in providing real-world evidence. In addition, there are knowledge gaps in understudied mental conditions owing to the poor infrastructure [103], fear of stigmatization, or inadequate surveillance/reporting systems [71]. This will impact the comprehensiveness and accuracy of the present data and further restricting the effectiveness of AI in clinical practice. To enhance AI performance, several key recommendations are essential. Conducting comprehensive population-based research is crucial to collect large set of data that ensures the representation of real-world data. These data sources can be derived from diverse and reliable sources, including EHR, government registries, online platforms, surveys, wearable devices, and environmental data. Effectively integrating and processing this varied data into deep learning models remains a challenge. Therefore, various transformation processes should be performed on this data to be in a more clear and transparent form before being used. Moreover, establishing standards for assessing AI performance is crucial. Finally, employing time-sensitive algorithms is imperative to ensure the timely delivery of clinical care.

17.10.3 Interpretability/Explainability

AI is often perceived as a 'black box' by clinicians [71]. Understanding the logic behind how AI makes decisions in clinical practice allows healthcare professionals to evaluate and validate the outputs effectively, promoting AI integration into workflows. To achieve this, enhancing technological literacy of doctors through training and reimbursement policies will deepen their understanding of AI's role in clinical practice which is essential for effective and safe use [104]. Moreover, engaging physicians in the development and validation process of AI tools will facilitate adoption in mental healthcare.

17.10.4 Accuracy and Validity

AI demonstrates inaccuracies in clinical practice An example was highlighted by Nicholas et al. revealing that some mHealth apps were significantly far from best-practice guidelines. Several studies have proposed approaches to improve data accuracy in mental healthcare. Coravos et al. highlight the importance of continuous assessment, verification, and validation of technologies used in mental health to ensure they align with real-world data [105]. Moreover, the American Psychiatric Association has released a remarkable initiative by releasing a comprehensive app evaluation model to test the usability and safety of mental health apps [106]. This model consists of fundamental questions to collect data about every app's background and access, safety and privacy, clinical basis, practicability, and treatment approaches [106]. Besides, Sendak et al. proposed a four-phase path to boost AI products deployment in clinical settings [107]. Also, collaborative work between clinicians, technology developers, and other relevant sectors is crucial for assessment of the outcomes of AI.

17.10.5 Equity in Artificial Intelligence Use

Certain groups, such as those residing in economically disadvantaged areas, people with cultural and language barriers. Extreme age groups, individuals with disabilities, and those with limited digital literacy [108], may face challenges in benefiting from AI-driven mental health applications compared to their counterparts. Addressing these barriers is crucial to reduce health disparities. Strategies must be implemented to address biases and to ensure truly inclusive and equitable AI use in mHealth. Personalized approaches to customize digital tools features to users' personal, medical, and cultural characteristics is imperative.

17.11 CLINICAL CHALLENGES

17.11.1 Heterogeneity of Mental Symptoms

The symptomatology of mental illnesses varies based on individuals' genetic, physiological, and environmental factors. Besides, clinical factors such as symptoms pattern, stage, and presence of comorbidities all contribute to dramatic variations in symptom presentation. Given the heterogeneous nature of mental disorders, current AI models often lack training on the full spectrum of symptom presentation and associated risk factors and, thereby limiting their diagnostic accuracy and effectiveness [109]. To handle the variability of mental symptomatology using AI, several recommendations can be followed. Firstly, enhancing data collection by assembling large datasets supported with diverse clinical predictors can increase AI efficiency in managing mental health conditions. Additionally, the results of ML algorithms in clinical processes should disclose the quality of the input data and potential biases in outputs for maintaining the reliability of AI in clinical decision making. Integrating the expertise of experienced psychiatrists is also beneficial to identify and correct erroneous clinical outcomes produced by AI systems, thereby ensuring accurate diagnoses Multimodal AI models, despite their challenges, can differentiate various inputs, and improve diagnostic accuracy [110].

17.11.2 Interoperability

Clinicians face several issues with digitalization in mental healthcare. They are concerned that incorporating AI into their daily practice will add new administrative tasks which will affect evidence-based practice and increase the workload. Trust issues also arise, as clinicians worry that AI technology may rely on inaccurate data, potentially leading to misdiagnoses or missed diagnoses. Additionally, clinicians believe that AI chatbots deprive patients of the therapeutic aspects of in-person therapy, such as motivation, self-advocacy, and social communication [111]. This underscores the positive impact of onsite visits for individuals with mental problems. To address these concerns, careful attention must be given to how AI integrates into the context of care to ensure it fulfils its intended role. Achieving this requires collaborative efforts by all parties, clinicians, patients, family members, AI developers, and regulators to ensure the successful implementation of AI technologies in clinical practice. Furthermore, ensuring ease of use and promptly addressing technical challenges will facilitate greater acceptance and utilization of AI technologies in mental healthcare. Finally, getting feedback from users is crucial for continuous improvement of AI in the clinical settings.

17.12 SUMMARY

AI provides different powerful techniques for mental health disorders diagnosis and treatment. The use of variable data sources, ML algorithms, speech pattern analysis, predictive analysis, NLP, and variable AI-powered therapeutic intervention revolutionized the current clinical mental health practice.

Customization of treatment regimens improve the outcome of different mental disorders, furthermore the patient's behavioral analysis can help in assessing the prognosis and treatment response for various disorders including: SUDs, depression, schizophrenia and personality and EDs. Additionally, various diagnostic modalities have been implemented to ML models to diagnose and differentiate anxiety and mood disorders, and the use of structurally designed AI-assisted psychotherapy robots can help in providing easily accessed CBT-based psychotherapy and emotional therapy.

Considerable ethical, technological and clinical challenges arise when AI is used in mental health. To guarantee that AI is applied responsibly, worries about algorithmic bias risk, data security, and privacy must be properly controlled, in addition to the capability of AI systems in interpretation of symptoms presentation and the related risk factors. Strong ethical frameworks and legal safeguards are necessary to protect patient rights and advance fair use of AI systems. Thus, future researches are required, and usage standards must be established to address the ethical and legal issues related to AI use in mental health and to evaluate the impact of AI on the doctor-patient relationship.

ACKNOWLEDGMENTS

We would like to express our sincere gratitude to Nourhan S. Sultan. for her invaluable assistance in conducting the plagiarism check for our research. Her expertise and support were crucial in ensuring the integrity and quality of our work. We deeply appreciate her time and effort.

REFERENCES

1. Topol, E. (2019a). *Deep Medicine: How Artificial Intelligence Can Make Healthcare Human Again*. New York: Hachette.
2. Guntuku, S. C., Yaden, D. B., Kern, M. L., Ungar, L. H., & Eichstaedt, J. C. (2017). Detecting depression and mental illness on social media: An integrative review. *Current Opinion in Behavioral Sciences, 18*, 43–49.
3. Miner, A. S., Milstein, A., & Hancock, J. T. (2017). Talking to machines about personal mental health problems. *JAMA, 318*(13), 1217–1218.
4. Fulmer, R., Joerin, A., Gentile, B., Lakerink, L., & Rauws, M. (2018). Using psychological artificial intelligence (Tess) to relieve symptoms of depression and anxiety: Randomized controlled trial. *JMIR Mental Health, 5*(4), e9782.
5. Naslund, J. A., Aschbrenner, K. A., Marsch, L. A., & Bartels, S. J. (2016). The future of mental health care: Peer-to-peer support and social media. *Epidemiology and Psychiatric Sciences, 25*(2), 113–122.

6. Mohr, D. C., Zhang, M., & Schueller, S. M. (2017). Personal sensing: Understanding mental health using ubiquitous sensors and machine learning. *Annual Review of Clinical Psychology, 13*(1), 23–47. https://doi.org/10.1146/annurev-clinpsy-032816-044949
7. Torous, J., & Roberts, L. W. (2017). The ethical use of mobile health technology in clinical psychiatry. *The Journal of Nervous and Mental Disease, 205*(1), 4–8.
8. Kessler, R. C., Berglund, P., Demler, O., Jin, R., Merikangas, K. R., & Walters, E. E. (2005). Lifetime prevalence and age-of-onset distributions of DSM-IV disorders in the national comorbidity survey replication. *Archives of General Psychiatry, 62*(6), 593–602.
9. Majeed, S., & Khan, A. Q. (2024). Artificial intelligence and mental health services: A systematic review. *Pakistan Postgraduate Medical Journal, 35*(1), 35–40.
10. Soffer, M. Sr, Charney, S. Sr, Landi, A., Nadkarni, I., & Klang, G. Jr (2024). Applications of large language models in psychiatry: A systematic review. *medRxiv*, 2024–03.
11. Omarov, B., Narynov, S., & Zhumanov, Z. (2023). Artificial intelligence-enabled chatbots in mental health: A systematic review. *Computers, Materials & Continua, 74*(3), 5105–5122.
12. Graham, S., Depp, C., Lee, E. E., Nebeker, C., Tu, X., Kim, H.-C., & Jeste, D. V. (2019). Artificial intelligence for mental health and mental illnesses: An overview. *Current Psychiatry Reports, 21*(11), 116.
13. Jin, K. W., Li, Q., Xie, Y., & Xiao, G. (2023). Artificial intelligence in mental healthcare: An overview and future perspectives. *The British Journal of Radiology, 96*(1150), 20230213.
14. Gültekin, M., & Şahin, M. (2024). The use of artificial intelligence in mental health services in Turkey: What do mental health professionals think? *Cyberpsychology: Journal of Psychosocial Research on Cyberspace, 18*(1).
15. Li, H., Zhang, R., Lee, Y.-C., Kraut, R. E., & Mohr, D. C. (2023). Systematic review and meta-analysis of AI-based conversational agents for promoting mental health and well-being. *NPJ Digital Medicine, 6*(1), 236.
16. Chen, X., & Genc, Y. (2022). A systematic review of artificial intelligence and mental health in the context of social media. In H. Degen & S. Ntoa (Eds.), *Artificial Intelligence in HCI* (Vol. 13336, pp. 353–368). Springer International Publishing.
17. Lee, T. C., Shah, N. U., Haack, A., & Baxter, S. L. (2020). Clinical implementation of predictive models embedded within electronic health record systems: A systematic review. *Informatics, 7*(3), 25.
18. Thieme, A., Belgrave, D., & Doherty, G. (2020). Machine learning in mental health: A systematic review of the HCI literature to support the development of effective and implementable ML systems. *ACM Transactions on Computer-Human Interaction, 27*(5), 1–53.
19. Das, A. K., & Naskar, R. (2024). A deep learning model for depression detection based on MFCC and CNN generated spectrogram features. *Biomedical Signal Processing and Control, 90*, 105898.
20. Koutsouleris, N., Dwyer, D. B., Degenhardt, F., Maj, C., Urquijo-Castro, M. F., Sanfelici, R., Popovic, D., Oeztuerk, O., Haas, S. S., & Weiske, J. (2021). Multimodal machine learning workflows for prediction of psychosis in patients with clinical high-risk syndromes and recent-onset depression. *JAMA Psychiatry, 78*(2), 195–209.
21. Kishimoto, T., Nakamura, H., Kano, Y., Eguchi, Y., Kitazawa, M., Liang, K., Kudo, K., Sento, A., Takamiya, A., & Horigome, T. (2022). Understanding psychiatric illness through natural language processing (UNDERPIN): Rationale, design, and methodology. *Frontiers in Psychiatry, 13*, 954703.
22. Fitzpatrick, K. K., Darcy, A., & Vierhile, M. (2017). Delivering cognitive behavior therapy to young adults with symptoms of depression and anxiety using a fully automated conversational agent (Woebot): A randomized controlled trial. *JMIR Mental Health, 4*(2), e7785.
23. Chekroud, A. M., Zotti, R. J., Shehzad, Z., Gueorguieva, R., Johnson, M. K., Trivedi, M. H., Cannon, T. D., Krystal, J. H., & Corlett, P. R. (2016). Cross-trial prediction of treatment outcome in depression: A machine learning approach. *The Lancet Psychiatry, 3*(3), 243–250.
24. Fiske, A., Henningsen, P., & Buyx, A. (2019a). Your robot therapist will see you now: Ethical implications of embodied artificial intelligence in psychiatry, psychology, and psychotherapy. *Journal of Medical Internet Research, 21*(5), e13216.
25. Thakkar, A., Gupta, A., & De Sousa, A. (2024). Artificial intelligence in positive mental health: A narrative review. *Frontiers in Digital Health, 6*, 1280235.
26. Reilly, J., Meurk, C., Heffernan, E., & Sara, G. (2023). Substance use disorder screening and brief intervention in routine clinical practice in specialist adult mental health services: A systematic review. *Australian & New Zealand Journal of Psychiatry, 57*(6), 793–810.
27. Acion, L., Kelmansky, D., van der Laan, M., Sahker, E., Jones, D., & Arndt, S. (2017). Use of a machine learning framework to predict substance use disorder treatment success. *PloS One, 12*(4), e0175383.
28. Kashyap, A., Callison-Burch, C., & Boland, M. R. (2023). A deep learning method to detect opioid prescription and opioid use disorder from electronic health records. *International Journal of Medical Informatics, 171*, 104979.
29. Goodman-Meza, D., Tang, A., Aryanfar, B., Vazquez, S., Gordon, A. J., Goto, M., Goetz, M. B., Shoptaw, S., & Bui, A. A. (2022). Natural language processing and machine learning to identify people who inject drugs in electronic health records. *Open Forum Infectious Diseases, 9*(9), ofac471.

30. DiGangi, J. A., Gomez, D., Mendoza, L., Jason, L. A., Keys, C. B., & Koenen, K. C. (2013). Pretrauma risk factors for posttraumatic stress disorder: A systematic review of the literature. *Clinical Psychology Review, 33*(6), 728–744.
31. Garg, S., Taylor, J., El Sherief, M., Kasson, E., Aledavood, T., Riordan, R., Kaiser, N., Cavazos-Rehg, P., & De Choudhury, M. (2021). Detecting risk level in individuals misusing fentanyl utilizing posts from an online community on Reddit. *Internet Interventions, 26*, 100467.
32. Tilala, M. H., Chenchala, P. K., Choppadandi, A., Kaur, J., Naguri, S., Saoji, R., & Devaguptapu, B. (2024). Ethical considerations in the use of artificial intelligence and machine learning in health care: A comprehensive review. *Cureus, 16*(6). e62443. 10.7759/cureus.62443
33. Bharat, C., Hickman, M., Barbieri, S., & Degenhardt, L. (2021). Big data and predictive modelling for the opioid crisis: Existing research and future potential. *The Lancet Digital Health, 3*(6), e397–e407.
34. Lauvsnes, A. D. F., Gråwe, R. W., & Langaas, M. (2022). Predicting relapse in substance use: Prospective modeling based on intensive longitudinal data on mental health, cognition, and craving. *Brain Sciences, 12*(7), 957.
35. Ahmed, A., Aziz, S., Alzubaidi, M., Schneider, J., Irshaidat, S., Serhan, H. A., … Househ, M. (2023). Wearable devices for anxiety & depression: A scoping review. *Computer Methods and Programs in Biomedicine Update, 3*, 100095.
36. Mbunge, E., Batani, J., Chitungo, I., Moyo, E., Musuka, G., Muchemwa, B., & Dzinamarira, T. (2024). Towards data-driven artificial intelligence models for monitoring, modelling and predicting illicit substance use. In R. Silhavy & P. Silhavy (Eds.), *Data Analytics in System Engineering* (935) (pp. 361–379). Cham: Springer Nature Switzerland.
37. Arif, M., Basri, A., Melibari, G., Sindi, T., Alghamdi, N., Altalhi, N., & Arif, M. (2020). Classification of anxiety disorders using machine learning methods: A literature review. *Insights of Biomedical Research, 4*(1), 95–110.
38. Yan, W.-J., Ruan, Q.-N., & Jiang, K. (2022). Challenges for artificial intelligence in recognizing mental disorders. *Diagnostics, 13*(1), 2.
39. Gorman, J. M. (2003). Treating generalized anxiety disorder. *Journal of Clinical Psychiatry, 64*(1), 24–29.
40. Goodwin, H., Yiend, J., & Hirsch, C. R. (2017). Generalized anxiety disorder, worry and attention to threat: A systematic review. *Clinical Psychology Review, 54*, 107–122.
41. Hood, B. M., Schacter, D., Wegner, D., & Gilbert, D. (2015). *Psychology: Second European Edition*. London: Macmillan Education UK.
42. Sau, A., & Bhakta, I. (2019). Screening of anxiety and depression among seafarers using machine learning technology. *Informatics in Medicine Unlocked, 16*, 100228.
43. Hilbert, K., Lueken, U., Muehlhan, M., & Beesdo-Baum, K. (2017). Separating generalized anxiety disorder from major depression using clinical, hormonal, and structural MRI data: A multimodal machine learning study. *Brain and Behavior, 7*(3), e00633.
44. Jacobson, N. C., & Feng, B. (2022). Digital phenotyping of generalized anxiety disorder: Using artificial intelligence to accurately predict symptom severity using wearable sensors in daily life. *Translational Psychiatry, 12*(1), 336.
45. Caballo, V. E., Salazar, I. C., Irurtia, M. J., Arias, B., & Nobre, L. (2013). The assessment of social anxiety through five self-report measures, LSAS-SR, SPAI, SPIN, SPS, and SIAS: A critical analysis of their factor structure. *Behavioral Psychology/Psicología Conductual, 21*(3), 423–448.
46. von Glischinski, M., Willutzki, U., Stangier, U., Hiller, W., Hoyer, J., Leibing, E., Leichsenring, F., & Hirschfeld, G. (2018). Liebowitz Social Anxiety Scale (LSAS): Optimal cut points for remission and response in a German sample. *Clinical Psychology & Psychotherapy, 25*(3), 465–473.
47. Fathi, S., Ahmadi, M., Birashk, B., & Dehnad, A. (2020). Development and use of a clinical decision support system for the diagnosis of social anxiety disorder. *Computer Methods and Programs in Biomedicine, 190*, 105354.
48. Lueken, U., Straube, B., Yang, Y., Hahn, T., Beesdo-Baum, K., Wittchen, H.-U., Konrad, C., Ströhle, A., Wittmann, A., & Gerlach, A. L. (2015). Separating depressive comorbidity from panic disorder: A combined functional magnetic resonance imaging and machine learning approach. *Journal of Affective Disorders, 184*, 182–192.
49. Sickel, A. E., Seacat, J. D., & Nabors, N. A. (2014). Mental health stigma update: A review of consequences. *Advances in Mental Health, 12*(3), 202–215.
50. Su, S., Wang, Y., Jiang, W., Zhao, W., Gao, R., Wu, Y., Tao, J., Su, Y., Zhang, J., & Li, K. (2022). Efficacy of artificial intelligence-assisted psychotherapy in patients with anxiety disorders: A prospective, national multicenter randomized controlled trial protocol. *Frontiers in Psychiatry, 12*, 799917.
51. Mehta, A., Niles, A. N., Vargas, J. H., Marafon, T., Couto, D. D., & Gross, J. J. (2021). Acceptability and effectiveness of artificial intelligence therapy for anxiety and depression (Youper): Longitudinal observational study. *Journal of Medical Internet Research, 23*(6), e26771.
52. Botella, C., Fernández-Álvarez, J., Guillén, V., García-Palacios, A., & Baños, R. (2017). Recent progress in virtual reality exposure therapy for phobias: A systematic review. *Current Psychiatry Reports, 19*(7), 42.
53. Parsons, T. D., & Rizzo, A. A. (2008). Affective outcomes of virtual reality exposure therapy for anxiety and specific phobias: A meta-analysis. *Journal of Behavior Therapy and Experimental Psychiatry, 39*(3), 250–261.

54. Kroenke, K., Spitzer, R. L., & Williams, J. B. (2003). The Patient Health Questionnaire-2: Validity of a two-item depression screener. *Medical Care, 41*(11), 1284–1292.
55. Schmaal, L., Marquand, A. F., Rhebergen, D., van Tol, M.-J., Ruhé, H. G., van der Wee, N. J., Veltman, D. J., & Penninx, B. W. (2015). Predicting the naturalistic course of major depressive disorder using clinical and multimodal neuroimaging information: A multivariate pattern recognition study. *Biological Psychiatry, 78*(4), 278–286.
56. Nemesure, M. D., Heinz, M. V., Huang, R., & Jacobson, N. C. (2021). Predictive modeling of depression and anxiety using electronic health records and a novel machine learning approach with artificial intelligence. *Scientific Reports, 11*(1), 1980.
57. Abd-Alrazaq, A., AlSaad, R., Aziz, S., Ahmed, A., Denecke, K., Househ, M., … Sheikh, J. (2023). Wearable artificial intelligence for anxiety and depression: Scoping review. *Journal of Medical Internet Research, 25*, e42672.
58. Lee, S., Kim, H., Park, M. J., & Jeon, H. J. (2021). Current advances in wearable devices and their sensors in patients with depression. *Frontiers in Psychiatry, 12*, 672347.
59. Ayers, J. W., Poliak, A., Dredze, M., Leas, E. C., Zhu, Z., Kelley, J. B., Faix, D. J., Goodman, A. M., Longhurst, C. A., & Hogarth, M. (2023). Comparing physician and artificial intelligence chatbot responses to patient questions posted to a public social media forum. *JAMA Internal Medicine, 183*(6), 589–596.
60. Doraiswamy, S., Abraham, A., Mamtani, R., & Cheema, S. (2020). Use of telehealth during the COVID-19 pandemic: Scoping review. *Journal of Medical Internet Research, 22*(12), e24087.
61. Cohen, S. E., Zantvoord, J. B., Wezenberg, B. N., Bockting, C. L., & van Wingen, G. A. (2021). Magnetic resonance imaging for individual prediction of treatment response in major depressive disorder: A systematic review and meta-analysis. *Translational Psychiatry, 11*(1), 168.
62. Phillips, M. L., & Kupfer, D. J. (2013). Bipolar disorder diagnosis: Challenges and future directions. *The Lancet, 381*(9878), 1663–1671.
63. Ferrari, A. J., Stockings, E., Khoo, J., Erskine, H. E., Degenhardt, L., Vos, T., & Whiteford, H. A. (2016). The prevalence and burden of bipolar disorder: Findings from the global burden of disease study 2013. *Bipolar Disorders, 18*(5), 440–450.
64. Hirschfeld, R. M. (2014). Differential diagnosis of bipolar disorder and major depressive disorder. *Journal of Affective Disorders, 169*, S12–S16.
65. Jan, Z., Ai-Ansari, N., Mousa, O., Abd-Alrazaq, A., Ahmed, A., Alam, T., & Househ, M. (2021). The role of machine learning in diagnosing bipolar disorder: Scoping review. *Journal of Medical Internet Research, 23*(11), e29749.
66. Sonkurt, H. O., Altınöz, A. E., Çimen, E., Köşger, F., & Öztürk, G. (2021). The role of cognitive functions in the diagnosis of bipolar disorder: A machine learning model. *International Journal of Medical Informatics, 145*, 104311.
67. Ezell, J. M., Ajayi, B. P., Parikh, T., Miller, K., Rains, A., & Scales, D. (2024). Drug use and artificial intelligence: Weighing concerns and possibilities for prevention. *American Journal of Preventive Medicine, 66*(3), 568–572.
68. Kondou, A., Haku, M., & Yasui, T. (2021). Development and psychometric testing of the mental health scale for childrearing fathers. *Healthcare, 9*(11), 1587.
69. Costantini, L., Pasquarella, C., Odone, A., Colucci, M. E., Costanza, A., Serafini, G., Aguglia, A., Murri, M. B., Brakoulias, V., & Amore, M. (2021). Screening for depression in primary care with patient health questionnaire-9 (PHQ-9): A systematic review. *Journal of Affective Disorders, 279*, 473–483.
70. Esteva, A., Robicquet, A., Ramsundar, B., Kuleshov, V., DePristo, M., Chou, K., Cui, C., Corrado, G., Thrun, S., & Dean, J. (2019). A guide to deep learning in healthcare. *Nature Medicine, 25*(1), 24–29.
71. Miotto, R., Wang, F., Wang, S., Jiang, X., & Dudley, J. T. (2018a). Deep learning for healthcare: Review, opportunities and challenges. *Briefings in Bioinformatics, 19*(6), 1236–1246.
72. Chanen, A. M., Nicol, K., Betts, J. K., & Thompson, K. N. (2020). Diagnosis and treatment of borderline personality disorder in young people. *Current Psychiatry Reports, 22*(5), 25.
73. Owen, M. J., Sawa, A., & Mortensen, P. B. (2016). Schizophrenia. *Lancet, 388*, 86–97.
74. Leucht, S., Cipriani, A., Spineli, L., Mavridis, D., Örey, D., Richter, F., Samara, M., Barbui, C., Engel, R. R., & Geddes, J. R. (2013). Comparative efficacy and tolerability of 15 antipsychotic drugs in schizophrenia: A multiple-treatments meta-analysis. *The Lancet, 382*(9896), 951–962.
75. Plis, S. M., Hjelm, D. R., Salakhutdinov, R., Allen, E. A., Bockholt, H. J., Long, J. D., Johnson, H. J., Paulsen, J. S., Turner, J. A., & Calhoun, V. D. (2014). Deep learning for neuroimaging: A validation study. *Frontiers in Neuroscience, 8*, 229.
76. Schnack, H. G., Nieuwenhuis, M., van Haren, N. E., Abramovic, L., Scheewe, T. W., Brouwer, R. M., Pol, H. E. H., & Kahn, R. S. (2014). Can structural MRI aid in clinical classification? A machine learning study in two independent samples of patients with schizophrenia, bipolar disorder and healthy subjects. *Neuroimage, 84*, 299–306.
77. Jackson, R. G., Patel, R., Jayatilleke, N., Kolliakou, A., Ball, M., Gorrell, G., Roberts, A., Dobson, R. J., & Stewart, R. (2017). Natural language processing to extract symptoms of severe mental illness from clinical text: The clinical record interactive search comprehensive data extraction (CRIS-CODE) project. *BMJ Open, 7*(1), e012012.

78. Zhang, J.-P., Lencz, T., Zhang, R. X., Nitta, M., Maayan, L., John, M., Robinson, D. G., Fleischhacker, W. W., Kahn, R. S., & Ophoff, R. A. (2016). Pharmacogenetic associations of antipsychotic drug-related weight gain: A systematic review and meta-analysis. *Schizophrenia Bulletin, 42*(6), 1418–1437.
79. Hollis, C., Falconer, C. J., Martin, J. L., Whittington, C., Stockton, S., Glazebrook, C., & Davies, E. B. (2017). Annual research review: Digital health interventions for children and young people with mental health problems – A systematic and meta-review. *Journal of Child Psychology and Psychiatry, 58*(4), 474–503.
80. Topol, E. J. (2019b). High-performance medicine: The convergence of human and artificial intelligence. *Nature Medicine, 25*(1), 44–56.
81. Zimmerman, M., & Morgan, T. A. (2013). The relationship between borderline personality disorder and bipolar disorder. *Dialogues in Clinical Neuroscience, 15*(2), 155–169.
82. Zucchetti, A., Nibbio, G., Altieri, L., Bertorni, L., Calzavara-Pinton, I., Invernizzi, E., Necchini, N., Cerati, C., Poddighe, L., & Bulgari, V. (2024). Artificial intelligence applications in mental health: The state of the art. *Italian Journal of Psychiatry*. https://www.italianjournalofpsychiatry.it/article/view/544
83. Göndöcs, D., & Dörfler, V. (2024). AI in medical diagnosis: AI prediction & human judgment. *Artificial Intelligence in Medicine, 149*, 102769.
84. Schmaal, L., Hibar, D. P., Sämann, P. G., Hall, G. B., Baune, B. T., Jahanshad, N., Cheung, J. W., van Erp, T. G., Bos, D., & Ikram, M. A. (2017). Cortical abnormalities in adults and adolescents with major depression based on brain scans from 20 cohorts worldwide in the ENIGMA major depressive disorder working group. *Molecular Psychiatry, 22*(6), 900–909.
85. Bleidorn, W., & Hopwood, C. J. (2019). Using machine learning to advance personality assessment and theory. *Personality and Social Psychology Review, 23*(2), 190–203.
86. Philippe, T. J., Sikder, N., Jackson, A., Koblanski, M. E., Liow, E., Pilarinos, A., & Vasarhelyi, K. (2022). Digital health interventions for delivery of mental health care: Systematic and comprehensive meta-review. *JMIR Mental Health, 9*(5), e35159.
87. Char, D. S., Shah, N. H., & Magnus, D. (2018). Implementing machine learning in health care—Addressing ethical challenges. *New England Journal of Medicine, 378*(11), 981–983.
88. Crone, C., Fochtmann, L. J., Attia, E., Boland, R., Escobar, J., Fornari, V., Golden, N., Guarda, A., Jackson-Triche, M., Manzo, L., Mascolo, M., Pierce, K., Riddle, M., Seritan, A., Uniacke, B., Zucker, N., Yager, J., Craig, T. J., Hong, S.-H., & Medicus, J. (2023). The American psychiatric association practice guideline for the treatment of patients with eating disorders. *American Journal of Psychiatry, 180*(2), 167–171.
89. Abbate-Daga, G., Taverna, A., & Martini, M. (2023). The oracle of Delphi 2.0: Considering artificial intelligence as a challenging tool for the treatment of eating disorders. *Eating and Weight Disorders-Studies on Anorexia, Bulimia and Obesity, 28*(1), 50.
90. Panero, M., Longo, P., De Bacco, C., Abbate-Daga, G., & Martini, M. (2022). Shame, guilt, and self-consciousness in anorexia nervosa. *Journal of Clinical Medicine, 11*(22), 6683.
91. Solans Noguero, D., Ramírez-Cifuentes, D., Ríssola, E. A., & Freire, A. (2023). Gender bias when using artificial intelligence to assess anorexia nervosa on social media: Data-driven study. *Journal of Medical Internet Research, 25*, e45184.
92. Huang, Z., & Cai, J. (2023). Progress in brain magnetic resonance imaging of individuals with Prader–Willi syndrome. *Journal of Clinical Medicine, 12*(3), 1054.
93. Kilpatrick, D. G., Resnick, H. S., Milanak, M. E., Miller, M. W., Keyes, K. M., & Friedman, M. J. (2013). National estimates of exposure to traumatic events and PTSD prevalence using DSM-IV and DSM-5 criteria: DSM-5 PTSD prevalence. *Journal of Traumatic Stress, 26*(5), 537–547.
94. McGorry, P. D., Hickie, I. B., Yung, A. R., Pantelis, C., & Jackson, H. J. (2006). Clinical staging of psychiatric disorders: A heuristic framework for choosing earlier, safer and more effective interventions. *Australian & New Zealand Journal of Psychiatry, 40*(8), 616–622.
95. Hahn, T., Nierenberg, A. A., & Whitfield-Gabrieli, S. (2017). Predictive analytics in mental health: Applications, guidelines, challenges and perspectives. *Molecular Psychiatry, 22*(1), 37–43.
96. Ramos-Lima, L. F., Waikamp, V., Antonelli-Salgado, T., Passos, I. C., & Freitas, L. H. M. (2020). The use of machine learning techniques in trauma-related disorders: A systematic review. *Journal of Psychiatric Research, 121*, 159–172.
97. Passos, I. C., Ballester, P. L., Barros, R. C., Librenza-Garcia, D., Mwangi, B., Birmaher, B., Brietzke, E., Hajek, T., Lopez Jaramillo, C., Mansur, R. B., Alda, M., Haarman, B. C. M., Isometsa, E., Lam, R. W., McIntyre, R. S., Minuzzi, L., Kessing, L. V., Yatham, L. N., Duffy, A., & Kapczinski, F. (2019). Machine learning and Big Data analytics in bipolar disorder: A position paper from the International Society for Bipolar Disorders Big Data Task Force. *Bipolar Disorders, 21*(7), 582–594.
98. Tylee, D. S., Chandler, S. D., Nievergelt, C. M., Liu, X., Pazol, J., Woelk, C. H., Lohr, J. B., Kremen, W. S., Baker, D. G., & Glatt, S. J. (2015). Blood-based gene-expression biomarkers of post-traumatic stress disorder among deployed marines: A pilot study. *Psychoneuroendocrinology, 51*, 472–494.
99. Chancellor, S., & De Choudhury, M. (2020). Methods in predictive techniques for mental health status on social media: A critical review. *NPJ Digital Medicine, 3*(1), 43.

100. Singh, S., & Sagar, R. (2022). Time to have effective regulation of the mental health apps market: Maximize gains and minimize harms. *Indian Journal of Psychological Medicine*, *44*(4), 399–404.
101. Nicholas, J., Larsen, M. E., Proudfoot, J., & Christensen, H. (2015). Mobile apps for bipolar disorder: A systematic review of features and content quality. *Journal of Medical Internet Research*, *17*(8), e198.
102. Borgesius, F. Z., Asghari, H., Bangma, N., & Hoepman, J.-H. (2023). The GDPR's rules on data breaches: Analysing their rationales and effects. *SCRIPTed*, *20*, 352.
103. Mental Health ATLAS 2020. (n.d.). Retrieved August 20, 2024, from https://www.who.int/publications/i/item/9789240036703
104. Torous, J., & Firth, J. (2018). Bridging the dichotomy of actual versus aspirational digital health. *World Psychiatry*, *17*(1), 108.
105. Coravos, A., Goldsack, J. C., Karlin, D. R., Nebeker, C., Perakslis, E., Zimmerman, N., & Erb, M. K. (2019). Digital medicine: A primer on measurement. *Digital Biomarkers*, *3*(2), 31–71.
106. The App Evaluation Model. (n.d.). Retrieved August 20, 2024, from https://www.psychiatry.org:443/psychiatrists/practice/mental-health-apps/the-app-evaluation-model
107. Sendak, M. P., D'Arcy, J., Kashyap, S., Gao, M., Nichols, M., Corey, K., Ratliff, W., & Balu, S. (2020). A path for translation of machine learning products into healthcare deliver. *EMJ Innovations*, *10*, 19–00172.
108. Torous, J., Myrick, K., & Aguilera, A. (2023). The need for a new generation of digital mental health tools to support more accessible, effective and equitable care. *World Psychiatry*, *22*(1), 1.
109. Lee, E. E., Torous, J., De Choudhury, M., Depp, C. A., Graham, S. A., Kim, H.-C., Paulus, M. P., Krystal, J. H., & Jeste, D. V. (2021). Artificial intelligence for mental health care: Clinical applications, barriers, facilitators, and artificial wisdom. *Biological Psychiatry: Cognitive Neuroscience and Neuroimaging*, *6*(9), 856–864.
110. Multimodal Artificial Intelligence: Opportunities and Challenges in HIV Clinical Care – National Institute of Mental Health (NIMH). (2024, January 29). https://www.nimh.nih.gov/funding/grant-writing-and-application-process/concept-clearances/2024/multimodal-artificial-intelligence-opportunities-and-challenges-in-hiv-clinical-care
111. Brown, J. E., & Halpern, J. (2021). AI chatbots cannot replace human interactions in the pursuit of more inclusive mental healthcare. *SSM-Mental Health*, *1*, 100017.

CHAPTER 18

Insights from Global and UAE AI Applications in Public Health and Epidemiology

Haifa Ben-Romdhane

18.1 INTRODUCTION

In a time of exponential technological progress, artificial intelligence (AI) is a game-changer in many industries, with public health and epidemiology emerging as particularly dynamic fields undergoing profound change [1]. The integration of AI into these disciplines is not merely a trend but a revolution, reshaping how health challenges are understood, managed, and mitigated on a global scale [2, 3]. This chapter investigates the pivotal role of AI in revolutionizing public health, highlighting its impact on disease prevention, outbreak prediction, and healthcare delivery.

Insights are brought from exemplary global applications and models that showcase the diverse and innovative ways AI is being harnessed to address public health challenges. Among these, the advancements made by the United Arab Emirates (UAE) provide compelling case studies. The UAE, with its forward-thinking approach and commitment to leveraging cutting-edge technologies, offers a rich tapestry of AI-driven initiatives that have not only enhanced health outcomes domestically but also serve as a model for international best practices [4–7].

Through a detailed examination of AI applications from around the world and in the UAE, including predictive analytics for disease surveillance, AI-powered diagnostic tools, and smart health infrastructure, lessons that extend beyond national borders are uncovered. These experiences illuminate the potential of AI to transform public health strategies, providing valuable insights into effective implementation, scalability, and the ethical considerations inherent in this technological evolution [8–11].

In this chapter, the revolutionary effect of AI on healthcare and public health is explored through a detailed examination of six key areas. The adopted approach (Figure 18.1) starts by tracing the evolution of AI from its early days with MYCIN to today's advanced systems, then examines its applications in disease surveillance and outbreak detection. Following this, the critical importance of timely data access and analysis is discussed, exploring innovative AI-driven initiatives like drug discovery and predictive analytics, and reflecting on key lessons learned from AI implementations. Finally, the significance of interdisciplinary collaboration and the need for scalable AI models are highlighted to enhance the effectiveness and sustainability of AI solutions in public health.

Through the exploration of these global applications and the UAE's trailblazing initiatives, this chapter attempts to provide a thorough understanding of how epidemiology and public health are changing as a result of AI. As these advancements are navigated through, the challenges and future directions of integrating AI into public health frameworks are also addressed, ensuring that the benefits of these innovations are realized in a sustainable, ethical, and equitable manner.

Due to its comprehensive and integrated exploration of AI across multiple facets of public health, this approach is considered to be novel, to the best

DOI: 10.1201/9781003533351-18

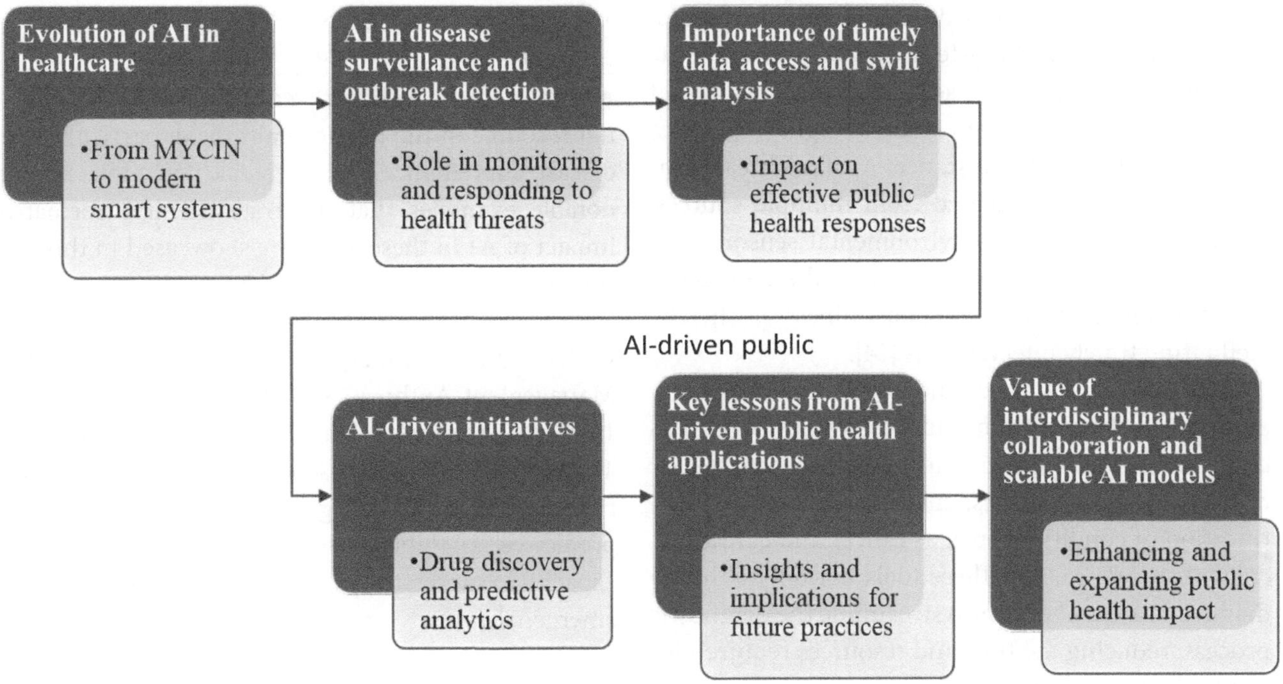

FIGURE 18.1 A summary workflow of the adopted approach.

knowledge of the author, as it uniquely synthesizes historical evolution, current applications, and strategic recommendations while incorporating both global and UAE-specific case studies, offering insights into the local and international advancements in smart healthcare systems. It uniquely combines historical evolution, contemporary applications, and strategic recommendations, providing a holistic view of AI's transformative impact. This multifaceted approach not only highlights the cutting-edge developments in AI but also addresses emerging challenges and strategic solutions tailored to the evolving landscape of public health.

18.2 THE EVOLUTION OF AI IN HEALTHCARE: FROM MYCIN TO MODERN SMART SYSTEMS

The evolution of AI in healthcare reflects a journey from early, exploratory research to sophisticated, transformative systems that are now integral to public health and epidemiology. This journey traces its roots back to the 1970s, a pivotal period when the nascent field of AI first intersected with medical science [12–14]. One of the seminal innovations from this era was MYCIN, an early AI program designed to assist clinicians in diagnosing and treating blood infections [15, 16].

Developed at Stanford University, MYCIN was a groundbreaking system that employed rule-based reasoning to analyze clinical data and recommend treatments for bacterial infections [15, 16]. Though MYCIN itself was never widely implemented in clinical practice, its development marked a significant milestone in demonstrating AI's potential to enhance medical decision-making [17, 18]. Through the utilization of a database of clinical knowledge and applying logical inference, MYCIN laid the groundwork for future AI applications in healthcare, illustrating how machine learning and expert systems could support and augment human expertise.

Looking at the current state, the landscape of smart healthcare systems has undergone remarkable evolution. Modern AI technologies now encompass a diverse array of tools and methodologies, from cutting edge natural language processing systems to machine learning methods [17]. These contemporary applications are designed to tackle a broad spectrum of public health challenges, significantly advancing epidemiological research and interventions [19–21].

One prominent area of development is in predictive analytics, where AI systems analyze vast amounts of health data to forecast disease outbreaks and detect patterns which human researchers might not notice right away [22, 23]. For instance, AI models can process real-time data gathered from multiple sources, including social media, environmental sensors, and electronic health records, to forecast the emergence and transmission of contagious illnesses, thereby facilitating timely interventions [24].

Additionally, AI-driven diagnostic tools have revolutionized the precision and effectiveness of illness detection. Modern imaging technologies, powered by deep learning algorithms, enable more precise identification of conditions such as cancer and cardiovascular diseases [25, 26]. These tools not only enhance diagnostic accuracy but also streamline the diagnostic process, reducing the time and resources required for comprehensive analysis.

Furthermore, AI's role in personalized medicine exemplifies its transformative impact. Through the analysis of genetic, environmental, and lifestyle data, AI systems are able to offer customized health recommendations and treatment plans, optimizing patient outcomes and minimizing adverse effects [27–29].

In the sector of public health management, AI-powered systems contribute to more effective resource allocation and policy planning. Through data-driven insights, health organizations can optimize the distribution of medical resources, plan vaccination campaigns, and design interventions that are both targeted and efficient [30, 31].

As these advancements are examined, it becomes evident that AI has evolved from a novel concept to a central pillar of modern healthcare. The legacy of early systems like MYCIN is evident in today's smart healthcare systems, which leverage AI to address pressing public health challenges with unprecedented precision and scale. This transformation not only highlights the potential of AI to revolutionize healthcare but also underscores the ongoing need for innovation and adaptation in the pursuit of better health outcomes worldwide.

After examining AI's revolutionary journey in healthcare, attention is focused on how important it is for disease surveillance and outbreak detection, areas where these developments are having a major influence.

18.3 AI IN DISEASE SURVEILLANCE AND OUTBREAK DETECTION

As stated above, AI has increasingly become an essential resource in the public health field, particularly in disease surveillance and outbreak detection. Several notable examples that illustrate the transformative impact of AI in these areas are showcased in this section, e.g., the Johns Hopkins University Coronavirus Resource Center's COVID-19 Dashboard [32], the UAE National Emergency Crisis and Disasters Management Authority's Al Hosn App for COVID-19 contact tracing [33], and Google Inc.'s Google Flu Trends [34]. Each of these examples underscores the power of AI in providing real-time data insights and predictive analytics, ultimately strengthening the capacity to successfully manage and respond to health emergencies.

18.3.1 The Johns Hopkins University COVID-19 Dashboard

The Johns Hopkins University COVID-19 Dashboard (https://coronavirus.jhu.edu/map.html) is a prime example of how AI and data visualization can be harnessed to provide real-time insights into a global health crisis. Launched in early 2020, this dashboard quickly became an essential tool for tracking the spread of COVID-19 worldwide.

Utilizing AI algorithms and data aggregation techniques, the dashboard compiled information from various sources; including government reports, health organizations, and news outlets, into a comprehensive, user-friendly interface.

The dashboard's ability to provide up-to-the-minute statistics on case numbers, deaths, and recoveries, broken down by country and region, proved AI's ability to manage enormous volumes of data and present it in an accessible format. Beyond simple tracking, the dashboard offered insights into trends and patterns, helping public health officials, researchers, and the public understand the trajectory of the pandemic and make informed decisions. The success of the Johns Hopkins University Dashboard highlights the crucial role of AI in synthesizing and disseminating information during an evolving public health emergency.

18.3.2 The Al Hosn App for COVID-19 Contact Tracing

In the UAE, the Al Hosn App (https://www.ncema.gov.ae/alhosn/index.html) emerged as a vital tool in managing COVID-19. This application leveraged AI to enhance contact tracing efforts, a critical component of controlling the spread of the virus. The app used Bluetooth technology to identify and log proximity data between users, allowing for the efficient tracing of potential exposure events.

AI played a significant role in analyzing this data, identifying patterns, and generating alerts for individuals who may have been in contact with confirmed cases. By means of combining real-time data with test results and medical records, the Al Hosn App facilitated targeted interventions and notifications, contributing to the UAE's successful containment strategies. The app exemplifies how AI-driven contact tracing can streamline the response to infectious diseases and support public health measures on a national scale.

18.3.3 Google Flu Trends

Google Flu Trends, though no longer operational, was an early and influential example of using AI for disease surveillance. Launched in 2008, this tool analyzed search query data to estimate flu activity levels across different regions. While examining the frequency and geographic distribution of flu-related search terms, Google Flu Trends provided estimates of the flu prevalence and outbreaks in real time, frequently weeks before they are reported using conventional methods.

The project's reliance on AI and machine learning algorithms to process and interpret search data demonstrated how digital behavior could be leveraged to predict and monitor disease trends. Although the project faced challenges related to data accuracy and model performance, it marked a significant step in using AI for epidemiological forecasting and highlighted the potential for digital tools to complement traditional surveillance systems. On August 9, 2015, Google Flu Trends discontinued providing up-to-date estimates. Current statistics are made available for stated research reasons, while historical estimates can still be downloaded.

A comparative summary of the three models, highlighting their respective purposes, functions, data sources, and impacts on public health is given in Table 18.1.

Through the exploration of these examples, AI has proven to be a powerful ally in disease surveillance and outbreak detection. These applications not only illustrate AI's ability to manage and examine big datasets in real time but also highlight its role in enhancing public health responses and shaping the understanding of disease dynamics. Through these case studies, valuable insights are gained into how AI can be effectively harnessed to address pressing health challenges and improve global health outcomes.

TABLE 18.1 Comparative Analysis Outlining the Three Studied AI-Driven Models in Disease Surveillance and Outbreak Detection

Feature	Johns Hopkins University COVID-19 Dashboard	UAE Al Hosn App	Google Flu Trends
Purpose	Real-time COVID-19 tracking and visualization	COVID-19 contact tracing and health status	Flu trend monitoring and prediction
Primary function	Provides global COVID-19 case data, visualizations, and updates	Tracks COVID-19 exposure and helps with contact tracing	Analyzes search queries to estimate flu activity
Data sources	Aggregates data from multiple global sources	Collects data from user interactions and health reports	Uses Google search data and trends
Real-time data	Yes, updated frequently with global data	Yes, updates based on user interactions and reports	Yes, based on search trends
Predictive analytics	No, primarily a tracking tool	No, focuses on exposure notification	Yes, estimates flu activity based on search patterns
Geographic coverage	Global	UAE-specific	Global
Impact on public health	Enhances understanding of pandemic spread and impact	Facilitates contact tracing and health management	Provides early warnings of flu outbreaks
User interaction	Web-based, accessible to the public	Mobile app, requires user participation	Web-based, utilizes search engine data

With a clear understanding of how AI enhances disease surveillance and outbreak detection, the crucial role of timely data access and rapid analysis in public health is, then, investigated, underscoring how these AI applications are shaping effective responses.

18.4 THE IMPORTANCE OF TIMELY DATA ACCESS AND FAST ANALYSIS IN PUBLIC HEALTH: INSIGHTS FROM AI APPLICATIONS

In the field of public health, having quick access to and analysis of data is crucial for effective decision-making and response to emerging health threats. The applications discussed; namely the Johns Hopkins University COVID-19 Dashboard, the Al Hosn App for COVID-19 contact tracing, and Google Flu Trends, demonstrate how timely data access and rapid analysis can profoundly impact public health strategies and outcomes.

18.4.1 Johns Hopkins University COVID-19 Dashboard

The Johns Hopkins University COVID-19 Dashboard exemplifies the critical role of timely data access in managing a global health crisis. The dashboard aggregated data from multiple sources around the world, including governmental health departments and international organizations, and presented it in real time. This immediate visibility into the evolving situation allowed for rapid assessment of the pandemic's progression.

Through the provision of up-to-date statistics on case counts, deaths, and recoveries, the dashboard enabled public health officials, policymakers, and the general public to track the pandemic's trajectory and make informed decisions. The ability to visualize trends and spikes in cases in real time facilitated proactive responses, such as implementing localized lockdowns or reallocating medical resources. The dashboard's success underscores how timely data and effective visualization can drive prompt and evidence-based decision-making during public health emergencies.

18.4.2 Al Hosn App for COVID-19 Contact Tracing

The Al Hosn App in the UAE emphasizes the significance of quick data analysis for contact tracing. Through the utilization of Bluetooth technology to log proximity data, the app was able to quickly identify and alert those who might have come into contact with COVID-19 [35, 36]. The app's AI-driven algorithms processed this data in real time to generate exposure alerts and risk assessments.

The immediacy of this data processing allowed for rapid isolation of potential cases and facilitated timely testing and quarantine measures. This proactive approach helped curb the dissemination of the virus by promptly recognizing and addressing potential hotspots. The Al Hosn App illustrates how rapid data analysis can enhance the effectiveness of contact tracing and support targeted public health interventions.

18.4.3 Google Flu Trends

Google Flu Trends demonstrated how real-time analysis of digital behavior could provide early warnings of flu outbreaks. Subsequent to analyzing search query data related to flu symptoms and treatments, the tool was able to estimate flu activity levels with a speed that traditional surveillance systems could not match. The capacity to quickly access and evaluate search data allowed for early detection of flu trends, often predicting outbreaks weeks before conventional reports.

This early warning capability enabled health authorities to prepare and respond more proactively to flu seasons, such as by ramping up vaccination campaigns and distributing resources to areas experiencing higher-than-normal flu activity. Although the project faced challenges, it highlighted the value of timely data access and rapid analysis in forecasting and mitigating public health threats.

A summary of how each application's capabilities in timely data access and fast analysis contribute to effective public health strategies and outcomes is tabulated in Table 18.2.

The effectiveness of these AI-driven applications underscores the crucial importance of timely access to data and the ability to analyze it quickly. Real-time data visibility and rapid analysis enable decision-making by public health officials based on knowledge, anticipate and respond to emerging threats, and implement targeted interventions with greater precision.

Having established the critical role of timely data access and fast analysis in public health, the focus is, subsequently, turned to AI-driven initiatives, examining how advancements in drug discovery and predictive analytics are further revolutionizing healthcare.

TABLE 18.2 Comparative Analysis Highlighting the Importance of Timely Data Access and Fast Analysis in Public Health for the Three Studied Applications

Feature	Johns Hopkins University COVID-19 Dashboard	UAE Al Hosn App	Google Flu Trends
Purpose	Global COVID-19 tracking and visualization	COVID-19 contact tracing and health status	Flu trend monitoring and prediction
Timeliness of data access	Real-time updates with frequent data refreshes	Near real-time updates based on user interactions and health reports	Near real-time updates based on search trends
Speed of data analysis	Rapid aggregation and visualization of global COVID-19 data	Quick analysis for contact tracing and exposure notifications	Fast analysis of search trends to estimate flu activity
Impact on decision-making	Enables swift understanding of pandemic trends and impacts	Facilitates immediate response to potential exposures	Provides early warnings for potential flu outbreaks
Data updating frequency	Constantly updated with global data inputs	Updated as users report data and interact with the app	Updated regularly based on search engine data
Role in public health response	Supports global and local decision-making by visualizing current data	Enhances individual and public health safety through contact tracing	Assists in predicting and preparing for flu season trends
Accessibility of information	Publicly accessible through a web interface	Mobile app requires user engagement	Publicly accessible via web-based trends analysis
Contribution to health outcomes	Improves situational awareness and planning	Enhances tracking and management of COVID-19 spread	Supports early intervention and resource allocation

18.5 EXPLORING AI-DRIVEN INITIATIVES: FROM DRUG DISCOVERY TO PREDICTIVE ANALYTICS

A range of AI-driven initiatives that exemplify how interdisciplinary collaboration and scalable solutions can address diverse public health challenges are explored. After examining IBM Watson for Drug Discovery [37–39], Singapore's National AI Strategy for Healthcare [40–42], South Korea's Infectious Disease Prediction Model [43–46], and Abu Dhabi Health Services Company (SEHA)'s AI Predictive Analytics System in the UAE [47, 48], insights into the transformative potential of AI across various facets of public health are gained, from drug discovery and personalized healthcare to infectious disease prediction and proactive resource management.

18.5.1 IBM Watson for Drug Discovery

IBM Watson for Drug Discovery is an important development in the field of pharmaceutical research. Leveraging AI and machine learning, this platform accelerates the process of finding new drugs through the examination of enormous volumes of biomedical data, such as scientific publications, clinical trial findings, and genetic information. Watson's AI algorithms detect links and patterns in the data that traditional research techniques might overlook, enabling them to forecast the efficacy of potential drug candidates and identify them.

The initiative exemplifies interdisciplinary collaboration by integrating knowledge from fields such as genomics, pharmacology, and bioinformatics. By cutting the time and expense of drug development dramatically, Watson for Drug Discovery not only enhances the efficiency of finding new treatments but also paves the way for personalized medicine approaches, where therapies can be tailored to individual genetic profiles.

18.5.2 Singapore's National AI Strategy for Healthcare

Singapore's National AI Strategy for Healthcare illustrates how a comprehensive national approach can harness AI to transform healthcare systems. This strategy encompasses a range of initiatives aimed at enhancing patient care, expediting the delivery of healthcare, and promoting medical research. Key components include AI-powered diagnostic tools, predictive analytics for patient management, and smart health infrastructure.

The strategy's emphasis on integrating AI into various aspects of healthcare demonstrates a commitment to leveraging technology for advantages that are both short- and long-term. By focusing on interoperability and data sharing across healthcare providers, Singapore is setting a precedent for how national strategies can drive innovation, enhance healthcare delivery, and support proactive health management.

18.5.3 South Korea's Infectious Disease Prediction Model

South Korea's Infectious Disease Prediction Model demonstrates how AI can be used to predict and control infectious disease outbreaks. Examining historical health data, environmental variables, and patterns of population mobility using machine learning algorithms, this model provides predictions on potential disease spread and identifies high-risk areas.

The model's ability to generate actionable insights supports timely interventions and resource allocation, enabling public health authorities to implement preventive measures before outbreaks reach critical levels. South Korea's approach highlights how AI can enhance epidemic preparedness and response by providing early warnings and guiding strategic planning.

18.5.4 SEHA's AI Predictive Analytics System in the UAE

In the UAE, SEHA's AI Predictive Analytics System demonstrates how AI can optimize healthcare delivery and resource allocation. This system uses predictive analytics to analyze patient data, hospital performance metrics, and healthcare utilization patterns. By forecasting demand for medical services and identifying potential bottlenecks, SEHA's system enables proactive planning and more efficient use of resources.

The system's impact on improving patient outcomes is evident through its ability to enhance care coordination, reduce wait times, and tailor healthcare services to meet community needs. SEHA's initiative underscores the importance of AI in enhancing healthcare infrastructure and ensuring that resources are allocated effectively to address evolving public health needs.

A comparison among the studied AI-driven initiatives, illustrating their diverse applications, collaborative efforts, scalability, and impacts on public health, is given in Table 18.3.

The exploration of these AI-driven initiatives reveals the multifaceted ways in which AI is revolutionizing public health. From accelerating drug discovery and advancing personalized healthcare to predicting infectious disease outbreaks and optimizing resource management, these models demonstrate the ability of

TABLE 18.3 Comparative Summary of the Studied AI-Driven Initiatives, Showcasing Their Impact across Different Facets of Public Health

Feature	IBM Watson for Drug Discovery	Singapore National AI Strategy for Healthcare	South Korea Infectious Disease Prediction Model	SEHA AI Predictive Analytics System (UAE)
Purpose	Accelerate drug discovery and development	Integrate AI into healthcare for improved patient outcomes	Predict and manage infectious disease outbreaks	Enhance predictive analytics for healthcare management
Focus area	Drug discovery and development	Broad healthcare AI applications, including predictive analytics	Infectious disease prediction and response	Proactive resource management and predictive healthcare
Interdisciplinary collaboration	Collaboration between data scientists, biologists, and chemists	Involves collaboration between tech companies, healthcare providers, and policymakers	Combines expertise in data science, epidemiology, and public health	Integration of AI with healthcare systems and data science
Scalability	Designed to scale with large datasets and diverse drug research	Scalable framework applied across various healthcare domains	Adaptable to different infectious disease contexts	Scalable for expanding healthcare needs and predictive models
Innovation in drug discovery	Uses AI to identify potential drug candidates and accelerate R&D	N/A	N/A	N/A
Predictive analytics	Supports drug discovery through data-driven insights	Employs AI for predictive analytics in healthcare	Provides predictive insights for disease management	Utilizes AI to forecast healthcare trends and resource needs
Impact on public health	Speeds up drug discovery, potentially leading to faster treatments	Improves healthcare outcomes through AI-driven insights	Enhances ability to predict and manage disease outbreaks	Improves healthcare planning and response through predictive models
Geographic focus	Global	Singapore	South Korea	UAE

AI to solve numerous health-related issues. They also highlight the importance of interdisciplinary collaboration and scalable solutions in achieving significant public health improvements. As these examples show, AI improves public health infrastructure resilience and patient outcomes in addition to increasing the efficacy and efficiency of healthcare systems.

After exploring the transformative impact of AI-driven initiatives in drug discovery and predictive analytics, the key lessons learned from these applications are being reflected on to better understand their broader implications for public health.

18.6 KEY LESSONS FROM AI-DRIVEN PUBLIC HEALTH APPLICATIONS

This section aims to elucidate the critical lessons learned from exemplary AI-driven applications and models in public health. Through an exploration of these innovative initiatives, the importance of ethical considerations, community engagement, and continuous learning and improvement is highlighted. Addressing these aspects ensures the efficient application of AI in public health.

18.6.1 Ethical Considerations in AI Implementation

18.6.1.1 Ethical AI Design and Development

Designing and developing AI systems ethically is essential to making sure that these technologies benefit all stakeholders without perpetuating biases or causing harm. Best practices for incorporating ethical principles into AI development, including fairness, accountability, and transparency are explored from the featured applications to illustrate how ethical concerns were addressed or could be improved.

18.6.1.2 Adherence to Privacy Standards

Protecting patient data and ensuring privacy are fundamental to maintaining public trust in AI-driven healthcare solutions. The significance of abiding by privacy regulations like the General Data Protection Regulation and the Health Insurance Portability and Accountability Act allowed to investigate how the selected models, like SEHA's AI Predictive Analytics System, manage and secure sensitive health information.

18.6.1.3 Transparency and Comprehensibility

For AI systems to be trusted and effectively utilized, their operations must be transparent, and their decision-making processes should be explainable. It is emphasized how important it is to communicate in a clear and concise manner how AI models generate predictions and choices. Insights drawn from applications like IBM Watson for Drug Discovery and the Johns Hopkins University COVID-19 Dashboard demonstrated practices that promote transparency.

18.6.2 Community Engagement

18.6.2.1 Involving Stakeholders in AI Development

Engaging communities and stakeholders in the AI development process is essential for creating systems that meet real-world needs and gain acceptance. Methods for involving public health officials, healthcare providers, and the general public are investigated in the development and application of AI solutions. Models such as Singapore's National AI Strategy demonstrated the incorporation of stakeholder feedback.

18.6.2.2 Building Public Trust and Understanding

Building trust in AI-driven public health initiatives requires effective communication and education. The study of strategies for informing the public about AI technologies, their benefits, and their limitations help to explain how initiatives like the Al Hosn App in the UAE effectively communicated their purpose and function to users.

18.6.2.3 Addressing Concerns and Misconceptions

Public misconceptions and concerns about AI can hinder its adoption and effectiveness. Examining how to proactively address and mitigate these concerns through transparent practices and educational outreach helped to draw lessons from South Korea's Infectious Disease Prediction Model.

18.6.3 Continuous Learning and Improvement

18.6.3.1 Iterative Development and Feedback Loops

AI systems in public health must evolve based on new data, research findings, and user feedback. The importance of iterative development and the establishment of feedback loops to continuously improve AI models is highlighted through the study of examples from Google Flu Trends and other initiatives to illustrate how ongoing learning is integrated into AI systems.

18.6.3.2 Evaluating Impact and Effectiveness

AI systems must be routinely evaluated to determine their impact and efficacy in real-world settings. Methods for evaluating AI applications, including performance metrics and user satisfaction surveys were explored. Insights from SEHA's AI Predictive Analytics System and similar models were used to highlight effective evaluation practices.

18.6.3.3 Adapting to Emerging Challenges

The dynamic nature of public health requires AI systems to be flexible and responsive to new challenges and shifting health needs. The current analytical approach showed that AI models can be updated and refined to respond to new diseases, changing health trends, and advances in technology. Case studies from across the discussed models provided examples of adaptability and resilience.

The chapter then looks at these important lessons, community involvement, ethical issues, and ongoing learning and development. Finally, it attempts to offer a thorough framework for the ethical and successful application of AI in public health. These observations aid in making sure AI-driven solutions are created and implemented in a way that is moral, open, and considerate of the requirements of all parties involved, ultimately enhancing public health outcomes and fostering trust in AI technologies.

The value of interdisciplinary collaboration and scalable AI models, which are crucial for advancing and expanding the impact of these innovations in public health, is highlighted.

18.7 THE VALUE OF INTERDISCIPLINARY COLLABORATION AND SCALABLE AI MODELS IN PUBLIC HEALTH

Through the examination of the various AI-driven public health initiatives, this section highlights how integrating expertise from multiple fields and ensuring that AI models can scale and adapt are essential for effective and sustainable health interventions.

18.7.1 Interdisciplinary Collaboration

18.7.1.1 Integrating Diverse Expertise

Effective AI solutions in public health often require the integration of knowledge from diverse disciplines, including data science, epidemiology, healthcare, computer science, and ethics. The exploration of examples, such as IBM Watson for Drug Discovery and Singapore's National AI Strategy, demonstrate that interdisciplinary teams made of researchers, clinicians, and technologists, contribute to the development and implementation of AI models by combining their expertise to address complex health challenges and lead to more robust and effective solutions.

18.7.1.2 Bridging the Gap between Research and Practice

It is imperative that the gap between scholarly research and real-world application be closed for translating AI innovations into real-world health solutions. The study of the stated examples and initiatives, e.g., South Korea's Infectious Disease Prediction Model and SEHA's AI Predictive Analytics System, illustrate how interdisciplinary collaboration facilitates this process, ensuring that AI research is aligned with the needs of healthcare providers and public health authorities, and how collaborative efforts help bridge this gap and enhance practical utility.

18.7.1.3 Fostering Collaborative Innovation

Innovation in AI-driven public health solutions often arises from collaborative efforts across different sectors. Insights from initiatives like Google Flu Trends and the Al Hosn App highlight how collaborative approaches can lead to the development of cutting-edge solutions. Partnerships between academic institutions, industry, government agencies, and non-profits are proven to drive innovation and address health challenges more effectively.

18.7.2 Scalability of AI Models

18.7.2.1 Designing Scalable Solutions

Scalability is a key consideration for AI models in public health, as they must be able to handle increasing data volumes and adapt to varying health contexts. The current analysis of scalable solutions from initiatives such as the Johns Hopkins University COVID-19 Dashboard, proved that these models, effectively, managed large-scale data aggregation and visualization including modular architecture, cloud computing, and data integration techniques.

18.7.2.2 Adapting to Regional and Global Needs

AI models must be adaptable to different geographic and demographic contexts to be effective across diverse populations. Case studies from Singapore's National AI Strategy and SEHA's AI Predictive Analytics System illustrate how adaptability ensures that AI solutions remain relevant and effective in various settings. These prove that AI systems can, indeed, be tailored to meet regional health needs and how they can be scaled to address global health threats.

18.7.2.3 Ensuring Flexibility for Emerging Threats

Public health is characterized by its dynamic nature, with new health threats emerging over time. Insights are drawn from applications such as South Korea's Infectious Disease Prediction Model, which adapted to changing epidemiological patterns. The examination of the importance of designing AI models with built-in flexibility to respond to these evolving challenges shows how iterative development, continuous monitoring, and adaptive learning mechanisms contribute to the resilience and effectiveness of AI models.

18.8 RECOMMENDATIONS AND CONCLUSIVE NOTES

AI has become a revolutionary force in the field of public health, providing hitherto unheard-of chances to improve drug discovery and disease monitoring, healthcare delivery, and outbreak prediction. The current study highlights the profound influence of AI on public health by approaching distinguished examples such as the Johns Hopkins University COVID-19 Dashboard, the Al Hosn App, Google Flu Trends, IBM Watson for Drug Discovery, Singapore's National AI Strategy, South Korea's Infectious Disease Prediction Model, and SEHA's AI Predictive Analytics System. Based on these models, public health systems can become more efficient and adaptable by utilizing AI's real-time data insights, predictive analytics, and proactive management capabilities. Key Findings from the adopted approach are illustrated in Figure 18.2.

Figure 18.2 highlights crucial aspects such as the importance of timely data access, ethical

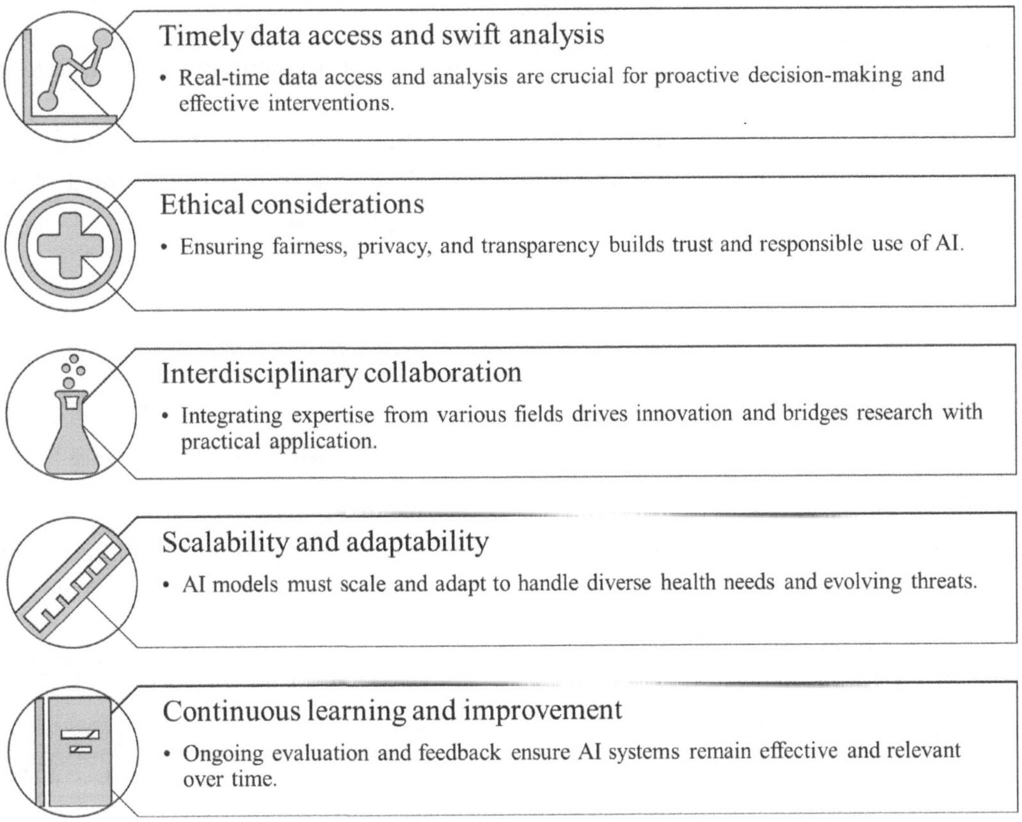

FIGURE 18.2 Key recommendations for optimizing AI in public health based on the analysis of global and UAE case studies.

considerations, interdisciplinary collaboration, scalability, and continuous improvement which, respectively, translate into the following recommendations.

- Real-time data analysis and access capabilities are crucial for effective public health management. In fact, AI-driven tools that provide up-to-date insights enable proactive decision-making and timely interventions, thereby improving health outcomes and resource allocation.
- The ethical AI implementation is essential for ensuring fairness, protecting privacy, and maintaining transparency. As a matter of fact, adherence to privacy standards and transparent communication with stakeholders are fundamental to fostering trust and guaranteeing the appropriate application of AI in public health.
- An interdisciplinary collaboration through the provision of effective AI solutions requires the integration of expertise from various fields, including data science, healthcare, and ethics. Collaborative efforts, indeed, drive innovation and help bridge the gap between research and practical application, leading to more robust and impactful solutions.
- The design of AI models to scale and their adaptation to different contexts and evolving health threats is highly recommended. Scalable solutions can, indeed, handle increasing data volumes and diverse health needs, while adaptable models remain effective in dynamic and changing environments, and
- An ongoing evaluation and iterative development to refine AI systems and ensure their relevance and effectiveness. In point of fact, feedback loops, performance metrics, and adaptability to emerging challenges are crucial to maintaining the efficacy of AI-driven public health tools.

These key recommendations serve as a foundation for understanding how these elements contribute to optimizing AI applications in the public health field.

Upon reviewing the main conclusions regarding AI's application to public health, the focus is, finally, shifted to the strategic recommendations designed to enhance these applications, as illustrated in Figure 18.3. The latter presents strategic recommendations for optimizing AI-driven public health solutions. It outlines key actions such as prioritizing ethical AI practices, fostering interdisciplinary collaboration, designing scalable and adaptable models, engaging and educating communities, and implementing continuous improvement processes, as follows.

- Prioritizing ethical AI practices through the development and enforcement of guidelines for ethical AI use, including fairness, accountability,

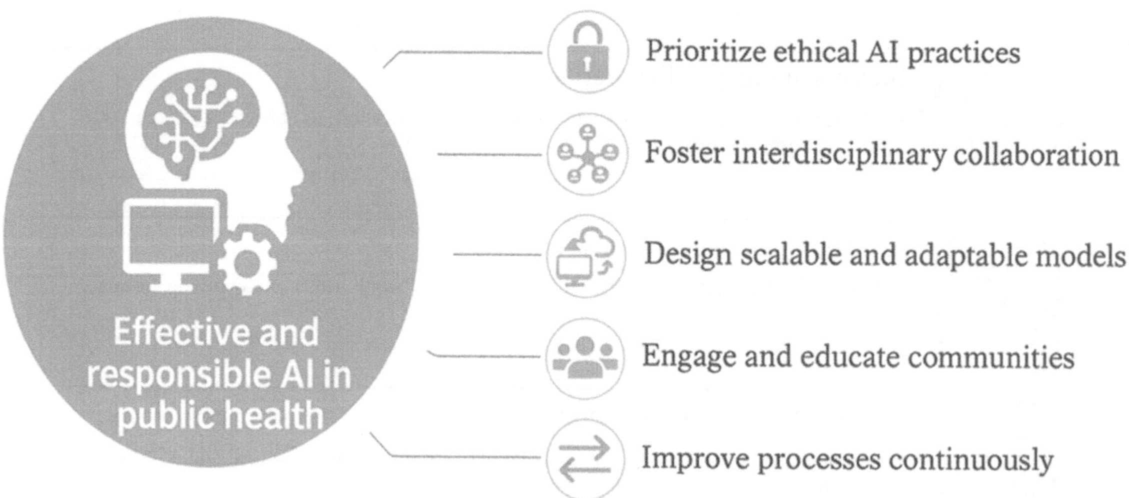

FIGURE 18.3 Strategic recommendations for enhancing AI-driven public health solutions based on the analysis of global and UAE case studies.

and transparency. This ensures that privacy standards are rigorously adhered to and that the goal of AI systems' design is clearly conveyed. This would make it possible to consult ethicists and privacy specialists while developing and implementing AI technology.

- Fostering interdisciplinary collaboration by encouraging partnerships among researchers, healthcare providers, technologists, and policymakers to drive the development of AI solutions. This facilitates cross-sector collaboration to integrate diverse expertise and address complex public health challenges effectively.
- Designing scalable and adaptable models through the implementation of AI models with scalable architecture and adaptable frameworks to handle large datasets and diverse health contexts. This involves investing in modular and cloud-based solutions that can be easily updated and customized to meet changing needs.
- Engaging and educating communities and stakeholders in the creation and application of AI systems to guarantee that the solutions meet their requirements and address their issues. In order to increase public trust and comprehension of AI technologies, this offers clear, transparent communication and educational materials.
- Implementing continuous improvement processes through the establishment of mechanisms for regular evaluation and feedback to refine AI systems and enhance their performance. Monitoring the effectiveness of AI applications and being prepared to adapt and update models based on new data, research findings, and emerging health threats are also essential.

While addressing these recommendations, public health authorities and AI developers can fully utilize AI to enhance healthcare delivery, enhance health outcomes and create enduring public health systems that can handle challenges in the present and future.

Continued research and development are expected to drive further advancements, ensuring that AI technologies remain responsive to emerging health challenges and contribute to more effective and equitable public health outcomes.

The study's exploration of interdisciplinary collaboration and scalable, adaptable AI models highlights the multifaceted approach needed to address complex public health challenges. By integrating diverse expertise and designing models that can scale and adapt, public health initiatives can leverage AI's full potential to improve health outcomes and respond effectively to evolving threats. These considerations ensure that AI-driven solutions are not only innovative and effective but also sustainable and able to meet the various and ever-changing demands of global health.

18.9 SUMMARY

Due to its comprehensive and integrated exploration of AI across multiple facets of public health, this analytical study is characterized by its significant novelty in synthesizing historical development, current applications, and strategic recommendations, while also incorporating both global and UAE-specific case studies. The adopted approach, however, acknowledges a number of limitations, including the focus on a select number of high-profile AI applications that may not fully represent the diversity of innovations in public health, and the rapid pace of technological change which may render some findings outdated. Additionally, the lessons and recommendations provided may not be universally applicable due to variations in local healthcare systems and regulatory environments. Ethical and privacy concerns, while addressed, require deeper exploration given their complexity and context-dependence. Lastly, the impact assessments are based on potentially incomplete data. To address these limitations, future research should expand the scope by including a wider range of AI applications and regional case studies, regularly update findings to reflect technological advancements, and explore how AI solutions can be adapted to different contexts. Enhanced focus on ethical issues and robust evaluation frameworks is expected to provide more nuanced insights and better assess the long-term effects of AI in public health.

REFERENCES

1. Panch, T., Szolovits, P., & Atun, R. (2018). Artificial intelligence, machine learning and health systems. *Journal of Global Health*, *8*(2), 020303.
2. Ibeneme, S., Okeibunor, J., Muneene, D., Husain, I., Bento, P., Gaju, C., Housseynou, B., Chibi, M.,

Karamagi, H., & Makubalo, L. (2021). Data revolution, health status transformation and the role of artificial intelligence for health and pandemic preparedness in the African context. *BMC Proceedings, 15*, 1–12.
3. Silcox, C., Zimlichmann, E., Huber, K., Rowen, N., Saunders, R., McClellan, M., Kahn, C. N. III, Salzberg, C. A., & Bates, D. W. (2024). The potential for artificial intelligence to transform healthcare: Perspectives from international health leaders. *NPJ Digital Medicine, 7*(1), 88.
4. Alhashmi, S. F., Salloum, S. A., & Abdallah, S. (2020, October). Critical success factors for implementing artificial intelligence (AI) projects in Dubai government United Arab Emirates (UAE) health sector: Applying the extended technology acceptance model (TAM). In: Hassanien, A., Shaalan, K., Tolba, M. (eds). *Proceedings of the International Conference on Advanced Intelligent Systems and Informatics. Advances in Intelligent Systems and Computing*, vol 1058, 393–405. Springer, Cham.
5. Almarzooqi, A. (2019). Towards an Artificial Intelligence (AI)-Driven Government in the United Arab Emirates (UAE): A Framework for Transforming and Augmenting Leadership Capabilities. Pepperdine University.
6. Khadragy, S., Elshaeer, M., Mouzaek, T., Shammass, D., Shwedeh, F., Aburayya, A., Jasri, A., & Aljasmi, S. (2022). Predicting diabetes in United Arab Emirates healthcare: Artificial intelligence and data mining case study. *South Eastern European Journal of Public Health*, 2022(5).
7. Sarirete, A., Balfagih, Z., Brahimi, T., Mousa, M. E. A., Lytras, M., & Visvizi, A. (2021). Artificial intelligence: Towards digital transformation of life, work, and education. Procedia Computer Science, 194, 1–8.
8. Hendawy, M., & Kumar, N. (2024). AI in the National AI Strategies of the Arab Region. *Arab Reform Initiative*. https://www.arab-reform.net/publication/ai-in-the-national-ai-strategies-of-the-arab-region/?tztc=1
9. Hjaltalin, I. T., & Sigurdarson, H. T. (2024). The strategic use of AI in the public sector: A public values analysis of national AI strategies. *Government Information Quarterly, 41*(1), 101914.
10. Mulukuntla, S., & Pamulaparthyvenkata, S. (2022). Realizing the potential of AI in improving health outcomes: Strategies for effective implementation. *ESP Journal of Engineering and Technology Advancements*, *2*(3), 32–40.
11. Periáñez, Á., Del Río, A. F., Nazarov, I., Jané, E., Hassan, M., Rastogi, A., & Tang, D. (2024). The digital transformation in health: How AI can improve the performance of health systems. Health Systems & Reform, 10(2). https://doi.org/10.1080/23288604.2024.2387138
12. Kononenko, I. (2001). Machine learning for medical diagnosis: History, state of the art and perspective. *Artificial Intelligence in Medicine, 23*(1), 89–109.
13. Szolovits, P. (2019). Artificial intelligence and medicine. In *Artificial Intelligence in Medicine* (pp. 1–19). Routledge.
14. Szolovits, P., Patil, R. S., & Schwartz, W. B. (1988). Artificial intelligence in medical diagnosis. *Annals of Internal Medicine, 108*(1), 80–87.
15. Shortliffe, E. (2012). *Computer-Based Medical Consultations: MYCIN* (Vol. 2). Elsevier.
16. Van Melle, W. (1978). MYCIN: A knowledge-based consultation program for infectious disease diagnosis. *International Journal of Man-Machine Studies, 10*(3), 313–322.
17. Patil, S., & Shankar, H. (2023). Transforming healthcare: Harnessing the power of AI in the modern era. *International Journal of Multidisciplinary Sciences and Arts, 2*(1), 60–70.
18. Victor, L. Y., Fagan, L. M., Bennett, S. W., Clancey, W. J., Scott, A. C., Hannigan, J. F., Blum, R. L., Buchanan, B. G., & Cohen, S. N. (1984). An evaluation of MYCIN's advice. *Buchanan and Shortliffe, 16*, 589–596.
19. Senbekov, M., Saliev, T., Bukeyeva, Z., Almabayeva, A., Zhanaliyeva, M., Aitenova, N., Toishibekov, Y., & Fakhradiyev, I. (2020). The recent progress and applications of digital technologies in healthcare: A review. *International Journal of Telemedicine and Applications, 2020*(1), 8830200.
20. Shaheen, M. Y. (2021). Applications of artificial intelligence (AI) in healthcare: A review. *ScienceOpen Preprints*. https://www.scienceopen.com/hosted-document?doi=10.14293/S2199-1006.1.SOR-.PPVRY8K.v1
21. Thangam, D., Malali, A. B., Subramaniyan, G., Mariappan, S., Mohan, S., & Park, J. Y. (2022). Relevance of artificial intelligence in modern healthcare. Edited by D. Jeya Mala, In *Integrating AI in IoT Analytics on the Cloud for Healthcare Applications* (pp. 67–88). IGI Global.
22. Eckhoff, P. A., & Tatem, A. J. (2015). Digital methods in epidemiology can transform disease control. *International Health, 7*(2), 77–78.
23. Mittal, S., Singh, P. K. K., Gochhait, S., & Kumar, S. (2024). AI-driven data integration to transform epidemiology. Edited by Ashish Khanna and Saikat Gochhait, In *Green AI-Powered Intelligent Systems for Disease Prognosis* (pp. 41–56). IGI Global Scientific Publishing.
24. Zhao, A. P., Li, S., Cao, Z., Hu, P. J.-H., Wang, J., Xiang, Y., Xie, D., & Lu, X. (2024). AI for science: Predicting infectious diseases. *Journal of Safety Science and Resilience*, 2(5), 130–146.
25. Asif, S., Wenhui, Y., ur-Rehman, S., ul-ain, Q., Amjad, K., Yueyang, Y., Jinhai, S., & Awais, M. (2024). Advancements and prospects of machine learning in medical diagnostics: Unveiling the future of diagnostic precision. *Archives of Computational Methods in Engineering, 2025*(32), 1–31.

26. Zhou, S. K., Greenspan, H., Davatzikos, C., Duncan, J. S., Van Ginneken, B., Madabhushi, A., Prince, J. L., Rueckert, D., & Summers, R. M. (2021). A review of deep learning in medical imaging: Imaging traits, technology trends, case studies with progress highlights, and future promises. *Proceedings of the IEEE*, *109*(5), 820–838.
27. Johnson, K. B., Wei, W., Weeraratne, D., Frisse, M. E., Misulis, K., Rhee, K., Zhao, J., & Snowdon, J. L. (2021). Precision medicine, AI, and the future of personalized health care. *Clinical and Translational Science*, *14*(1), 86–93.
28. Schork, N. J. (2019). Artificial intelligence and personalized medicine. *Precision Medicine in Cancer Therapy*, *2019*(178), 265–283.
29. Udegbe, F. C., Ebulue, O. R., Ebulue, C. C., & Ekesiobi, C. S. (2024). AI's impact on personalized medicine: Tailoring treatments for improved health outcomes. *Engineering Science & Technology Journal*, *5*(4), 1386–1394.
30. Olorunsogo, T. O., Anyanwu, A., Abrahams, T., Ehimuan, B., & Reis, O. (2024). Emerging technologies in public health campaigns: Artificial intelligence and big data. *International Journal of Science and Research Archive*, *11*(1), 478–487.
31. Rocha, T. A. H., de Almeida, D. G., Kozhumam, A. S., da Silva, N. C., Thomaz, E. B. A. F., de Sousa Queiroz, R. C., de Andrade, L., Staton, C., & Vissoci, J. R. N. (2021). Microplanning for designing vaccination campaigns in low-resource settings: A geospatial artificial intelligence-based framework. *Vaccine*, *39*(42), 6276–6282.
32. Dong, E., Ratcliff, J., Goyea, T. D., Katz, A., Lau, R., Ng, T. K., Garcia, B., Bolt, E., Prata, S., Zhang, D., Murray, R. C., Blake, M. R., Du, H., Ganjkhanloo, F., Ahmadi, F., Williams, J., Choudhury, S., Gardner, L. M. (2022). The Johns Hopkins University Center for Systems Science and Engineering COVID-19 Dashboard: Data collection process, challenges faced, and lessons learned. *Lancet Infect Dis*, *22*(12):e370–e376.
33. Al-Hosani, F., Al-Mazrouei, S., Al-Memari, S., Al-Yafei, Z., Paulo, M. S., & Koornneef, E. (2021). A review of COVID-19 mass testing in the United Arab Emirates. *Frontiers in Public Health*, *9*, 661134.
34. Santillana, M., Zhang, D. W., Althouse, B. M., & Ayers, J. W. (2014). What can digital disease detection learn from (an external revision to) Google Flu Trends? *American Journal of Preventive Medicine*, *47*(3), 341–347.
35. Alfayez, Z. H., Al-Sinayyid, N. M., & AL-Ameri, S. I. (2021). Mobile Applications developed by Arab countries in response to covid-19: A review. *Journal of Information System and Technology Management*, *6*(22), 200–211.
36. Obaid, W., Farag, M. M., & Hamid, A. K. (2022). *Smart Information Recognition on COVID-19 APPs for User Health Identification. 2022 Advances in Science and Engineering Technology International Conferences (ASET)*, Dubai, United Arab Emirates, 1–4.
37. Agrawal, P. (2018). Artificial intelligence in drug discovery and development. *Journal of Pharmacovigilance*, *6*(2), 1000e173.
38. Chen, Y., Argentinis, J. E., & Weber, G. (2016). IBM Watson: How cognitive computing can be applied to big data challenges in life sciences research. *Clinical Therapeutics*, *38*(4), 688–701.
39. Martin, R. L., Iraola, D. M., Louie, E., Pierce, D., Tagtow, B., Labrie, J. J., & Abrahamson, P. G. (2018). Hybrid natural language processing for high-performance patent and literature mining in IBM Watson for drug discovery. *IBM Journal of Research and Development*, *62*(6), 8–1.
40. Chik, W. B., & Sugumaran, H. (2024). Legal governance of artificial intelligence in healthcare devices and services in Singapore. In Solaiman, B., Cohen, I.G., editors. *Research Handbook on Health, AI and the Law* (pp. 373–391). Edward Elgar Publishing.
41. Liu, S., Ko, Q. S., Heng, K. Q. A., Ngiam, K. Y., & Feng, M. (2020). Healthcare transformation in Singapore with artificial intelligence. *Frontiers in Digital Health*, *2*, 592121.
42. Ta, A. W. A., Goh, H. L., Ang, C., Koh, L. Y., Poon, K., & Miller, S. M. (2022). Two Singapore public healthcare AI applications for national screening programs and other examples. *Health Care Science*, *1*(2), 41–57.
43. Jee, Y., Kim, Y.-J., Oh, J., Kim, Y.-J., Ha, E.-H., & Jo, I. (2022). A COVID-19 mortality prediction model for Korean patients using nationwide Korean disease control and prevention agency database. *Scientific Reports*, *12*(1), 3311.
44. Mijwil, M. M., Al-Mistarehi, A.-H., & Aggarwal, K. (2021). The effectiveness of utilising modern artificial intelligence techniques and initiatives to combat COVID-19 in South Korea: A narrative review. *Asian Journal of Applied Sciences*, *9*(5), 343–352.
45. Song, H., Han, K., Park, J., Liu, Z., Goo, T., Krishnamurthy, A., & Park, T. (2024). K-track-covid: Interactive web-based dashboard for analyzing geographical and temporal spread of Covid-19 in South Korea. *Frontiers in Public Health*, *12*, 1347862.
46. Yoo, K. J., Kwon, S., Choi, Y., & Bishai, D. M. (2021). Systematic assessment of South Korea's capabilities to control COVID-19. *Health Policy*, *125*(5), 568–576.
47. Deeb, A., Lakinska, B., Goyal, R., Al Braiki, A., & Al Hashemi, H. (2023). Bringing health information exchange to the Middle East and North Africa: The case of Malaffi in Abu Dhabi. In *Health Information Exchange* (pp. 665–680). Academic Press.
48. Haneya, H., AlKaf, D., Bajammal, F., & Brahimi, T. (2021). A meta-analysis of artificial intelligence applications for tracking COVID-19: The case of the UAE. *Procedia Computer Science*, *194*, 180–189.

CHAPTER 19

Healthcare Robotics

Rehabilitation and Assistance Devices

Rahma Mahmoud Saad, Ismail A. Ibrahim, Rowida Hesham Ali, Tasneem Mahmoud Eltantawy, Noor Razzaq Hayat, Ahmed Tharwat Al-Namrouti, and Doaa S. R. Khafaga

19.1 INTRODUCTION

Robots are physically embodied systems with the ability to interact with their environment and sense it. Their sensory capacities, degree of autonomy, and visual appearance can all differ[59]. They are resistant to psychological wear and tear, have remarkable precision when performing repetitive tasks, and can handle materials that pose a risk to humans [1]. Given that healthcare accounts for over 10% of the global domestic product [2] digital innovation is becoming more and more crucial to lowering costs and enhancing results. One digital innovation field that has the potential to significantly affect healthcare is robotics. It is not a novel idea to employ robotic technology in healthcare. One of the first known uses of robotics technologies in healthcare was to determine the course of a brain biopsy, thanks to technological advancements in the 1980s [3].

Over time, technology has advanced quickly, which has improved robot capabilities. Nowadays, well-defined functions like productivity, pharmacy automation, and hospital logistics are handled by robotic systems [4, 5].We predict that future developments in technology will further enhance robot capabilities, enabling them to function more independently and collaborate closely with people. Automation and robotics are being utilized in healthcare at an ever-increasing rate [6]. Despite notable advancements within the domain of robotics [7–14], a key obstacle to positive attitudes among stakeholders, including clinicians, is the widespread lack of exposure to and knowledge of robots in healthcare [11, 15]. Depending on their use, robots are accepted in society in very different ways. Robots in healthcare should concentrate on helping patients and professionals in order to gain more acceptability. It is wise to exercise caution when considering tasks that seem to replace human labor [16]. Although a number of experimental studies have shown the potential benefits of robots in healthcare, little is known about how long-term adoption will be affected; large-scale studies in real-world settings are still required to reveal the effects on the economy, society, and health [17]. It's hard to predict with certainty where robotics in healthcare will go in the future because the field is still in its infancy. However, in order to highlight the main ideas and potential paths for this discipline, this paper builds on earlier research [18, 19] and presents examples from both recent and historical studies as well as commercial applications. The causes, difficulties, applications, and important suggestions for implementing robots in healthcare that are outlined here can serve as a roadmap for the creation and application of robotic solutions in the field in the future.

19.2 TECHNOLOGICAL COMPONENTS AND INNOVATIONS

Robotics encompasses a wide range of technological disciplines, including computer science, electrical engineering, mechanical engineering, sensors, electronics,

DOI: 10.1201/9781003533351-19

actuators, and artificial intelligence (AI) [20]. It's mainly composed of various items as: Hardware components which are the components that form the mechanical structure of the robot that include: the chassis form the body, wheels allow it to move, ultrasonic sensors or cameras give it information about the surroundings as distances, obstacles, and topography to be able to see and react, actuators enable the robot to perform actions, energy sources provide the robot with energy needed [21]. The control unit, the robot's brain, includes microcontrollers, processors, and software, managing sensor data, decision-making, actuator commands, and communication [21]. As well as communication interferences like Wi-Fi and Bluetooth that allow robots to interact with humans [21]. Thus, it is a complex field that incorporates several branches of engineering and requires the application of advanced mathematical principles [20].

19.3 CLASSIFICATION OF HEALTHCARE ROBOTS

Robots were generally classified by the International Organization for Standardization into industrial, medical, and service robots while the International Federation of Robotics classified robots into industrial and service robots only, identifying medical robots as service robots [22, 23].

Advancing toward our objective, medical robots are classified into robotics to assist doctors/surgeons for surgery, exploration, diagnosis, etc., and robotics to assist people, which are also classified into assistive technologies and rehabilitation robotics [24] (Figure 19.1). For more detailed categorization of healthcare/medical robots, we will take a vast look into literature; many classifications were used. Some [25] are related to [25]:

- Size; macro, micro, and biological robots.
- Application; surgical, rehabilitation, and hospital service robots.
- Department; neurosurgery, orthopedic, laparoscopic robots, etc.

However, over time, the need for more innovative classification increases, and interesting proposals are made. For example, a proposal in conceptualized healthcare robots based on service roles as three types [26]:

1. Functional tools.
2. Service assistants.
3. Social colleagues.

Each type of them is described in this classification according to its appearance, presence, social capability, AI, and autonomy capability.

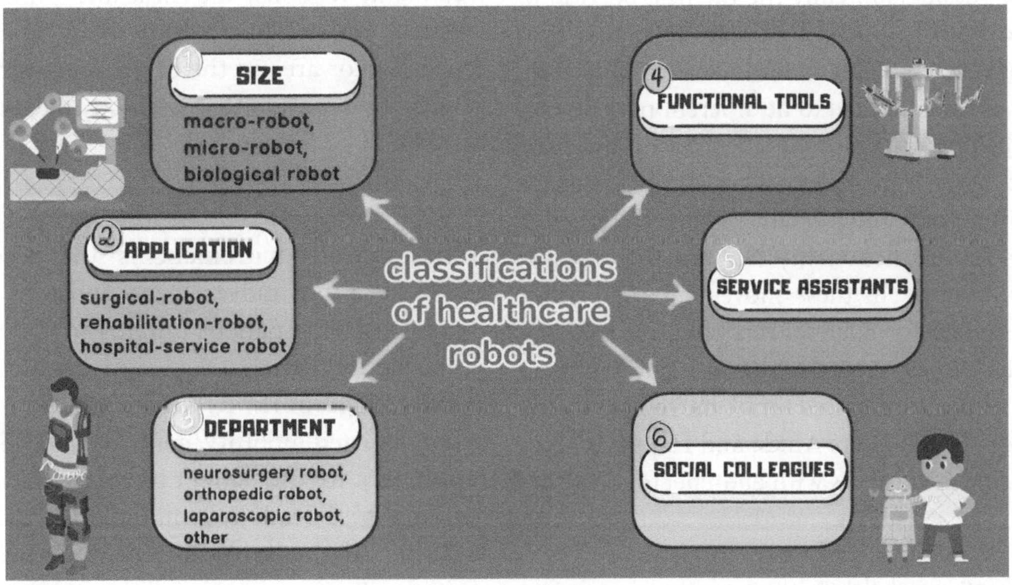

FIGURE 19.1 Different classifications of healthcare robots.

19.4 TYPES OF REHABILITATION ROBOTS

Today, patients recuperating from a variety of neurological and coordination diseases, including stroke, brain injury, spinal cord injury, cerebral palsy in children, orthopedic surgery, etc., can benefit from the use of rehabilitation robots. Research has demonstrated that using rehabilitation robots in physical therapy is more effective and long-lasting than traditional therapy approaches [27].

Rehabilitation devices can operate in three modes: passive, where the affected limb is limited to a specific range of motion without external assistance or resistance; active, where actuators direct the limb along a predetermined path; and interactive, where forces are applied in response to the limb's movement to enhance assistance [28]. The majority of robotics now employed in healthcare are categorized as either active or interactive; however, they can also function in a passive manner. Furthermore, these devices are classified as either end-effector devices or exoskeletal devices depending on their method of interaction with the user [29].

19.4.1 According to Robot Structure

19.4.1.1 End-Effector Devices

End-effector robots were the focus of early rehabilitation robot research. Robots with end-effectors grasp the injured limb of the patient at a single location while producing pressures at the point of interaction. Human limb joints are not compatible with those of end-effector robots [30]. Only the furthest portion of the afflicted limb is guided by robots [31]. This specific type of robot is easier to build, less complicated, and can be easily modified to fit different patient arm lengths [30]. Robots that use end-effectors are also used to improve lower limb functioning. This is because training lower limb functionality is typically easier than upper limb functionality because there is less complexity involved in these movements [29]. End-effectors designed for bilateral rehabilitative training have been developed recently. This is a synchronized mimicking of the unimpaired limb's movement by the afflicted limb [32]. Rutgers Ankle and Haptic Walker are two examples of robots with end-effectors [33].

19.4.1.2 Exoskeletal Devices

Robots with an exoskeleton-like structure can interact with a user's targeted joints directly and provide structure and design obstacles. They have to fit the contours of the limb, adjust to clients with different body sizes, and fasten to the limb in several places [29]. Patients could receive treatment from exoskeletons without the need for a therapist, which could result in more frequent treatments and possibly lower costs. An exoskeleton can also use precise quantitative data to evaluate the state of the patient. By using specially designed virtual games for the exoskeleton, therapy sessions can be made more fun, and patient participation in activities can be encouraged [30]. Robots that use exoskeletons are distinguished by having limbs enclosed in splints or bionic structures. They also regulate the movement of the limbs and determine the necessary torque for every joint [32]. Compared to traditional manual therapy, exoskeleton robots are better at diagnosing and treating injuries. Robots such as Lokomat, LOPES, and BLEEX are examples of exoskeletons [33].

19.4.2 According to Robot Treatment Approach

Two categories of rehabilitation robots can be distinguished according to the therapeutic approaches they employ.

19.4.2.1 Continuous Passive Movement

Continuous passive movement (CPM) is the first tactic. Since the limb is manipulated and controlled by the robot, CPM requires no voluntary effort from the person receiving medical care. As a result of CPM treatment, tendons, muscles, and joints can move without restriction. Additionally, the associated sensorimotor area in the brain is activated by CPM, producing a motion that is exactly like typical movements [32].

19.4.2.2 Active-Assisted Movement

Active-assisted movement is the second choice when the robot moves in response to input from the patient. The patient intends to use the limb, as indicated by this electrical signal, which may be an electromyogram (EMG). Rehabilitation robots with active-assisted mobility, compared to CPM rehabilitation, require the patient to voluntarily exert dedication, which leads to a notable increase in motor function. The majority of recent studies, therefore, concentrate on robot-assisted active-assisted rehabilitation [32].

19.5 APPLICATIONS IN REHABILITATION

For humans, rehabilitation is a tedious long-term process which requires great coherence. As a consequence of this, utilizing technology-based tools has shown drastically superior outcomes. Rehabilitation robotics can be addressed using two main domains: supporting therapy for individuals recovering their physical, cognitive, social, and functional ability, or assisting individuals suffering from long-term disability to accommodate and execute their daily activities of living as independently as possible [34]. Applications of rehabilitation (Figure 19.2) robots involve variety of diseases and conditions for instance: cerebral palsy, stroke, amyotrophic lateral sclerosis, and spinal cord injury.

19.5.1 Exoskeletons Robots

The robotic exoskeleton is a patient dependent/active orthosis that can be described as wearable robot. It assists in cases of disabilities involving mobility and manipulation defects, it can be used for upper or lower limbs. The upper limb exoskeleton design is all about dexterity and precision of manipulation whilst lower limb is all about supporting balance while walking and standing. Its control depends on acquisition and identification of human intent prior to any actual movement through human-robot interaction by attaching sensors of device to human body [27].

19.5.2 Robotic Prostheses

A type of rehabilitation robot called a robotic prosthesis is used to replace an amputated limb with similar human-robot interaction to that of the exoskeleton. It is directly attached to the human body and imitates the functions of the original limb through everyday activities. Crucial considerations in prostheses design include anthropoid appearance, personal physical measurements, identification of user intent and simulate movement, force and grip of actual human limb [27].

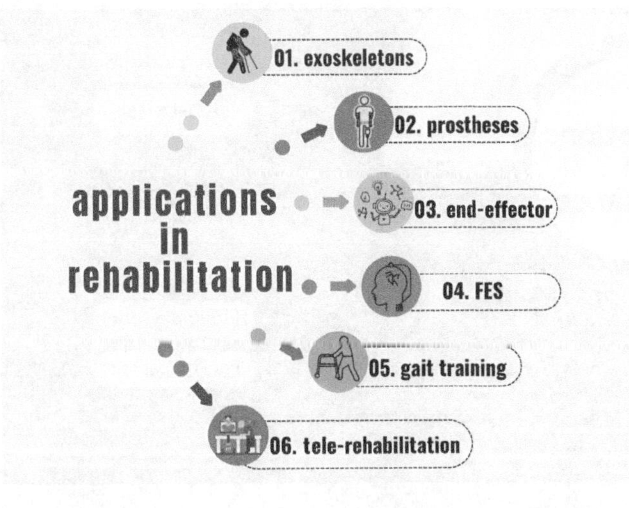

FIGURE 19.2 Applications of robotic technologies in rehabilitation.

19.5.3 End-Effector Robot

End-effector robots produce motion of limb through distal joint, human body joints work as an articulated chain so moving distal joints will mobilize proximal joints. End-effector robots apply this principle using robotic foot plates for the lower limb or handles for upper limb. Unlike exoskeleton that gets dressed all over the limb it only connects to distal parts of limb to achieve similar functions [35].

19.5.4 Gait Training Robot

Gait training significantly stimulates recovery of mobility, manual gait training is exhausting, time consuming and ineffective compared to gait training with robots as it achieves more regulated and meticulous motion. It decreases physical therapist's workload in addition to providing assessment in an objective and quantitative manner. Structure of gait training robot include end-effector robot that is a footplate imitating stance and swing phases of gait and robotic exoskeleton to help direct other proximal joint through physiological movement pattern, ensuring the user receives proprioceptive feedback while loading [36].

19.5.5 Robotic Systems with Functional Electrical Stimulation "FES"

This is a hybrid robotic system combines passive exoskeleton and EMG-triggered functional electrical stimulation commonly used for post stroke patients aiming to restore upper limb motor functionality, it works even in case of severely impaired patients. It showed more improvement in dexterity and hand functions recovery compared to advanced conventional therapy [37].

19.5.6 Tele-Rehabilitation Robots

Telerehabilitation is the process of delivering rehabilitation services to remote patient in their home environment using communication technologies like online video conferencing. Robots can improve telerehabilitation efficiency by delivering consistent, objective, and reliable assessment of motor functions of patients from their homes to the therapist. This makes rehabilitation more accessible for patients and reduces bias of subjective assessment and allows remote supervision by therapist [29].

19.6 TYPES AND APPLICATIONS IN ASSISTANCE

Assistive robotics encompasses a wide range of devices, including manipulative and mobile assistive robots that help patients with motor function impairments [38] to social assistance robots that support cognitive rehabilitation and education [7], or wearable robots that enhance the human body physically [7, 39].

They mainly help people with daily activities [40] including clothing, running, or walking[1] through physical or social interaction [41] when these tasks are limited because of an injury or disability so these technologies help to improve the individual's quality of life by providing independence and autonomy [42].

There are many different categories of assistive devices as mentioned by Khasnabis et al. [35] (Figure 19.3).

19.6.1 Mobility Devices

For example, wheelchairs, tricycles, crutches, walking frames/walkers, and Co-bots.

They may provide unique functionality to meet the demands of the user. A wheelchair with trunk/head supports, for instance, can be necessary for a person with cerebral palsy in order to guarantee that they can sit comfortably [34, 43].

19.6.2 Orthotics and Prosthetics

For example, hand/leg splints, spinal braces, and artificial legs or hands.

Typically, they are specially designed tools that replace, support, or adjust bodily parts. Professionals with training in prosthetics and orthotics design, produce, and install them at specialized workshops or centers [43]

19.6.3 Positioning Devices

For example, standing frames, wedges, chairs, and special seats.

These devices assist individuals with physical disabilities who frequently struggle to maintain comfortable sitting, standing, or laying postures during functional tasks, helping to prevent improper

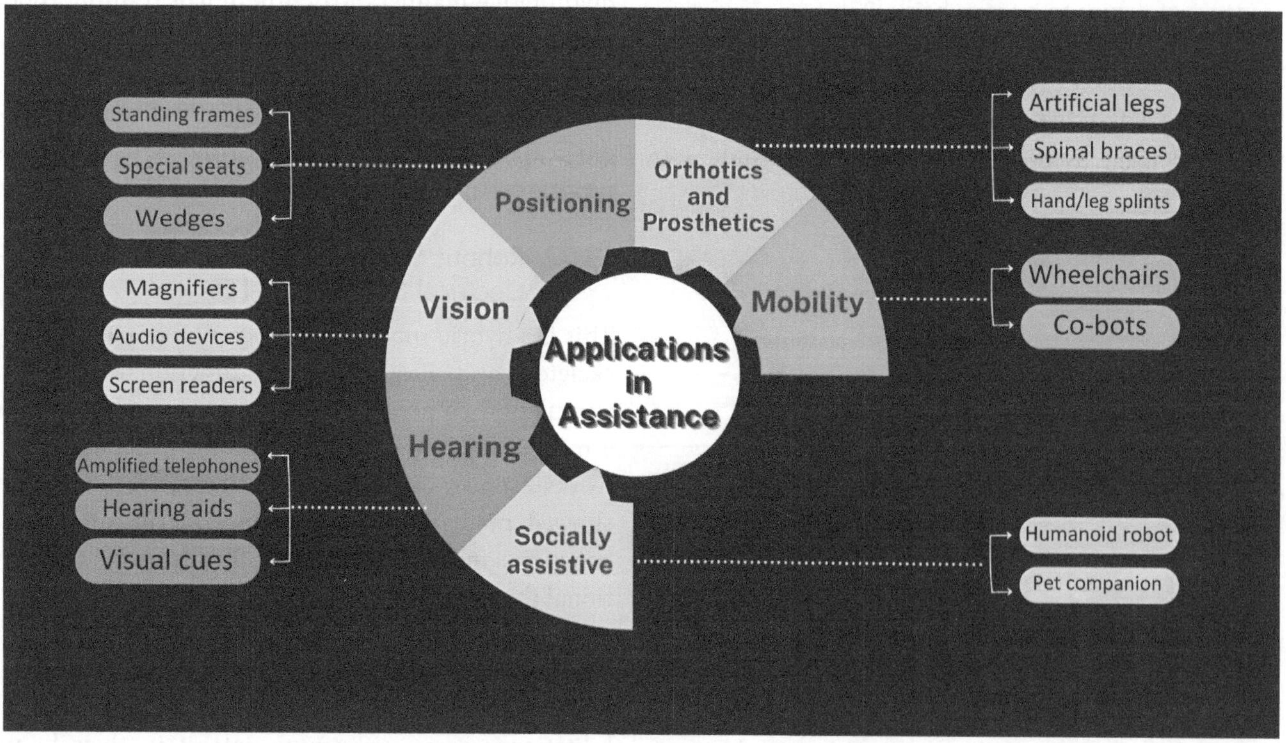

FIGURE 19.3 Applications of healthcare robots in assistive devices.

positioning that could lead to the development of deformities [43].

19.6.4 Vision Devices

For example, screen readers for computers, magnifiers, eyeglasses or spectacles, audio devices like radios and talking books, and braille systems for reading and writing.

To enhance involvement and independence of blind or low vision persons, those variety of tools—from basic to sophisticated—can be employed [43].

19.6.5 Hearing Devices

For example, hearing aids, visual systems to provide cues as light-linked Doorbells or vibration Alerts, amplified telephones, and telecommunication devices.

Communication and social interaction are hampered by hearing loss, which may have an impact on speech and language development among other developmental domains. Aside from that, it can restrict access to jobs and education, which can result in social isolation and discrimination, so hearing devices improve the patient's quality of life significantly [43].

19.6.6 Socially Assistive Robots

For example, humanoid robots, pet companion-bots, and telepresence robots.

These devices provide emotional support to patients and may decrease mental illnesses such as depression, anxiety, autism, and dementia through engaging in play and conversations [44]. Also, they offer motivation for recovery and support for exercise during the patient's rehabilitation phase [44].

19.7 IMPORTANCE OF REHABILITATION AND ASSISTIVE DEVICES

Sophisticated diagnosis, treatment, and rehabilitation systems have been made possible by advancements in information technology, electronics, radio engineering, and robotics as well as in rehabilitation. A key component of rehabilitation for a long time has been mechanotherapy, especially when it comes to regaining motor function. Robotic mechanotherapy is a new approach to rehabilitation that uses customized robotic devices to teach the upper and lower limbs via feedback mechanisms. It has recently gained popularity. Compared to traditional physical therapy, this method has a number of benefits, such as longer sessions, precise and repetitive movements, reliable training regimens, and the capacity to assess and demonstrate to patients the efficacy of exercises. The efficacy of physical rehabilitation, treatment, and diagnostic procedures is greatly increased by these current robotic systems [44].

In Southern Punjab, a study examined the function of assistive technology in the rehabilitation of children with physical disabilities. 170 senior and junior special education teachers, instructors of physical education, and physical education teachers who worked with students who were physically disabled made up the sample. According to the study's findings, children with physical disabilities experience less disability when they use assistive technology [45].

Another advancement improved the quality and results of rehabilitation for patients with myocardial infarction: the design and development of a myocardial infarction rehabilitation aid system. The system gathers and transmits vital sign data and basic patient information using clever wearable device technologies. A cloud module allows for efficient management of system data, assisting families in understanding the patient's state, and a communication module gives clinicians real-time access and rehabilitation guidance [46].

By creating modern technologies and helpful technologies, rehabilitation engineering has significantly improved the quality of life for people with impairments. Given that the population over 75 is more likely to experience long-term diseases and disabilities, this is becoming more and more crucial. Furthermore, patients suffering from diverse neuromuscular disorders want prompt assistance in order to avert serious or irreversible harm. Human aid has become the primary emphasis of robotics and mechatronics, with a shift from industrial applications to rehabilitation and functional enhancements. In an effort to aid people in recovering from or improving their cognitive and physical abilities, researchers are driven to create innovative rehabilitative and assistive technology [47].

Within the field of rehabilitation robots, the research paper "Robotics for rehabilitation of hand movement in stroke survivors" explores into the latest developments in hand rehabilitation robotics, emphasizing training paradigms, mechanical design, and usability to facilitate successful hand recovery [48] "Design and control of a diagnosis and treatment-focused robotic platform for the wrist and forearm rehabilitation: DIAGNOBOT," which introduces a robotic platform capable of carrying out therapeutic exercises involving certain wrist and forearm movements, is another noteworthy work. Tests conducted on persons in good health show that the platform can carry out these workouts efficiently [49].

The HomeRehab robot, intended for use at home to support rehabilitation therapies in a three-dimensional setting, was developed in accordance with the "Development of a robotic device for post-stroke home tele-rehabilitation" paper. Though it is lighter, smaller, and less expensive than hospital systems, the gadget provides comparable performance. Initial usability studies demonstrate that it is more effective than the Robotherapist 2D system [50].

A customizable hand exoskeleton intended for "Mirror Therapy" is described in the paper "Hand exoskeleton for rehabilitation therapies with integrated optical force sensor." This device has an integrated force sensor that serves as a human-machine interface and can move a handicapped hand with success. The technical feasibility of the exoskeleton has been validated by preliminary clinical trials and research involving healthy participants [51].

Mobility impairments have a major influence on independent living and quality of life and are common in our aging population. Various mobility aid robots have been developed, characterized by actuation, kinematic structure, sensors, and human-machine interfaces. These robots help with walking, mobility, navigation, and transitions from sit to stand and stand to sit [52]. There is a growing need for service robots that can do intricate activities like dressing help due to the aging population. These multi-controller robots, which are often bimanual or single manipulators, help with tasks such as clothing assembly [53].

Furthermore, assistive robotic systems offer educational assistance by identifying and classifying dysgraphia in pupils in order to customize rehabilitation plans [54]. Processing information and patient data, helping with everyday living activities, bringing and fetching things, telepresence and communication, monitoring, safety, navigation, and other sophisticated aid functions are all included in the scope of assistive robotic systems [55].

Robots detect people and their surroundings using a range of sensors. For tracking motion and kinematic evaluation, research is concentrated on technologies such as depth cameras, laser rangefinders, infrared sensors, and stereo vision devices. While force and torque sensors track mechanical stress and interaction forces, proprioception sensors record the movements and positioning states of the robots [56]. The fifth class of rehabilitation robots is made for workouts that demand adaptation and response from patients. By providing dynamic surroundings, these robots help patients recover more effectively [57].

19.8 BENEFITS AND OUTCOMES

With the beginning of progress and development in the new era, it promises to bring about innovation in AI and healthcare robotics. Make diagnosis easier, enhance patient care and make processes more streamlined. These new technologies enhance patient outcomes and increase accuracy and efficiency of medical field practices in addition to reducing duty and difficulty on healthcare professionals [58]. Also, never diminish that rehabilitation robotics assist elderly and disabled patients with important daily activities, mobility, social interactions, and medications reminders aiming to improve a person's physical or cognitive abilities, independence, and quality of life [34, 59]. Patients who recover from stroke or spinal injuries benefit from neurorehabilitation as brain-machine interfaces and virtual reality integration as they restore the patient's communication, provide cognitive trainings, enhance engagement and task specific training by allowing patients to experience real world actions [60]. Therefore, we could summarize robotics benefits in two categories: physical and non-physical as improving upper and lower limbs control, pain relief, increasing cognitive functions, decreasing the perception of weariness while doing daily activities, enhancing quality of life, providing social support, feeling hopeful, and assisting patient with mobility, hearing and vision [25, 61].

19.9 DISADVANTAGES AND COMPLICATIONS

While rehabilitation and assistive robots in healthcare offer significant benefits, they also have certain disadvantages and complications related to their system, application, or interaction with the human body. End-effector robots are unsuitable for pronation and supination as they provide limited rotational movement, also overstretching or exceeding the physiological range of motion can result in traumatic joint injuries [62, 63]. High pressures and Interaction forces at the human-robot interface, especially in exoskeleton-type devices, can raise the risk of pain and soft tissue injuries like abrasions because of the greater contact area and even can compress tissues and obstruct blood flow leading to injuries such as bruises or pressure ulcers, particularly over bony prominences, also the restraining energy exerted by the robot can damage skin and soft tissues [62]. Additionally, fall risks are increased by gait rehabilitation robots because of their extra weight and interference with the patient's ability to interact with the ground [62]. A misalignment of the exoskeleton joints with the human skeleton might result in unwanted stresses, compromise safety and comfort, and even induce musculoskeletal damage [62]. Furthermore, implementing sensors such as bend or strain sensors in hand orthosis poses challenges, and individuals with disabilities may face problems in responding promptly to avert potential harm caused by significant faults and even the utilization of these devices is complicated by problems such as the failure of voice recognition, the possibility of spool rotational limit failure, and the erroneous beginning of modes by patients [64]. Finally, it's challenging to personalize assistance levels according to each patient's interests, needs, and specific abilities [32].

19.10 CHALLENGES AND LIMITATIONS

Even with a wealth of evidence supporting its effectiveness, telerehabilitation and robotics-based rehabilitation are not yet widely used as therapy modalities in most settings [29].

19.10.1 Cost

One of the main obstacles to the promotion of rehabilitation robots is cost. The price/performance ratio is not very good due to the relatively modest profit for customers and centers and high expenses for development [32].

19.10.2 Difficulty to Be Personalization

When a patient receives robotic assistance, it is best to do so in a manner that makes communication between him or her and the robot easier. Depending on the patient's residual abilities, the level of assistance offered should be modified [32].

19.10.3 Safety

The security of these gadgets would be an additional essential concern. To guarantee a fail-safe performance if it becomes necessary, programmers must make sure the necessary procedures are in place. There should be no major consequences for the patient from breakdowns in systems [65].

Price, security, effectiveness, and both patient and practitioner familiarity with use are among the challenges that remain to be tackled before robotic therapies become more accessible and user-friendly. Much study has additionally been conducted to examine the ability of low-cost rehabilitation systems to provide useful data for remote evaluation and change them to make such devices easier to purchase [29].

19.11 CLINICAL AND ETHICAL CONSIDERATIONS

The use of robotics technology in healthcare increases ethical considerations and they must be addressed in order to ensure patient's safety and the safe implantation of these technologies [21].

19.11.1 Data Security and Patient Privacy

Ensuring the safety and privacy of the large amount of sensitive patient data is crucial as any breach will lead to unfavorable consequences for patient trust and data integrity and this will be done by implanting encryption methods during transmission and storage of data [58].

19.11.2 Bias and Equity

Continuous monitoring of AI algorithms for bias is important to limit disparities in diagnosis and treatment among populations [58].

19.11.3 Accountability and Transparency

Determining who is accountable for AI systems mistakes is challenging but crucial in their ethical implantation in healthcare thus there must be a clear responsibility [58].

19.11.4 Robots Design

From a system's perspective, the robot design must prevent it from causing harm to its user in case all other safety measures are ineffective [34].

19.11.5 Informed Consent

Transparent communication is necessary when discussing healthcare robots as the healthcare provider has to discuss all the advantages, disadvantages and dangers with the patient to let him give his permission [60].

19.12 FUTURE DIRECTIONS AND TRENDS

Based on current trends and developing technology, future research in the healthcare fields of rehabilitation and assistive device robots is predicted to enhance significantly by advancement of technology like AI, robotics, engineering. Alongside the advancement of healthcare robots, more personalized rehabilitation robots that adapt to the needs and advancement of their patients are on the horizon. Robotic rehabilitation technology that patients may use at home will provide regular and convenient therapy [60]. Our goal from improving these techniques is to improve the lives of those in recovery, promote resilience, and enable individuals to overcome physical challenges with technological aid [60]. Additionally, certain features will need to be prioritized in the future as develop exercises that practice everyday life functions using actual things, not only virtual ones, obtain more advanced sensory input from the patient, optimize the range of motion, and boost the ratio of cost to benefit [66]. All innovations will lead to the development of more personalized and adaptive tools that can better meet individual needs and subsequently improve the quality of life and integration of individuals with physical and cognitive impairments into society.

19.13 CONCLUSION

With the hope that in the coming decades, robots could assist in addressing major difficulties, and offer efficiency. The implementation of robotics in healthcare is a relatively recent and interesting field. Only a small number of robots are in widespread usage and are currently offered for sale. With numerous other uses being developed, service and surgery robots are currently the most well-known types of healthcare robots in use. Although the exact nature of these robots' integration into the healthcare system is unknown, it is certain that they will play a significant role in facilitating the provision healthcare. It's challenging to integrate modern robotic technology into healthcare. However, a significant and effective technological disruption can be achieved with the right approach. Before robots are used in healthcare and made commercially available, prototype experimental systems must be improved to make them user-friendly, safe, and dependable. The robot must have a suitable appearance and behavior that complements the intended application. It is necessary to consider specific metrics for system performance and cost-effectiveness. It is necessary to conduct feasibility, impact, and economic assessments to validate the acceptability and usability. Acceptance of robots is higher when they assist hospital staff in improving care delivery, as opposed to displacing or isolating them. In the end, widespread robot adoption is unlikely to occur until positive long-term effects are demonstrated.

REFERENCES

1. Siciliano, B., & Khatib, O. (Eds.). (2016). *Springer Handbook of Robotics* (2nd ed.). Switzerland: Springer International Publishing.
2. World Health Organization. (2019). *Global Spending on Health: A World in Transition* (p. 49). World Health Organization. https://iris.who.int/handle/10665/330357. License: CC BY-NC-SA 3.0 IGO.
3. Kwoh, Y. S., Hou, J., Jonckheere, E. A., & Hayati, S. (1988). A robot with improved absolute positioning accuracy for CT guided stereotactic brain surgery. *IEEE Transactions on Biomedical Engineering, 35*(2), 153–160.
4. Kaiser, M. S., Al Mamun, S., Mahmud, M., & Tania, M. H. (2021). Healthcare robots to combat COVID-19. *COVID-19: Prediction, Decision-Making, and Its Impacts*, 83–97.
5. Kyrarini, M., Lygerakis, F., Rajavenkatanarayanan, A., Sevastopoulos, C., Nambiappan, H. R., Chaitanya, K. K., … Makedon, F. (2021). A survey of robots in healthcare. *Technologies, 9*(1), 8.
6. Müller, C. (2019). International Federation of Robotics Press Conference. IFR Press Conference, September 18, 2019, Shanghai.
7. Abdi, J., Al-Hindawi, A., Ng, T., & Vizcaychipi, M. P. (2018). Scoping review on the use of socially assistive robot technology in elderly care. *BMJ Open, 8*(2), e018815.
8. Avgousti, S., Christoforou, E. G., Panayides, A. S., Voskarides, S., Novales, C., Nouaille, L., & Vieyres, P. (2016). Medical telerobotic systems: Current status and future trends. *Biomedical Engineering Online, 15*(1), 44.
9. Conti, D., Di Nuovo, A., & Di Nuovo, S. (2019, December). Social robots for older users: A possibility to support assessment and social interventions. In *CEUR Workshop Proceedings, City Campus*, (Vol. 2524). Tilburg University, Netherlands.

10. Dickstein-Fischer, L. A., Crone-Todd, D. E., Chapman, I. M., Fathima, A. T., & Fischer, G. S. (2018). Socially assistive robots: Current status and future prospects for autism interventions. *Innovation and Entrepreneurship in Health, 5*, 15–25.
11. Dolic, Z., Castro, M. C. R., & Moarcas, M. C. A. (2019). *Robots in healthcare: A solution or a problem? Workshop proceedings*. Brussels: European Parliament, Panel for the Future of Science and Technology (STOA), Committee on Environment, Public Health and Food Safety. PE 638.391. https://www.europarl.europa.eu/RegData/etudes/STUD/2019/638391/EPRS_STU(2019)638391_EN.pdf
12. Khan, Z. H., Siddique, A., & Lee, C. W. (2020). Robotics utilization for healthcare digitization in global COVID-19 management. *International Journal of Environmental Research and Public Health, 17*(11), 3819.
13. Mois, G., & Beer, J. M. (2020). Robotics to support aging in place. In R. Pak, E. J. de Visser, & E. Rovira (Eds.), *Living with Robots, US*, (pp. 49–74). Cambridge, MA: Academic Press.
14. Shi, D., Zhang, W., Zhang, W., & Ding, X. (2019). A review on lower limb rehabilitation exoskeleton robots. *Chinese Journal of Mechanical Engineering, 32*(1), 1–11.
15. Whelan, S., Murphy, K., Barrett, E., Krusche, C., Santorelli, A., & Casey, D. (2018). Factors affecting the acceptability of social robots by older adults including people with dementia or cognitive impairment: A literature review. *International Journal of Social Robotics, 10*(5), 643–668.
16. Irujo Aizcorbe, J. (2020). *A review on human-robot collaboration and its application in the health care sector* (Master's thesis). Hochschule Niederrhein, University of Applied Sciences, Germany.
17. Silvera-Tawil, D., & Roberts-Yates, C. (2018). Socially-assistive robots to enhance learning for secondary students with intellectual disabilities and autism. In *2018 27th IEEE International Symposium on Robot and Human Interactive Communication (RO-MAN)* (pp. 838–843). Nanjing: IEEE, US.
18. Cresswell, K., Cunningham-Burley, S., & Sheikh, A. (2018). Health care robotics: Qualitative exploration of key challenges and future directions. *Journal of Medical Internet Research, 20*(7), e10410.
19. Riek, L. D. (2017). Healthcare robotics. *Communications of the ACM, 60*(11), 68–78.
20. Nwazor, N. O., & Orakwue, S. I. (2023). Enhancing healthcare through automation and robotics. In E. O. Nwaichi (Ed.), *Modernity in Health and Disease Diagnosis: The Account from STEM Women*, 59–67. Cham: Springer Nature Switzerland.
21. Kebede, G. A., Gelaw, A. A., Andualem, H., & Hailu, A. T. (2024). Review of the characteristics of mobile robots for health care application. *International Journal of Intelligent Robotics and Applications, 8*(2), 480–502.
22. Duan, F., Li, W., & Tan, Y. (2023). Overview of robots. In *Intelligent Robot: Implementation and Applications* (pp. 3–30). Singapore: Springer Nature Singapore.
23. Haidegger, T. (2021). Taxonomy and standards in robotics. In M. H. Ang, O. Khatib, & B. Siciliano (Eds.), *Encyclopedia of Robotics* (pp. 1–10). Berlin, Heidelberg: Springer.
24. Boubaker, O. (2020). Medical robotics. *Control Theory in Biomedical Engineering*, 153–204.
25. Guo, Y., Yang, Y., Liu, Y., Li, Q., Cao, F., Feng, M., … Kang, Y. (2021). Development status and multilevel classification strategy of medical robots. *Electronics, 10*(11), 1278.
26. Huang, R. (2022, August). Classification of healthcare robots. In *International Conference on Well-Being in the Information Society* (pp. 115–123). Cham: Springer International Publishing.
27. Mohebbi, A. (2020). Human-robot interaction in rehabilitation and assistance: A review. *Current Robotics Reports, 1*(3), 131–144.
28. Morales, R., Badesa, F. J., García-Aracil, N., Sabater, J. M., & Pérez-Vidal, C. (2011). Pneumatic robotic systems for upper limb rehabilitation. *Medical & Biological Engineering & Computing, 49*, 1145–1156.
29. Laut, J., Porfiri, M., & Raghavan, P. (2016). The present and future of robotic technology in rehabilitation. *Current Physical Medicine and Rehabilitation Reports, 4*, 312–319
30. Lo, H. S., & Xie, S. Q. (2012). Exoskeleton robots for upper-limb rehabilitation: State of the art and future prospects. *Medical Engineering & Physics, 34*(3), 261–268.
31. Mazzoleni, S., Duret, C., Grosmaire, A. G., & Battini, E. (2017). Combining upper limb robotic rehabilitation with other therapeutic approaches after stroke: Current status, rationale, and challenges. *BioMed Research International, 2017*(1), 8905637.
32. Qian, Z., & Bi, Z. (2015). Recent development of rehabilitation robots. *Advances in Mechanical Engineering, 7*(2), 563062.
33. Zhang, X., Yue, Z., & Wang, J. (2017). Robotics in lower-limb rehabilitation after stroke. *Behavioural Neurology, 2017*(1), 3731802
34. Van der Loos, H. M., Reinkensmeyer, D. J., & Guglielmelli, E. (2016). Rehabilitation and health care robotics. *Springer Handbook of Robotics* (pp. 1685–1728). Cham: Springer International Publishing.
35. Molteni, F., Gasperini, G., Cannaviello, G., & Guanziroli, E. (2018). Exoskeleton and end-effector robots for upper and lower limbs rehabilitation: Narrative review. *PM&R, 10*(9), S174–S188.
36. Warutkar, V., Dadgal, R., & Mangulkar, U. R. (2022). Use of robotics in gait rehabilitation following stroke: A review. *Cureus, 14*(11), e31075–e31075.
37. Ambrosini, E., Gasperini, G., Zajc, J., Immick, N., Augsten, A., Rossini, M., & Krakow, K. (2021). A robotic system with EMG-triggered functional

electrical stimulation for restoring arm functions in stroke survivors. *Neurorehabilitation and Neural Repair, 35*(4), 334–345.
38. Campeau-Lecours, A., Lamontagne, H., Latour, S., Fauteux, P., Maheu, V., Boucher, F., … L'Ecuyer, L. J. C. (2019). Kinova modular robot arms for service robotics applications. In *Rapid Automation: Concepts, Methodologies, Tools, and Applications. International Journal of Robotics Applications and Technologies (IJRAT), 5(2)* (pp. 693–719). IGI Global Scientific Publishing.
39. Kumar, V., Hote, Y. V., & Jain, S. (2019, October). Review of exoskeleton: History, design and control. In *2019 3rd International Conference on Recent Developments in Control, Automation & Power Engineering (RDCAPE), Noida, India* (pp. 677–682). IEEE.
40. Kim, J., Gu, G. M., & Heo, P. (2016). Robotics for healthcare. *Biomedical Engineering: Frontier Research and Converging Technologies, 9*, 489–509.
41. Fosch-Villaronga, E., & Drukarch, H. (2021). On healthcare robots: Concepts, definitions, and considerations for healthcare robot governance. *arXiv Preprint Arxiv:2106.03468.*
42. Abarca, V. E., & Elias, D. A. (2023). A review of parallel robots: Rehabilitation, assistance, and humanoid applications for neck, shoulder, wrist, hip, and ankle joints. *Robotics, 12*(5), 131.
43. Khasnabis, C., Motsch, K. H., Achu, K., Al Jubah, K., Brodtkorb, S., Chervin, P., … Lander, T. (2010). *Community-Based Rehabilitation: CBR Guidelines*. Geneva: World Health Organization.
44. Rastorguyeva, I., Efimova, A., Khatuntseva, S., & Kara, S. (2022). Modern rehabilitación tecnologías in the field of health. *Scientific Papers of Berdiansk State Pedagogical University Series Pedagogical Sciences, 1(3)*:370–378
45. Ashfaq, M., Ali, H. H., & Aftab, M. J. (2022). Role of assistive devices in the rehabilitation of physically handicapped children in Southern Punjab. *Pakistan Journal of Social Research, 4*(4), 502–514.
46. Chen, B. (2023, November). Design and development of a myocardial infarction rehabilitation assistance system based on intelligent wearable devices. In *International Conference on Cognitive Based Information Processing and Applications, vol 3, CIPA 2023. Lecture Notes on Data Engineering and Communications Technologies, vol 198.* (pp. 201–211). Springer Nature Singapore.
47. Aggogeri, F. (2020). Special collection on advances in rehabilitation engineering with robotics and mechatronic devices. *Advances in Mechanical Engineering, 12*(8), 1687814020949086.
48. Aggogeri, F., Mikolajczyk, T., & O'Kane, J. (2019). Robotics for rehabilitation of hand movement in stroke survivors. *Advances in Mechanical Engineering, 11*(4), 1687814019841921.
49. Aktan, M. E., & Akdoğan, E. (2018). Design and control of a diagnosis and treatment aimed robotic platform for wrist and forearm rehabilitation: DIAGNOBOT. *Advances in Mechanical Engineering, 10*(1), 1687814017749705.
50. Díaz, I., Catalan, J. M., Badesa, F. J., Justo, X., Lledo, L. D., Ugartemendia, A., … García-Aracil, N. (2018). Development of a robotic device for post-stroke home tele-rehabilitation. *Advances in Mechanical Engineering, 10*(1), 1687814017752302.
51. Han, J., Lian, S., Guo, B., Li, X., & You, A. (2017). Active rehabilitation training system for upper limb based on virtual reality. *Advances in Mechanical Engineering, 9*(12), 1687814017743388.
52. Geravand, M., Werner, C., Hauer, K., & Peer, A. (2023). A survey on rollator-type mobility assistance robots. *Handbook of Human-Machine Systems*, 165–179.
53. Nocentini, O., Kim, J., Bashir, Z. M., & Cavallo, F. (2022). Learning-based control approaches for service robots on cloth manipulation and dressing assistance: A comprehensive review. *Journal of NeuroEngineering and Rehabilitation, 19*(1), 117.
54. Gouraguine, S., Riad, M., Rafik, M., Qbadou, M., & Mansouri, K. (2023, May). A humanoid robot assistant for the classification of students according to their type of dysgraphia. In *2023 3rd International Conference on Innovative Research in Applied Science, Engineering and Technology (IRASET), Mohammedia, Morocco* (pp. 1–5). IEEE.
55. Ohneberg, C., Stöbich, N., Warmbein, A., Rathgeber, I., Mehler-Klamt, A. C., Fischer, U., & Eberl, I. (2023). Assistive robotic systems in nursing care: A scoping review. *BMC Nursing, 22*(1), 72.
56. Zheng, Y., Zhong, P., Liu, K., Yang, K., & Yue, Q. (2019). Human motion capture system based 3D reconstruction on rehabilitation assistance stability of lower limb exoskeleton robot climbing upstairs posture. *IEEE Sensors Journal, 20*(20), 11778–11786.
57. Michmizos, K. P., & Krebs, H. I. (2017). Pediatric robotic rehabilitation: Current knowledge and future trends in treating children with sensorimotor impairments. *NeuroRehabilitation, 41*(1), 69–76.
58. Elendu, C., Amaechi, D. C., Elendu, T. C., Jingwa, K. A., Okoye, O. K., Okah, M. J., … Alimi, H. A. (2023). Ethical implications of AI and robotics in healthcare: A review. *Medicine, 102*(50), e36671.
59. Ness, S., Xuan, T. R., & Oguntibeju, O. O. (2024). Influence of AI: Robotics in healthcare. *Asian Journal of Research in Computer Science, 17*(5), 222–237.
60. Kaluvakuri, S. (2022). Revolutionizing healthcare: The impact of robotics on health services. *Malaysian Journal of Medical and Biological Research, 9*(2), 41–50.
61. Soljacic, F., Law, T., Chita-Tegmark, M., & Scheutz, M. (2024). Robots in healthcare as envisioned by care professionals. *Intelligent Service Robotics, 17*(3), 685–701.

62. Bessler, J., Prange-Lasonder, G. B., Schaake, L., Saenz, J. F., Bidard, C., Fassi, I., ... Buurke, J. H. (2021). Safety assessment of rehabilitation robots: A review identifying safety skills and current knowledge gaps. *Frontiers in Robotics and AI*, *8*, 602878.
63. Qassim, H. M., & Wan Hasan, W. Z. (2020). A review on upper limb rehabilitation robots. *Applied Sciences*, *10*(19), 6976.
64. Chu, C. Y., & Patterson, R. M. (2018). Soft robotic devices for hand rehabilitation and assistance: A narrative review. *Journal of Neuroengineering and Rehabilitation*, *15*(1), 14.
65. Suppiah, R., Kim, N., Abidi, K., & Sharma, A. (2023). A comprehensive review of motor movement challenges and rehabilitative robotics. *Smart Health*, *29*, 100402.
66. Fazekas, G., & Tavaszi, I. (2019). The future role of robots in neuro-rehabilitation. *Expert Review of Neurotherapeutics*, *19*(6), 471–3.

CHAPTER 20

Artificial Intelligence for Population Health and Wellness

Diana Malaeb, Ahmad El Ouweini, Samar Younes, Malik Sallam, Sami El Khatib, Muna Barakat, May G. Mohamed, Naglaa M. Marzuuk, and Souheil Hallit

20.1 INTRODUCTION

Population health management (PHM) is an evolving field aimed at improving overall community health through data-driven strategies and preventive care. By leveraging technology and healthcare analytics, PHM helps identify health trends, predict potential risks, and create intervention plans that benefit both individuals and larger populations.

In PHM, artificial intelligence (AI) has become a potent instrument that provides fresh approaches to effectively evaluate and control health hazards. Large volumes of both organized and unstructured data may be processed by AI algorithms, allowing medical professionals to identify trends and make wise choices. These advancements allow for better allocation of resources and more precise interventions that improve patient outcomes.

However, integrating AI into PHM comes with notable challenges. Issues related to data security, ethical concerns, and the need for unbiased algorithms must be carefully managed. Ensuring transparency and accountability in AI applications will be essential for building trust among healthcare professionals and the populations they serve.

This chapter will examine the transformative role of AI in PHM, with a particular focus on its applications in patient risk assessment, personalized interventions, and disease prevention. These applications will be illustrated through real-world examples of AI-driven public health initiatives. Additionally, the chapter will discuss the difficulties in implementing AI in PHM, such as uncertainties in data, privacy concerns, and possible algorithmic biases. Finally, it will propose strategies to mitigate these challenges and optimize the impact of AI on PHM.

20.2 ARTIFICIAL INTELLIGENCE: A GAME-CHANGER FOR POPULATION HEALTH MANAGEMENT

PHM involves analyzing and facilitating healthcare services along with their delivery [1]. This approach focuses on improving the clinical outcomes of defined groups, typically within a specific patient population or geographical region [2]. Unlike conventional models that primarily treat diseases at an individual level, PHM takes a broader perspective by addressing key health determinants, prioritizing preventive care, and managing chronic conditions [3]. Additionally, PHM aims to minimize health disparities, reduce healthcare costs, and enhance both the efficiency and effectiveness of care delivery [4]. To achieve these goals, PHM utilizes data analytics to recognize high-risk populations and deploy targeted health interventions [5].

The growing interest in PHM is driven by multiple factors, including the rising prevalence of multimorbidity and disability due to aging and increased life expectancy, escalating healthcare costs amid financial constraints, widening health disparities, and shifting patient expectations [6, 7]. As a result, a greater emphasis on prevention and a deeper understanding of socioeconomic determinants of health are necessary to improve population health and reduce inequalities.

DOI: 10.1201/9781003533351-20

TABLE 20.1 Steps of PHM and Their Aims

Step	Aim
Defining and identifying the population	To identify vulnerable individuals and populations and offer them appropriate care even before they develop an illness or condition or its complications [7] To enable more effective interventions to be designed by enhancing the understanding of the interface between individual health and its broader determinants
Health assessment and segmentation	To collect the information needed to comprehend the variations within the population, the type of needs, and the services needed to maximize the population's health To divide people into comparatively uniform groups (segments) based on the kind and degree of care required
Risk stratification and impactibility	To determine the likelihood that unfavourable things may happen and give a person a risk status or score To determine how likely at-risk individuals or patients are to respond or be affected by various actions or treatment options
Tailored service delivery	To specify the care programs or actions that will proactively close the care gaps, adjusting them to fit the various risk groups
Evaluation and improvement	To monitor and evaluate the intervention implemented

Adopting proactive strategies like PHM can help achieve these goals [8]. The essential components of PHM can be structured into a sequence of five key steps applicable across various healthcare systems [9]. These steps are illustrated in Table 20.1.

By identifying subgroups with shared characteristics and needs, PHM can greatly assist primary healthcare providers (PCPs) in moving away from a standardized, one-size-fits-all approach toward more targeted and personalized strategies that address the specific requirements of different population groups or clusters with similar health conditions. It also facilitates the shift from passive, reactive care to a proactive model by recognizing individuals at varying risk levels and anticipating their healthcare needs. Furthermore, PHM broadens the scope beyond clinical concerns, incorporating psychosocial factors and social determinants of health into care strategies. Additionally, it supports PCPs in transitioning from fragmented and poorly coordinated care to a more integrated and well-coordinated system that connects seamlessly with secondary and tertiary healthcare services [7, 9].

Despite its well-documented advantages, PHM encounters several major challenges that can limit its effectiveness. The integration of data from various sources, including as claims data, electronic health records (EHRs), and socioeconomic determinants of health, is a major problem that is frequently hampered by interoperability barriers among healthcare systems [10, 11]. Furthermore, handling substantial amounts of private and sensitive patient data presents serious privacy and security issues, requiring close compliance with laws like the Health Insurance Portability and Accountability Act of 1996 (HIPAA) [12, 13].

Resource limitations, including inadequate financial investment, technological infrastructure, and a shortage of skilled personnel, further complicate the implementation and sustainability of comprehensive PHM programs, particularly for smaller healthcare organizations [14]. There are logistical issues as a result of the rising frequency of chronic diseases like diabetes and heart disease, which call for coordinated, long-term care across several providers [14, 15]. Addressing health disparities and incorporating social determinants of health into care models is essential yet difficult, as these factors are often complex and challenging to translate into actionable interventions [4].

Moreover, engaging patients in their own healthcare remains a critical challenge, particularly among populations with low health literacy or restricted access to medical resources [16]. PHM programs must also navigate a complex regulatory framework, balancing patient privacy protections with the need for data sharing and collaboration [17]. Measuring the return on investment and assessing the effectiveness of PHM initiatives is another challenge, given the numerous variables involved in long-term population health outcomes [18]. Finally, cultural and organizational resistance within healthcare institutions can create obstacles to the successful adoption of PHM strategies [19].

AI has the potential to significantly influence PHM in today's rapidly advancing healthcare landscape.

It can revolutionize how healthcare systems predict, prevent, and manage diseases [20]. AI is an innovative approach that leverages advanced algorithms and data analytics to forecast health outcomes, optimize interventions, and extract meaningful insights [21]. It entails creating computer programs that can carry out tasks that normally call for human intelligence. These programs can examine large datasets, such as insurance claims, EHRs, remote monitoring data, public health data, and social determinants of health, to find trends and use processed data to forecast results or make defensible decisions [22].

Machine learning (ML), which allows computers to learn from data and improve over time, and deep learning, which uses artificial neural networks to mimic human brain function, are two important AI technologies [23]. In addition, AI's capabilities extend to natural language processing, which facilitates the extraction and analysis of critical information from research papers, clinical documentation, and textual data [24]. By leveraging these techniques, AI uncovers hidden patterns within large, complex datasets, ultimately enhancing population health through data-driven decision-making and proactive interventions [25]. With AI integration, healthcare can shift from a reactive to a predictive model, identifying and addressing potential health risks before they escalate, therefore encouraging a healthcare system that is more patient-centred and efficient [26].

20.3 UNLOCKING THE POTENTIAL: HOW AI IS REVOLUTIONIZING POPULATION HEALTH

Through better diagnostics, more productivity, and the ability to treat patients proactively, AI is revolutionizing the healthcare industry. Predictive analytics, blockchain technology, and AI-driven tools are combining to shape the future of healthcare by guaranteeing improved results and efficient use of resources.

20.3.1 AI-Driven Diagnostic Tools: Advantages over Traditional Methods

AI minimizes human errors, improves diagnostic precision, and accelerates the decision-making process. AI-powered diagnostics reduce healthcare costs by streamlining workflows and eliminating redundant procedures. AI provides clinicians with real-time insights, assisting in detecting anomalies like fractures, tumors, and other medical conditions. ML algorithms analyze imaging data to generate accurate quantitative measurements, leading to better clinical outcomes.

20.3.2 AI in Medical Imaging and Healthcare Operations

AI enhances disease detection by analyzing complex imaging data, improving early diagnosis and intervention [27]. AI predicts patient admission rates, optimizes hospital resource allocation, and improves treatment efficiency. AI accelerates the identification of new therapeutic compounds, streamlining drug development and research processes.

20.3.3 Population Health Management and Technological Integration

PHM leverages AI to identify and address health risks at both individual and population levels. Health Information Exchange and EHRs: Seamless data sharing allows for the identification of high-risk patients, prediction of disease outbreaks, and targeted healthcare interventions. Integration of IT systems enhances patient-centered approaches, care coordination, and overall healthcare efficiency.

20.3.4 Predictive Analytics and Remote Monitoring

AI-driven predictive analytics identify individuals at risk for diseases, facilitating preventive care strategies. Smart devices improve data collection, enable remote monitoring, and contribute to early diagnosis and improved patient outcomes.

20.3.5 Blockchain for Healthcare Security and Privacy

Blockchain provides a decentralized and secure platform for managing medical records, ensuring patient privacy. Secure data-sharing capabilities across healthcare systems facilitate better coordination among medical professionals.

20.3.6 The Future of AI in Healthcare

Healthcare delivery is changing as a result of the confluence of EHRs, telemedicine, wearable technologies, AI-driven analytics, and blockchain. This will enhance patient outcomes, increase efficiency, and promote proactive health management [28].

20.4 ARTIFICIAL INTELLIGENCE APPLICATIONS IN HEALTHCARE

20.4.1 AI-Powered Patient Risk Prediction

AI-powered patient risk prediction uses sophisticated models to analyze extensive healthcare data and predict the probability of adverse health events [29]. A statistical technique called logistic regression forecasts the probability of a binary outcome, such as contracting an illness. In order to determine risk and guide focused interventions, it examines variables such test results, medical histories, and demographics. Random Forest is a ML method that uses multiple decision trees to predict outcomes. It's good at identifying complex patterns in healthcare data and predicting risk factors.

Gradient Boosting Machines is a ML method that builds multiple models that learn from each other's errors to improve accuracy. The model is effective at capturing complex patterns in healthcare data and predicting patient risk. ML techniques that can forecast patient risk include Support Vector Machines (SVM). By looking at their traits, they determine the best strategy to distinguish between patients who are at risk and those who are not. SVM excels in producing precise predictions and analyzing complicated data.

Convolutional neural networks (CNNs) and recurrent neural networks are two types of neural networks that are useful for forecasting patient risk and analyzing sequential healthcare data. In unstructured data, such as pictures, sensor readings, and medical records, they are able to spot intricate patterns.

Time series forecasting methods, such as Long Short Term Memory and Autoregressive Integrated Moving Average, examine trends in medical data over time to estimate patient risk. In order to forecast future health occurrences, disease development, and treatment outcomes, they utilize historical data.

20.5 ADVANTAGES OF AI-DRIVEN RISK PREDICTION

Early disease identification, personalized medication, better diagnostic precision, effective resource use, and preventive health measures are all made possible by risk prediction models in the healthcare industry. Early disease identification finds illnesses early, when they can be treated most successfully [29].

This approach proposes a medication adherence solution using an augmented reality (AR-headset) and Medication Coach Intelligent Agent. The evaluation showed that participants felt comfortable using the AR-headset for medication management tasks like taking pills and putting pills into pill dispensers, which are considered a useful tool for helping patients maintain self-management and medication adherence, potentially delaying or preventing the need for professional assistance. The drawbacks and limitations of the current AR-headset technology, such as short battery life and limited field of view. Additionally, possible drawbacks of using technology to replace human interaction, such as heightened loneliness and the possibility of receiving inaccurate medication information [30].

Wheeze Detection Using Convolutional Neural Networks for Respiratory Health Monitoring. Researchers developed a novel CNN architecture, termed WDCNN, for the accurate detection of wheezes in lung sound recordings. The approach, which leverages spectrograms and Mel-frequency cepstral coefficients as input features, consistently outperformed traditional ML methods in rigorous evaluations. Notably, using a lung sound dataset that was made publically available, WDCNN obtained a state-of-the-art area under the curve score of 0.96.

To assess the robustness of their method, researchers conducted extensive experiments under various conditions, including simulated lung sound shifts and the introduction of external noise. WDCNN demonstrated superior performance compared to baseline techniques, highlighting its resilience to real-world challenges.

The findings underscore the exceptional generalization capability and reliability of WDCNN, making it a promising candidate for deployment in both clinical settings and home-based applications. WDCNN's ability to accurately detect wheezes offers a valuable tool for respiratory health monitoring and early disease detection [31].

20.5.1 Multiple Disease Prediction System Tool

An innovative healthcare breakthrough called the numerous Disease Prediction System (MDPS) uses AI to determine a person's risk of developing numerous diseases before any symptoms appear. By shifting the focus from reactive treatment to proactive prevention, the MDPS aims to revolutionize healthcare by enabling early interventions and improved patient outcomes [32].

This system integrates data from various sources, such as genetic profiles, lifestyle habits, EHRs, and environmental influences, to predict the likelihood of multiple diseases. By identifying potential health risks in advance, MDPS represents a major advancement toward a future where diseases are not just managed but anticipated and prevented, significantly enhancing healthcare practices [32].

Studies have indicated that ML methods can enhance the early diagnosis and detection of diseases such as concomitant diabetes and breast cancer. Scientists have effectively identified patterns linked to various diseases by analyzing mammography pictures using techniques like Random Forest, deep CNNs, and generative adversarial networks. These models have demonstrated encouraging accuracy in differentiating between cancer patients and healthy people, highlighting the revolutionary potential of ML in contemporary healthcare [32].

20.5.2 AI-Powered Precision Medicine for Acute Kidney Injury

Critical care nephrology stands to benefit greatly from AI, which offers better clinical decision-making, risk prediction, individualized treatment, and greater early diagnosis of acute kidney injury (AKI). AI can find subphenotypes for focused therapies, and ML models can predict AKI up to 24–48 hours before changes in serum creatinine. Clinical notes can be automated and patient education materials can be produced using generative AI and large language models. However, thorough validation, ethical considerations, and data quality must all be addressed for deployment to be successful. Maximizing the advantages of AI in critical care nephrology requires constant training, education, and cooperation between nephrologists, intensivists, and AI specialists [33].

20.5.3 A Novel AI-Driven System for Brain Tumor Segmentation and Localization

The AI4Neurosurgery system is a novel AI-based system that significantly improves the accuracy and efficiency of brain tumor identification and localization. Its multi-modal approach, combining automated and interactive tools, offers a comprehensive solution for tumor segmentation and surgical planning. Preliminary testing indicates high usability and effectiveness. The system's features, including automatic segmentation and segment editing, have the potential to enhance patient care and surgical outcomes. In some situations, manual segmentation may still be required even if the system is generally easy to use. The AI4N seeks to improve brain tumour treatment's efficacy and efficiency [34].

AI in the medical field or healthcare creates and retains enormous amounts of private and sensitive medical data, necessitating strict data security and privacy regulations. Security breaches, insufficient access control, inadequate data encryption, and inadequate data retention policies are among the risks. Secure methods and procedures must be used by healthcare providers to ensure that patient data is protected from abuse or unauthorized access. Legal requirements like the HIPAA must be followed in order to protect patient data. Healthcare providers need to make sure patient privacy is maintained and that the data used to train and implement deep learning models is secure. Data integration continues to be challenged by technical and policy issues, with many healthcare organizations either reluctant to share patient data because of privacy and security concerns or lacking the necessary technical support [35].

20.6 PRECISION PREVENTION: AI-DRIVEN STRATEGIES FOR POPULATION HEALTH

The future of healthcare is becoming increasingly dependent on the efficient management of population health, shifting the focus from individual patient care to addressing the health needs of entire communities. In this area, AI is proven to be a game-changer, allowing for more accurate and proactive preventative measures. With its capacity to process and analyze vast amounts of data, AI can uncover hidden patterns and risk factors, allowing healthcare providers to stratify populations based on their likelihood of developing specific diseases or facing adverse health events. This data-driven approach facilitates more targeted interventions, ultimately enhancing overall health outcomes and improving preventative care efforts [36].

20.6.1 Keyways AI Can Aid in Precision

AI is capable of analyzing large datasets to find people who are more likely to have a particular disease or other health issue. This enables healthcare providers to target preventive interventions more effectively, reducing the likelihood of costly complications and hospitalizations. AI can assist in customizing interventions to meet the needs of each patient, guaranteeing that they receive the best care and treatment possible. Risks of unneeded operations and unfavourable outcomes can be decreased in this way. Early disease detection by AI-powered technologies enables more economical treatments. AI may, for instance, examine medical photos to find early indications of diseases like cancer.

20.6.2 Precise Risk Stratification

Enhancing preventative care and improving general health outcomes within a defined population are key goals of PHM. A key component of PHM is risk stratification, which groups people according to how likely they are to develop particular medical disorders. Age, gender, and family medical history are examples of demographic factors that have historically been used in conjunction with clinical records and basic risk-scoring models. However, these conventional approaches often lack the accuracy and depth required to fully optimize preventative care strategies [37].

AI offers a groundbreaking enhancement to risk stratification by leveraging advanced algorithms to analyze extensive datasets. Unlike traditional methods, AI can integrate information from diverse sources, including EHRs, genomic data, and lifestyle behaviors. By processing these multidimensional data points, AI can identify personalized risk factors and complex patterns that could go unnoticed otherwise, resulting in more accurate and successful health solutions [37].

20.6.3 AI-Driven Risk Stratification: A Framework

The application of AI in risk stratification typically involves the following steps (Figure 20.1):

i. Include all patients assigned to the health centre, not just those who have visited for care.

ii. Use the Uniform Data System High-Risk Conditions or a similar list to classify patients based on the number of health conditions they have.

iii. Group patients based on their number of conditions. This can create categories like "highly complex," "high-risk," "rising-risk," and "low-risk."

iv. Develop care models that match the specific needs of each risk group. For example, highly complex patients may require more intensive care, while low-risk patients may benefit from preventive care.

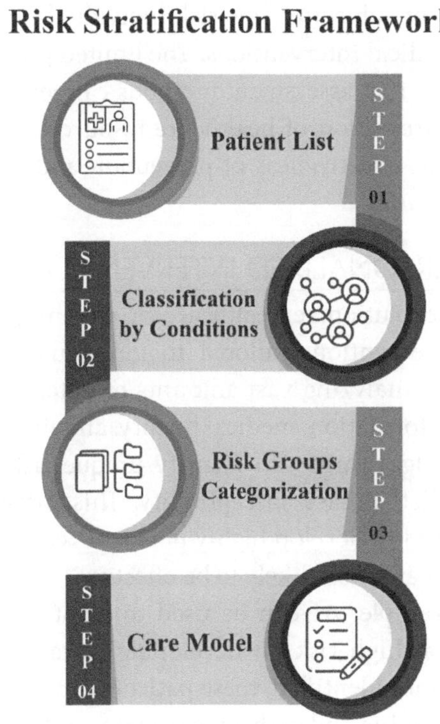

FIGURE 20.1 A brief summary for stepwise approach to implement AI in risk stratification.

By following these steps, healthcare centres can effectively stratify patients and provide targeted care to improve health outcomes [37].

20.6.4 Benefits of AI-Driven Risk Stratification

AI can integrate and analyze a wide range of data sources to precisely identify individuals who are at a high risk of developing specific diseases or health conditions [38]. Customized Interventions: AI-driven models tailor interventions to meet individual needs, enhancing the efficient and effective use of healthcare resources. AI helps healthcare providers allocate resources strategically, ensuring that high-risk populations receive appropriate care while minimizing waste. AI algorithms can be continuously refined and updated using new data and feedback, ensuring that risk assessment and interventions remain effective and up to date.

20.6.5 Limitations of Traditional Risk Stratification

Conventional methods primarily rely on easily accessible demographic and clinical data, overlooking other critical factors that influence health outcomes [38]. Basic risk scoring models may generate imprecise predictions, potentially overlooking individuals at high risk or subjecting low-risk individuals to unnecessary medical interventions. The limited precision of traditional risk assessment methods can lead to inefficient distribution of healthcare resources, reducing the overall effectiveness of prevention and intervention programs.

20.7 PERSONALIZED INTERVENTIONS

AI can revolutionize healthcare by enabling personalized interventions tailored to individual patients' needs. By analyzing vast amounts of data, including genetic information, medical history, and lifestyle factors, AI algorithms can identify unique risk factors and predict disease susceptibility. This information can be used to develop highly personalized treatment plans that are more likely to be effective.

For example, AI can be used to identify patients who are at high risk for developing certain types of cancer. Once identified, these patients can receive targeted screening and early intervention strategies, such as genetic testing or preventive surgeries. Additionally, AI can help determine the most effective treatment options for patients with chronic diseases, such as diabetes or heart disease. By considering a patient's individual characteristics, AI can recommend personalized treatment plans that are more likely to improve outcomes and reduce the risk of adverse events.

Moreover, personalized medicine can be enhanced through active patient engagement. By involving patients in their care decisions, it can lead to improved adherence to treatment plans and better health outcomes using AI-powered tools such as communication channels [36].

AI-driven patient engagement includes providing personalized health information: using AI to create personalized health reports and dashboards that patients can access easily; facilitating patient-provider communication: using AI-powered chatbots or virtual assistants to connect patients with healthcare providers and answer their questions; empowering patients to make informed decisions: patients can be empowered to make educated decisions by being informed about their treatment options and being encouraged to take part in the decision-making process; and tracking patient progress and outcomes: using AI to monitor patient progress and provide personalized feedback on their treatment plan.

20.8 EARLY DETECTION AND PREVENTION

20.8.1 AI in Early Disease Detection and Prevention

AI-powered tools (Figure 20.2) can play a crucial role in early detection and prevention of diseases. By analyzing medical images, EHRs, and other data, AI algorithms can identify early signs of diseases, even before symptoms appear. This early detection can enable timely interventions, potentially preventing the progression of diseases and reducing the need for more costly treatments [38].

20.8.2 Predicting Disease Risk with AI

AI improves risk prediction and illness detection by evaluating lifestyle and genetic factors, analyzing medical pictures, and facilitating early treatments to improve patient outcomes. AI can evaluate mammograms to find minor indications of breast cancer that human radiologists might miss [36]. Diabetic retinopathy screening – AI-driven retinal image analysis enables early detection of diabetic retinopathy, a major cause of blindness in diabetes patients [36]. AI assesses genetic and lifestyle factors to predict the likelihood of developing certain diseases [36]. AI-driven insights support early interventions, such as lifestyle modifications and screenings, to lower disease risk [36]. By leveraging AI for early diagnosis and prevention, healthcare providers can improve patient outcomes and reduce complications associated with chronic diseases.

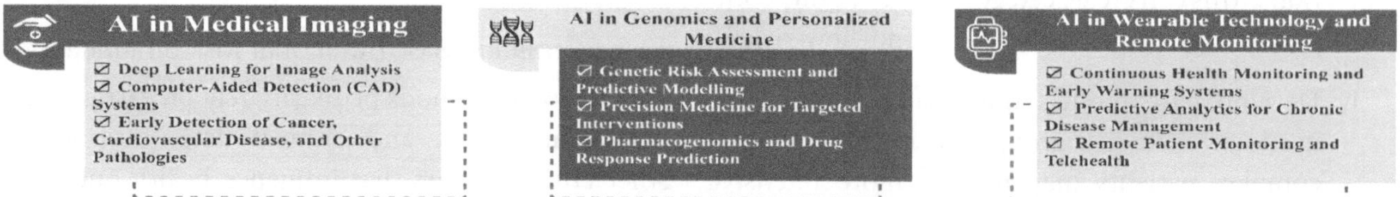

FIGURE 20.2 AI tools and their aims.

20.8.3 AI-Enabled Health Monitoring with Wearable Devices

AI enhances wearable devices by predicting health risks and monitoring health status, enabling early detection and timely interventions for improved outcomes. Efficient data analysis and AI and ML enable better processing of health data, offering real-time monitoring of vital signs, activity levels, and emotional states. AI improves data accuracy, reducing errors and refining health analytics for more precise monitoring.

AI-powered wearables contribute to disease prognosis, treatment feedback, and personalized health management. Wearables track heart rate [39], blood sugar levels, and sleep patterns [40], providing early warnings for conditions like atrial fibrillation, diabetes, and sleep apnea. Applications including fall detection, gait analysis, and sleep and mental health monitoring are supported by AI techniques like ML and deep learning [41].

AI-driven insights enable early interventions, such as medication adjustments, lifestyle modifications, or seeking medical care, reducing the risk of severe health complications [39–41]. Healthcare professionals and individuals can attain more accurate health monitoring, better illness prevention, and enhanced general well-being by combining AI with wearable technology.

20.9 AI – WEARABLE DEVICES IN CARDIOVASCULAR DISEASES

Wearable technologies are becoming essential tools in managing cardiovascular conditions, particularly hypertension and cardiac abnormalities. Frequent measurements enable detection of masked hypertension and blood pressure variability without disrupting daily routines. Continuous monitoring improves cardiovascular prognosis by allowing for timely treatment adjustments and better patient adherence.

Studies indicate that smartwatches with photoplethysmography technology can measure blood pressure as accurately as traditional devices, making them suitable for routine use. ECG analysis is enhanced by AI and ML, which makes it possible to identify arrhythmias and patterns of heart rate variability early. Ensures reliable ECG data even during physical activity, enhancing real-world usability. Monitoring blood pressure, temperature, activity levels, and respiration rate is made easier by AI-driven electrode designs and sensors. Data accessibility is improved for patients and healthcare practitioners through integration with cellphones and cloud platforms. AI-powered wearables promote early intervention, supporting proactive cardiovascular management. By integrating AI-driven wearables into clinical practice, healthcare providers can enhance cardiac care, improve patient outcomes, and advance preventive health strategies.

20.9.1 AI in Wearables: Challenges and Benefits

AI-enabled wearables offer significant potential for enhancing healthcare, but they also come with substantial challenges. Privacy and security risks – these devices collect sensitive health data (e.g., heart rate, sleep patterns, and location information), making them vulnerable to data breaches. Implementing robust encryption, data anonymization, and strict access controls is crucial to protect user information from hackers and malicious actors.

AI algorithms must be rigorously tested and validated to prevent false positives or false negatives, which could lead to incorrect medical decisions. AI models must account for diverse human physiology and behavior, ensuring that wearables perform accurately across different populations.

20.9.2 Opportunities and Benefits

Users receive instant feedback, empowering them to make healthier lifestyle choices and detect early signs of illness. Data collected from wearables can contribute to medical research and public health initiatives, enabling more effective disease prevention and management. Despite the challenges, AI-enabled wearables hold tremendous promise for personalized healthcare, early disease detection, and large-scale health studies. Addressing privacy, accuracy, and inclusivity concerns will be key to their widespread adoption and effectiveness.

20.10 CHALLENGES AND CONSIDERATIONS: RESPONSIBLE IMPLEMENTATION OF AI IN POPULATION HEALTH

The scope of PHM will undoubtedly be redefined by AI, the integration of this revolutionary technology into healthcare systems demands a careful assessment of various challenges [42]. These include the crucial

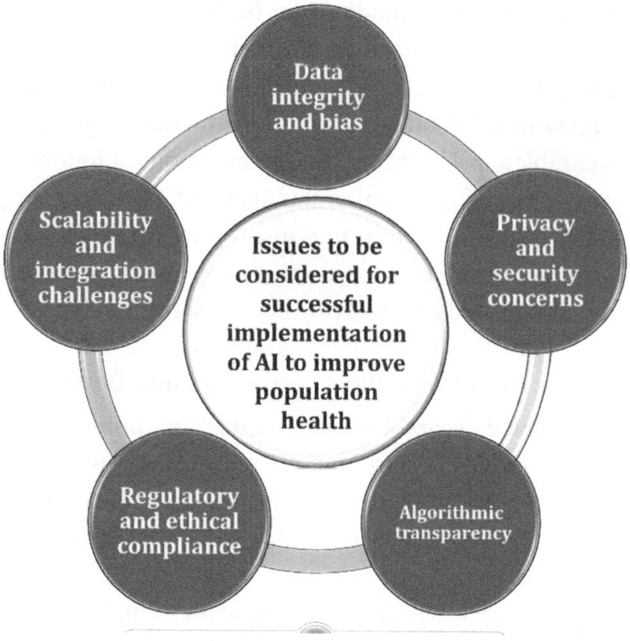

FIGURE 20.3 A schematic representation of the considerations before implementing artificial intelligence (AI) to improve population health.

concerns of algorithmic bias, fairness, data privacy, security, and the vital function of human oversight [43–45]. This section seeks to offer a thorough examination of these issues and worries, stressing the difficulties involved in implementing AI responsibly in health programs and the factors that need to be taken into account in order to reduce the hazards it presents. An overview of these considerations is highlighted in Figure 20.3.

20.10.1 Data Integrity and Bias

The reliability of AI technologies in healthcare is critically dependent on the integrity and diversity of the data used for development and training [22]. PHM relies on extensive data sources, including EHRs, wearable device outputs, and social determinants of health [46]. However, these data sources vary widely in accuracy, completeness, and formatting, presenting significant challenges in maintaining data integrity [47]. The variability in PHM data has profound implications beyond technical concerns, raising serious ethical issues [48].

A notable study by Obermeyer et al. exposed racial bias in AI-driven healthcare algorithms. An AI system deployed across multiple US health systems showed bias in selecting patients for high-risk care management programs [49]. The bias arose because the AI algorithm relied on healthcare costs as a proxy for health needs, unintentionally prioritizing populations with historically higher healthcare spending over those with greater medical needs [49]. Such data-driven biases distort AI decision-making, disproportionately affecting vulnerable populations and undermining both equity and trust in AI-driven PHM.

The direct consequences of AI biases are severe; these technologies can exacerbate existing disparities in healthcare access and outcomes. Abràmoff et al. proposed three key strategies to mitigate AI bias [10]: Ensuring collaboration between AI developers, regulators, and healthcare providers. Assessing AI models to identify and rectify biases before deployment. Implementing targeted interventions at every stage of AI development and application [50].

By embedding these strategies into AI development, healthcare stakeholders can reduce bias, enhance fairness, and improve health outcomes for diverse populations. Prioritizing health equity in PHM ensures that AI tools serve as instruments for reducing, rather than reinforcing, disparities in healthcare delivery [51].

20.10.2 Privacy and Security Concerns

There are serious privacy issues when integrating AI into healthcare systems because it necessitates maintaining enormous volumes of private information [42, 52]. As AI-driven technologies increasingly handle the collection, storage, and analysis of health information, the risks of unauthorized access and data breaches grow [44, 45, 53, 54]. The proliferation of health data touchpoints complicates the tracking of information flow, challenges consent processes, and risks infringing on patient autonomy [55]. Addressing these privacy challenges in PHM requires strict adherence to robust data protection standards [56, 57].

For instance, important principles like data minimization are enforced by the General Data Protection Regulation in Europe, which requires that just the bare minimum of personal data be processed, and purpose limitation, which limits the use of data to specified, explicit, and lawful purposes that the data subjects have consented to [58]. These regulations also strengthen individuals' rights over their data, ensuring greater control and transparency [58].

Additionally, safeguarding patient data in an AI-driven healthcare environment necessitates the implementation of strong cybersecurity measures [52]. These measures are essential to prevent potential data breaches and enhance the resilience of healthcare systems against cyber threats [59, 60]. Transparent data governance frameworks are also critical, as they establish clear guidelines for data access, processing, and sharing, Building trust between patients, medical professionals, and other interested parties [61, 62]. These frameworks further support AI accountability by making decision-making processes auditable and explainable [63].

Ultimately, embedding stringent privacy safeguards and governance principles into AI deployment in healthcare will help maximize AI's benefits while minimizing risks to privacy. This will enhance trust among all users, contributing to a more secure and equitable PHM [42, 62, 64].

20.10.3 Algorithmic Transparency

Many AI algorithms are opaque, sometimes referred to as "black boxes," which makes it difficult to integrate them into clinical settings where choices can have a big impact on patient care [42, 65, 66]. In healthcare, transparency is not just advantageous but essential for ensuring accountability and maintaining trust [67]. This is particularly crucial when AI-driven decisions directly affect patient outcomes, necessitating a level of openness that allows healthcare professionals to scrutinize and understand the decision-making process [68, 69].

The field of Explainable AI (XAI) has gained traction, emphasizing the development of AI models that are not only technically advanced but also interpretable by human experts [70]. A notable example of XAI's potential is in the management of diabetic retinopathy, where AI systems have been created to predict and diagnose the condition [71–73]. Research, such as that by Quellec et al., highlights how enhancing transparency in AI algorithms fosters trust among medical professionals while also empowering patients, ultimately promoting a more collaborative and informed decision-making process [74]. XAI adds value by offering insights into the reasoning behind diagnostic conclusions, thereby increasing the acceptance and trustworthiness of AI tools in clinical applications, including PHM [75].

However, achieving AI algorithmic transparency remains a complex challenge, primarily due to the inherent trade-off between model performance and explainability [76]. More sophisticated AI models may provide greater accuracy but often do so at the cost of interpretability, creating a dilemma for both developers and healthcare practitioners [77]. Thus, maintaining a balance between these two factors remains a core challenge in AI implementation within healthcare [78]. Continuous innovation in AI development methodologies is necessary to ensure that both accuracy and transparency are upheld [79]. Ultimately, a key goal is to guarantee that AI deployment in PHM aligns with ethical standards and transparency requirements, making healthcare applications both effective and trustworthy [62].

20.10.4 Regulatory and Ethical Compliance

The varying healthcare regulatory environments across different regions present a significant challenge to integrating AI in PHM [36, 80]. As AI technology continues to advance, constantly learning and adapting, regulatory frameworks must be flexible and forward-thinking to effectively address these evolving challenges [81]. From an ethical perspective, the use of AI in healthcare should strictly adhere to fundamental principles, including beneficence, non-maleficence, respect for patient choice, and fairness in the allocation of health resources [75]. This means that AI tools should not only enhance healthcare efficiency but must also be developed in a way that safeguards care quality and does not exacerbate existing health disparities [82]. Additionally, these AI-driven tools should empower patients to make informed decisions, reinforcing the ethical principle of respecting individual autonomy [83].

Furthermore, fairness dictates that AI implementation in healthcare should be inclusive and equitable, preventing biases or discrimination that could result in disparities in healthcare access [84]. This is particularly crucial in PHM, where the primary objective is to improve health outcomes at both community and societal levels [21]. Therefore, the establishment of comprehensive ethical guidelines tailored specifically for AI in PHM is imperative. These rules would give healthcare workers and AI developers explicit direction, guaranteeing that AI solutions are morally good, adhere to legal requirements, and reflect societal values [85].

20.10.5 Scalability and Integration Challenges

Expanding AI solutions from pilot projects to widespread clinical use presents significant challenges. Integrating these technologies into existing healthcare systems often requires substantial workflow modifications, extensive staff training, and updates to institutional policies [86]. Additionally, the variability of healthcare systems across different regions further complicates scalability, making it difficult to implement a universal solution effectively [87, 88].

Successfully transitioning pilot AI applications into real-world PHM settings demands considerable financial investment and a multidisciplinary approach involving clinicians, IT specialists, data scientists, and organizational leaders [20, 89]. Therefore, effectively integrating and scaling AI tools in healthcare requires adapting these technologies to the unique needs of different health systems while fostering an environment of continuous learning and innovation [90]. Collaboration across disciplines is crucial to ensuring that AI enhances, rather than disrupts, healthcare delivery [91].

Given these complexities, the responsible implementation of AI in PHM is a multifaceted challenge that involves technical, ethical, and practical considerations. Addressing these issues necessitates a unified effort from policymakers, healthcare professionals, AI developers, and the broader public. Only through this collective and engaged approach can AI fully achieve its transformative potential in healthcare.

20.11 SUMMARY

This chapter explores the transformative potential of AI in PHM. PHM aims to improve health outcomes by identifying at-risk individuals and implementing targeted interventions. Traditionally, risk stratification relied on basic demographic and clinical data, leading to imprecise interventions. AI provides a solution by identifying complicated risk variables and enabling more accurate risk stratification through the analysis of large datasets, such as genetic data, lifestyle factors, and EHRs.

AI can revolutionize healthcare through personalized medicine and early disease detection. By analyzing individual data, AI can create personalized treatment plans and identify early signs of disease before symptoms appear. This allows for timely intervention, potentially preventing disease progression and reducing healthcare costs.

However, there are obstacles to integrating AI in PHM, including algorithmic bias, privacy issues, and data availability and quality. Healthcare organizations must invest in cybersecurity measures, establish strong data governance frameworks, and guarantee algorithmic transparency in order to address these problems. Overcoming these obstacles will enable healthcare systems to fully utilize AI's promise to enhance population health outcomes, build a more just and effective healthcare system, and eventually improve everyone's health.

The future of AI in healthcare suggests XAI solutions for more transparency, federated learning for cooperative AI model training without revealing sensitive data, and combining AI and the Internet of Things (IoT) for remote patient monitoring and chronic illness management. AI has the ability to transform healthcare delivery by overcoming these obstacles and providing more transparent and private solutions for better population health outcomes.

REFERENCES

1. Smith L. www.snhu.edu. 2021. What is Population Health Management? Available from: https://www.snhu.edu/about-us/newsroom/health/what-is-population-health-management
2. Steenkamer B, Baan C, Putters K, Van Oers H, Drewes H. Population health management guiding principles to stimulate collaboration and improve pharmaceutical care. JHOM. 2018 Apr 6;32(2):224–45.
3. Mitri J, Gabbay R. Understanding Population Health Through Diabetes Population Management. Endocrinol Metabolism Clinics North Am. 2016 Dec;45(4):933–42.
4. NHS England. www.england.nhs.uk. 2023. NHS England Populationhealthmanagement.Availablefrom: https://www.england.nhs.uk/long-read/population-health-management/
5. Van Ede AFTM, Minderhout RN, Stein KV, Bruijnzeels MA. How to successfully implement population health management: A scoping review. BMC Health Serv Res. 2023 Aug 25;23(1):910.
6. Khan MohdR, Malik MA, Akhtar SN, Yadav S, Patel R. Multimorbidity and its associated risk factors among older adults in India. BMC Public Health. 2022 Dec;22(1):746.
7. www.who.int [Internet]. Population health management in primary health care: A proactive approach to improve health and well-being: Primary health care policy paper series. Available from: https://www.who.int/europe/publications/i/item/WHO-EURO-2023-7497-47264-69316

8. Chelak K, Chakole S. The role of social determinants of health in promoting health equality: A narrative review. Cureus [Internet]. 2023 Jan 5 [cited 2025 Sep 19]; Available from: https://www.cureus.com/articles/119651-the-role-of-social-determinants-of-health-in-promoting-health-equality-a-narrative-review
9. Van Weel C, Kidd MR. Why strengthening primary health care is essential to achieving universal health coverage. CMAJ. 2018 Apr 16;190(15):E463–6.
10. Ehrenstein V, Kharrazi H, Lehmann H, Taylor CO. Obtaining data from electronic health records [Internet]. Agency for Healthcare Research and Quality (US); 2020. (www.ncbi.nlm.nih.gov). Available from: https://www.ncbi.nlm.nih.gov/books/NBK551878/
11. Tsai CH, Eghdam A, Davoody N, Wright G, Flowerday S, Koch S. Effects of electronic health record implementation and barriers to adoption and use: A scoping review and qualitative analysis of the content. Life. 2020 Dec 4;10(12):327.
12. Edemekong PF, Pavan Annamaraju, Afzal M, Haydel MJ. Nih.gov. StatPearls Publishing; 2024. Health Insurance Portability and Accountability Act (HIPAA) Compliance. Available from: http://www.ncbi.nlm.nih.gov/books/NBK500019/
13. McGraw D, Mandl KD. Privacy protections to encourage use of health-relevant digital data in a learning health system. NPJ Digit Med. 2021 Jan 4;4(1):2.
14. Cowie J, Nicoll A, Dimova ED, Campbell P, Duncan EA. The barriers and facilitators influencing the sustainability of hospital-based interventions: A systematic review. BMC Health Serv Res. 2020 Dec;20(1):588.
15. Ghasemiardekani M, Willetts G, Hood K, Cross W. The effectiveness of chronic disease management planning on self-management among patients with diabetes at general practice settings in Australia: A scoping review. BMC Prim Care. 2024 Mar 1;25(1):75.
16. Krist AH, Tong ST, Aycock RA, Longo DR. Engaging patients in decision-making and behavior change to promote prevention. Stud Health Technol Inform. 2017;240:284–302.
17. Househ M, Grainger R, Petersen C, Bamidis P, Merolli M. Balancing between privacy and patient needs for health information in the age of participatory health and social media: A scoping review. Yearb Med Inform. 2018 Aug;27(01):029–36.
18. Turner HC, Hori Y, Revill P, Rattanavipapong W, Arai K, Nonvignon J, et al. Analyses of the return on investment of public health interventions: A scoping review and recommendations for future studies. BMJ Glob Health. 2023 Aug;8(8):e012798.
19. Mengstie MM, Biks GA, Cherlin EJ, Curry LA. Organizational culture and barriers to change in University of Gondar Comprehensive Specialized Hospital Cardiac Unit. BMC Health Serv Res. 2023 Mar 28;23(1):296.
20. Alowais SA, Alghamdi SS, Alsuhebany N, Alqahtani T, Alshaya AI, Almohareb SN, et al. Revolutionizing healthcare: The role of artificial intelligence in clinical practice. BMC Med Educ. 2023 Sep 22;23(1):689.
21. Ali O, Abdelbaki W, Shrestha A, Elbasi E, Alryalat MAA, Dwivedi YK. A systematic literature review of artificial intelligence in the healthcare sector: Benefits, challenges, methodologies, and functionalities. J Innovation Knowledge. 2023 Jan;8(1):100333.
22. Collins C, Dennehy D, Conboy K, Mikalef P. Artificial intelligence in information systems research: A systematic literature review and research agenda. Int J Inf Manag. 2021 Oct;60:102383.
23. Taye MM. Understanding of machine learning with deep learning: Architectures, workflow, applications and future directions. Computers. 2023 Apr 25;12(5):91.
24. Sim J ah, Huang X, Horan MR, Stewart CM, Robison LL, Hudson MM, et al. Natural language processing with machine learning methods to analyze unstructured patient-reported outcomes derived from electronic health records: A systematic review. Artificial Intelligence in Medicine. 2023 Dec;146:102701.
25. Olawade DB, Wada OJ, David-Olawade AC, Kunonga E, Abaire O, Ling J. Using artificial intelligence to improve public health: A narrative review. Front Public Health. 2023 Oct 26;11:1196397.
26. Adarsh DK. inHealth Strategies. 2024 [cited 2025 Jan 1]. Guide to AI in population health management: Revolutionizing healthcare. Available from: https://inhealth4change.com/guide-to-ai-in-population-health-management-revolutionizing-healthcare/
27. Lainjo B. A meta-study on optimizing healthcare performance with artificial intelligence and machine learning. J Autonom Intell. 2024 Mar 7;7(5):1535.
28. Tak A. Population health management: Leveraging it for better patient outcome's. IJHS. 2024 Apr 7;7(2):53–67.
29. Selvaraj S. Utilizing artificial intelligence for patient risk predictions: Empowering doctors with data-driven insights. IJSR. 2024 Jul 5;13(7):371–6.
30. Koch F. Artificial intelligence in health: First International Workshop, AIH 2018, Stockholm, Sweden, July 13-14, 2018, Revised Selected Papers. Cham: Springer International Publishing AG; 2019. 1 p. (Lecture Notes in Computer Science Ser).
31. Gama J, Lopes Cardoso H, Oliveira E, Vale Z, editors. Progress in Artificial Intelligence: 18th EPIA Conference on Artificial Intelligence, EPIA 2017, Porto, Portugal, September 5–8, 2017, Proceedings. 1st ed. 2017. Cham: Springer International Publishing : Imprint: Springer; 2017. 1 p. (Lecture Notes in Artificial Intelligence).
32. Kallepalli R, Pasumarthi N, Javvadi H, Kodithala R, Swaroop S. Multi-disease prediction system using machine learning. Int Res J Modern Eng Technol Sci [Internet]. 2024 Feb 22 [cited 2025 Sep 19]; Available from: https://www.irjmets.com/uploadedfiles/paper//issue_2_february_2024/49550/final/fin_irjmets1708595435.pdf

33. Cheungpasitporn W, Thongprayoon C, Kashani KB. Advances in critical care nephrology through artificial intelligence. Curr Opin Crit Care. 2024 Dec;30(6):533–41.
34. Reddy Kalli VD. Creating an AI-powered platform for neurosurgery alongside a usability examination: Progressing towards minimally invasive robotics. JAIGS. 2024 Apr 19;3(1):256–68.
35. Li YH, Li YL, Wei MY, Li GY. Innovation and challenges of artificial intelligence technology in personalized healthcare. Sci Rep. 2024 Aug 16;14(1):18994.
36. Davenport T, Kalakota R. The potential for artificial intelligence in healthcare. Future Healthcare J. 2019 Jun;6(2):94–8.
37. https://www.nachc.org [Internet]. 2025 [cited 2025 Jan 1]. Population health management risk stratification. Available from: https://www-nachc-org.webpkgcache.com/doc/-/s/www.nachc.org/wp-content/uploads/2023/07/Action-Guide_Risk-Stratification.pdf
38. Mohsin SN, Gapizov A, Ekhator C, Ain NU, Ahmad S, Khan M, et al. The role of artificial intelligence in prediction, risk stratification, and personalized treatment planning for congenital heart diseases. Cureus [Internet]. 2023 Aug 30 [cited 2025 Sep 19]; Available from: https://www.cureus.com/articles/181414-the-role-of-artificial-intelligence-in-prediction-risk-stratification-and-personalized-treatment-planning-for-congenital-heart-diseases
39. Cleveland Clinic. Cleveland Clinic. 2022. Heart Rate Monitors: How They Work and Accuracy. Available from: https://my.clevelandclinic.org/health/diagnostics/23429-heart-rate-monitor
40. De Zambotti M, Cellini N, Goldstone A, Colrain IM, Baker FC. Wearable sleep technology in clinical and research settings. Med Sci Sports Exercise. 2019 Jul;51(7):1538–57.
41. Bhaltadak V, Ghewade B, Yelne S. A comprehensive review on advancements in wearable technologies: Revolutionizing cardiovascular medicine. Cureus [Internet]. 2024 May 29 [cited 2025 Sep 19]; Available from: https://www.cureus.com/articles/244512-a-comprehensive-review-on-advancements-in-wearable-technologies-revolutionizing-cardiovascular-medicine
42. Sallam M. ChatGPT utility in healthcare education, research, and practice: Systematic review on the promising perspectives and valid concerns. Healthcare. 2023 Mar 19;11(6):887.
43. https://jjournals.ju.edu.jo/ [Internet]. Jordan Medical Journal; 2024. View of Ensuring Security and Privacy in Healthcare Systems: A Review Exploring Challenges, Solutions, Future Trends, and the Practical Applications of Artificial Intelligence.
44. Ali G, M. Mijwil M. Cybersecurity for sustainable smart healthcare: State of the art, taxonomy, mechanisms, and essential roles. Mesopotamian J CyberSecurity. 2024 May 23;4(2):20–62.
45. Mijwil M, Mohammad Aljanabi, Ahmed Hussein Ali. ChatGPT: Exploring the role of cybersecurity in the protection of medical information. Mesopotamian J CyberSecurity. 2023 Feb 1;2023:18–21.
46. Nasir K, Javed Z, Khan SU, Jones SL, Andrieni J. Big data and digital solutions: Laying the Foundation for Cardiovascular Population Management^{CME}. Methodist DeBakey Cardiovascular J. 2020 Oct 1;16(4):272.
47. Syed R, Eden R, Makasi T, Chukwudi I, Mamudu A, Kamalpour M, et al. Digital Health Data Quality Issues: Systematic review. J Med Internet Res. 2023 Mar 31;25:e42615.
48. Bovens D, Van Baarle E, Ziesemer K, Molewijk B. The ethical dimension of personal health monitoring in the armed forces: a scoping review. BMC Med Ethics. 2024 Aug 10;25(1):88.
49. Obermeyer Z, Powers B, Vogeli C, Mullainathan S. Dissecting racial bias in an algorithm used to manage the health of populations. Science. 2019 Oct 25;366(6464):447–53.
50. Abràmoff MD, Tarver ME, Loyo-Berrios N, Trujillo S, Char D, Obermeyer Z, et al. Considerations for addressing bias in artificial intelligence for health equity. NPJ Digit Med. 2023 Sep 12;6(1):170.
51. Gurevich E, El Hassan B, El Morr C. Equity within AI systems: What can health leaders expect? Healthc Manage Forum. 2023 Mar;36(2):119–24.
52. Williamson SM, Prybutok V. Balancing privacy and progress: A review of privacy challenges, systemic oversight, and patient perceptions in AI-driven healthcare. Applied Sci. 2024 Jan 12;14(2):675.
53. Bhagat SV, Kanyal D. Navigating the future: The transformative impact of artificial intelligence on hospital management: A comprehensive review. Cureus [Internet]. 2024 Feb 20 [cited 2025 Sep 19]; Available from: https://www.cureus.com/articles/219582-navigating-the-future-the-transformative-impact-of-artificial-intelligence-on-hospital-management–a-comprehensive-review
54. Nemec Zlatolas L, Welzer T, Lhotska L. Data breaches in healthcare: Security mechanisms for attack mitigation. Cluster Comput. 2024 Oct;27(7):8639–54.
55. Naik N, Hameed BMZ, Shetty DK, Swain D, Shah M, Paul R, et al. Legal and ethical consideration in artificial intelligence in healthcare: Who takes responsibility? Front Surg. 2022 Mar 14;9:862322.
56. Al Kuwaiti A, Nazer K, Al-Reedy A, Al-Shehri S, Al-Muhanna A, Subbarayalu AV, et al. A review of the role of artificial intelligence in healthcare. JPM. 2023 Jun 5;13(6):951.
57. Murdoch B. Privacy and artificial intelligence: challenges for protecting health information in a new era. BMC Med Ethics. 2021 Dec;22(1):122.
58. Hoofnagle CJ, Van Der Sloot B, Borgesius FZ. The European Union general data protection regulation: What it is and what it means. Inf Commun Technol Law. 2019 Jan 2;28(1):65–98.

59. Ibrahim A, Thiruvady D, Schneider JG, Abdelrazek M. The challenges of leveraging threat intelligence to stop data breaches. Front Comput Sci. 2020 Aug 28;2:36.
60. Murthy R. Exploring the use of PHM for Software System Security and Resilience. In: 2023 IEEE International Conference on Prognostics and Health Management (ICPHM) [Internet]. Montreal, QC, Canada: IEEE; 2023 [cited 2025 Sep 19]. p. 67–72. Available from: https://ieeexplore.ieee.org/document/10193932/
61. Liakh O. Accountability through sustainability data governance: Reconfiguring reporting to better account for the digital acceleration. Sustainability. 2021 Dec 14;13(24):13814.
62. Mennella C, Maniscalco U, De Pietro G, Esposito M. Ethical and regulatory challenges of AI technologies in healthcare: A narrative review. Heliyon. 2024 Feb;10(4):e26297.
63. Janssen M, Brous P, Estevez E, Barbosa LS, Janowski T. Data governance: Organizing data for trustworthy artificial intelligence. Gov Inf Q. 2020 Jul;37(3):101493.
64. Wang C, Zhang J, Lassi N, Zhang X. Privacy protection in using artificial intelligence for healthcare: Chinese regulation in comparative perspective. Healthcare. 2022 Sep 27;10(10):1878.
65. Karalis VD. The integration of artificial intelligence into clinical practice. Applied Biosci. 2024 Jan 1;3(1):14–44.
66. The Lancet Digital Health. ChatGPT: friend or foe? The Lancet Digital Health. 2023 Mar;5(3):e102.
67. Abor PA, Tetteh CK. Accountability and transparency: Is this possible in hospital governance? Cogent Business Manag. 2023 Dec 11;10(3):2266188.
68. Bekbolatova M, Mayer J, Ong CW, Toma M. Transformative potential of AI in healthcare: Definitions, applications, and navigating the ethical landscape and public perspectives. Healthcare. 2024 Jan 5;12(2):125.
69. Wang W, Wang Y, Chen L, Ma R, Zhang M. Justice at the forefront: Cultivating felt accountability towards Artificial Intelligence among healthcare professionals. Social Sci Med. 2024 Apr;347:116717.
70. Ali S, Abuhmed T, El-Sappagh S, Muhammad K, Alonso-Moral JM, Confalonieri R, et al. Explainable artificial intelligence (XAI): What we know and what is left to attain trustworthy artificial intelligence. Information Fusion. 2023 Nov;99:101805.
71. Grzybowski A, Brona P, Lim G, Ruamviboonsuk P, Tan GSW, Abramoff M, et al. Artificial intelligence for diabetic retinopathy screening: A review. Eye. 2020 Mar 1;34(3):451–60.
72. Kong M, Song SJ. Artificial intelligence applications in diabetic retinopathy: What we have now and what to expect in the future. Endocrinol Metab. 2024 Jun 30;39(3):416–24.
73. Raman R, Dasgupta D, Ramasamy K, George R, Mohan V, Ting D. Using artificial intelligence for diabetic retinopathy screening: Policy implications. Indian J Ophthalmol. 2021 Nov;69(11):2993–8.
74. Quellec G, Al Hajj H, Lamard M, Conze PH, Massin P, Cochener B. ExplAIn: Explanatory artificial intelligence for diabetic retinopathy diagnosis. Medical Image Analysis. 2021 Aug;72:102118.
75. Islam MR, Ahmed MU, Barua S, Begum S. A systematic review of explainable artificial intelligence in terms of different application domains and tasks. Applied Sci. 2022 Jan 27;12(3):1353.
76. Morandini S, Fraboni F, Balatti E, Hackmann A, Brendel H, Puzzo G, et al. Assessing the transparency and explainability of AI algorithms in planning and scheduling tools: A review of the literature. Human Interaction & Emerging Technologies (IHIET 2023): Artificial Intelligence & Future Applications 2023 [cited 2025 Sep 20]. Available from: https://openaccess.cms-conferences.org/publications/book/978-1-958651-87-2/article/978-1-958651-87-2_65
77. Hassija V, Chamola V, Mahapatra A, Singal A, Goel D, Huang K, et al. Interpreting black-box models: A review on explainable artificial intelligence. Cogn Comput. 2024 Jan;16(1):45–74.
78. MacDonald S, Steven K, Trzaskowski M. Interpretable AI in Healthcare: Enhancing Fairness, Safety, and Trust. In: Raz M, Nguyen TC, Loh E, editors. *Artificial Intelligence in Medicine* [Internet]. Singapore: Springer Nature Singapore; 2022 [cited 2025 Sep 20]. p. 241–58. Available from: https://link.springer.com/10.1007/978-981-19-1223-8_11
79. Aldoseri A, Al-Khalifa KN, Hamouda AM. Re-thinking data strategy and integration for artificial intelligence: Concepts, opportunities, and challenges. Applied Sci. 2023 Jun 13;13(12):7082.
80. Wubineh BZ, Deriba FG, Woldeyohannis MM. Exploring the opportunities and challenges of implementing artificial intelligence in healthcare: A systematic literature review. Urologic Oncology: Seminars and Original Investigations. 2024 Mar;42(3):48–56.
81. Bajwa J, Munir U, Nori A, Williams B. Artificial intelligence in healthcare: transforming the practice of medicine. Future Healthcare J. 2021 Jul;8(2):e188–94.
82. Siala H, Wang Y. SHIFTing artificial intelligence to be responsible in healthcare: A systematic review. Social Science & Medicine. 2022 Mar;296:114782.
83. Funer F, Wiesing U. Physician's autonomy in the face of AI support: walking the ethical tightrope. Front Med. 2024 Mar 28;11:1324963.
84. Ueda D, Kakinuma T, Fujita S, Kamagata K, Fushimi Y, Ito R, et al. Fairness of artificial intelligence in healthcare: review and recommendations. JPN J Radiol. 2024 Jan;42(1):3–15.
85. Solanki P, Grundy J, Hussain W. Operationalising ethics in artificial intelligence for healthcare: A framework for AI developers. AI Ethics. 2023 Feb;3(1): 223–40.

86. Ahmed MI, Spooner B, Isherwood J, Lane M, Orrock E, Dennison A. A systematic review of the barriers to the implementation of artificial intelligence in healthcare. Cureus [Internet]. 2023 Oct 4 [cited 2025 Sep 20]; Available from: https://www.cureus.com/articles/170025-a-systematic-review-of-the-barriers-to-the-implementation-of-artificial-intelligence-in-healthcare
87. Haefner N, Parida V, Gassmann O, Wincent J. Implementing and scaling artificial intelligence: A review, framework, and research agenda. Technological Forecasting and Social Change. 2023 Dec;197:122878.
88. Maleki Varnosfaderani S, Forouzanfar M. The Role of AI in Hospitals and Clinics: Transforming Healthcare in the 21st Century. Bioengineering. 2024 Mar 29;11(4):337.
89. Yelne S, Chaudhary M, Dod K, Sayyad A, Sharma R. Harnessing the Power of AI: A Comprehensive Review of Its Impact and Challenges in Nursing Science and Healthcare. Cureus [Internet]. 2023 Nov 22 [cited 2025 Sep 20]; Available from: https://www.cureus.com/articles/206741-harnessing-the-power-of-ai-a-comprehensive-review-of-its-impact-and-challenges-in-nursing-science-and-healthcare
90. Roppelt JS, Kanbach DK, Kraus S. Artificial intelligence in healthcare institutions: A systematic literature review on influencing factors. Technology in Society. 2024 Mar;76:102443.
91. Lämmermann L, Hofmann P, Urbach N. Managing artificial intelligence applications in healthcare: Promoting information processing among stakeholders. International Journal of Information Management. 2024 Apr;75:102728.

CHAPTER 21

AI-Powered Clinical Trials

Optimizing Research Process

Omar Alomari, Rumeysa Yegin, and Habiba Eyvazova

21.1 INTRODUCTION

Artificial intelligence (AI) is a new technical discipline that is widely accepted and adopted in many fields such as medicine, media, and e-commerce [1]. AI in the medical field is often implemented on computational tools that mimic the mechanisms supporting deep learning, interaction, and adaptability of human intelligence [2]. AI research conducted in recent years reveals that the medicine has a more promising output-input ratio compared to other areas. The combination of AI and medicine is changing the traditional medical model [3]. AI such as machine learning (ML), is acquiring increasing attention in clinical study and is engaging the clinical research landscape [4].

When a clinician wants to diagnose a patient with a particular disease/condition, thanks to AI, the duration needed to diagnose an issue can be significantly reduced and accuracy of diagnosis can be greatly increased [3]. AI also has the ability to influence efficiency and design of CTs. For instance, numerous applications are possible for AI in extracting and organizing data from electronic health records (EHRs) and medical statements. AI can also be used to track and enhance patient adherence during CT scans. Tools like smartphone notifications and reminders, missed clinic appointment tracking and electronic tracking of medications can help control this [4].

AI is also used to assist with image analysis recently. It contributes to more accurate examination of computed tomography scans and X-rays, which are medical imaging methods, and to diagnose faster and more effectively [5]. The application of AI and ML, especially through deep learning methods, brings numerous advantages to processing medical images by enabling the automation of tasks that were historically processed manually, such as organs, tissues segmentation. Today, algorithms outperform radiologists at identifying cancerous tumors and are helping researchers figure out how to create cohorts for costly CTs [6]. Additionally, using automated AI tools will minimize human interaction, reduce human-caused changeability, and reduce the required number of patients with higher statistical power [7].

There is a wealth of research arguing that AI is capable of doing fundamental healthcare activities, such as disease diagnosis, just as well as or even better than humans [6]. The use of AI in healthcare is through constructing algorithms that are able to examine big quantities of medical data and find trends that humans cannot easily notice. This can contribute to recognizing the early warning indicators of diseases or conditions and allow for earlier diagnosis, treatment, and prognosis assessment [8].

CTs are research studies that test new medical interventions, treatments, or devices on patients in order to assess their efficacy, safety, and possible adverse effects. These trials are essential for expanding our understanding of medicine and enhancing patient care. By systematically testing new therapies in controlled environments, researchers can determine the effective methods for avoiding, diagnosing, and treating various diseases and conditions. CTs also help ensure that novel therapies are safe and effective before they become widely available to the public. Through rigorous scientific methodology, CTs contribute to evidence-based medicine, leading to improved health results and the development of innovative therapeutic solutions.

DOI: 10.1201/9781003533351-21

The usage of AI in CTs has revolutionized the way research is conducted, offering numerous benefits that enhance the efficiency and accuracy of these studies. AI technologies, including ML algorithms and natural language processing (NLP), are employed to quickly evaluate large volumes of data, identify patterns, and predict results. This enables researchers to streamline patient acquisition through the selection of appropriate candidates more effectively and ensuring diverse and representative study populations. AI also assists in monitoring patient adherence and managing large datasets, lowering the possibility of human mistake and enhancing the reliability of trial results. Furthermore, AI-driven predictive modeling can accelerate the process of finding new drugs by identifying promising compounds and optimizing trial designs. Overall, the integration of AI into CTs promises to accelerate medical advancements and bring innovative treatments to patients faster.

The main aim of this book chapter is to investigate the transformative effect of AI on CTs, highlighting its potential to enhance different aspects of the research process. The chapter will delve into how AI technologies are revolutionizing patient recruitment, data analysis, and trial monitoring, ultimately raising the level of effectiveness, accuracy, and reliability of clinical studies. By examining the current applications and latest advancements, this chapter aims to provide a comprehensive understanding of the advantages and difficulties involved in incorporating AI into CTs, as well as its future implications for medical studies and patient care.

21.2 THE ROLE OF AI IN ADDRESSING CHALLENGES IN CTS

AI is playing a transformative role in addressing several key challenges inherent in CTs, such as data constraints and management, labor-intensive processes, remote patient monitoring needs, and the complexities of traditional trial methodologies. One of the primary challenges in CTs is managing the enormous volumes of data generated from diverse sources, including patient records, lab findings, imaging data, and more [9, 10]. This data, often heterogeneous and unstructured, poses significant hurdles in terms of integration, consistency, and analysis. AI, particularly ML algorithms, excels at processing and analyzing large datasets, enabling researchers to streamline data integration, ensure data consistency, and extract actionable insights with greater accuracy (Figure 21.1). Predictive analytics, powered by AI, can identify potential issues in real-time, allowing for proactive interventions and enhancing overall trial quality.

Labor-intensive processes represent another major challenge in CTs. Tasks such as patient recruitment, data entry, monitoring, and compliance checks are traditionally manual, time-consuming, and prone to human error [11]. AI technologies, including NLP and robotic process automation (RPA), offer robust solutions to these problems [12–14]. NLP can efficiently sift through medical records to identify suitable candidates for trials, drastically reducing the time and effort required for patient recruitment [12, 13]. RPA can also automate repetitive tasks like data entry, ensuring higher accuracy and releasing priceless personnel resources to enable more complex tasks [12–14]. As shown in Figure 21.1, these AI-driven automations not only enhance efficiency but also significantly reduce the likelihood of errors, thus improving the reliability of trial outcomes.

The need for remote patient monitoring has become more pronounced, especially in light of the COVID-19 pandemic, which emphasized the crucial importance of minimizing in-person visits to ensure patient safety. AI-driven wearable devices and telehealth platforms have emerged as vital tools in this context. These technologies enable constant observation of patients' vital signs, medication regimen adherence, and symptom progression from the comfort of their homes [15]. The data collected is analyzed in real-time using AI, providing healthcare professionals with timely insights and enabling prompt interventions. This not only improves patient compliance but also guarantees that any adverse events are detected and addressed swiftly (Figure 21.1).

The traditional complexities of CTs, including their length, cost, and the extensive coordination required among multiple stakeholders, are also being mitigated by AI. Designing an effective CT involves simulating various scenarios to identify the optimal approach. As shown in Figure 21.1, AI can streamline this process by using historical data and ML models to predict patient outcomes and refine inclusion criteria, thereby optimizing trial design and reducing trial duration [16–18]. Moreover, AI-powered project management tools enable smooth communication and teamwork

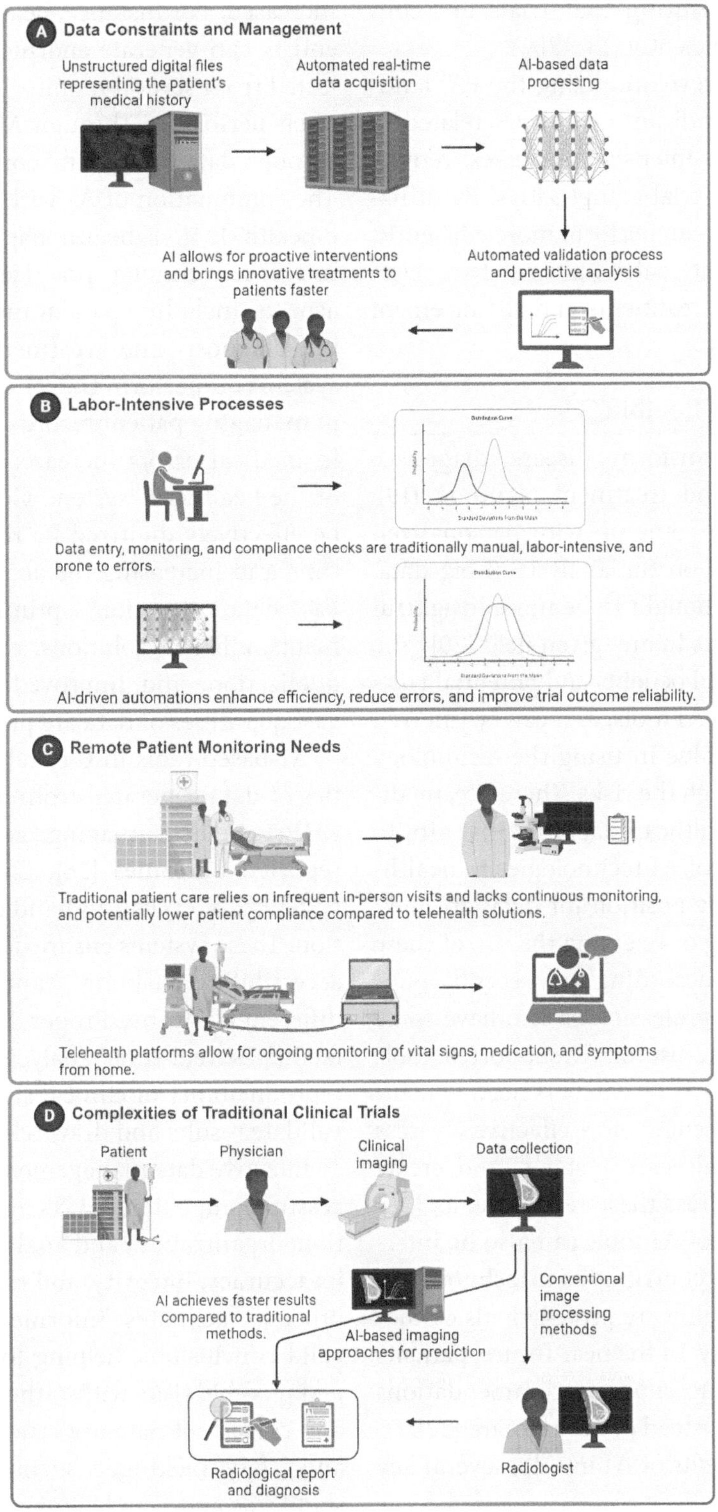

FIGURE 21.1 The role of AI in addressing challenges in CTs. AI is revolutionizing clinical research by improving data management through advanced analytics and predictive models, which in turn improves integration, consistency, and real-time problem identification (A); AI technologies like NLP and RPA streamline labor-intensive processes in CTs, increasing efficiency and accuracy in tasks like patient recruitment and data entry (B); AI-enabled wearable devices and telehealth platforms facilitate remote patient monitoring, provide real-time information that increases patient safety and compliance, and enable rapid interventions (C); AI reduces the complexity, cost, and duration of CTs by optimizing design through predictive modeling and improving stakeholder coordination with advanced project management tools (D).

among stakeholders, ensuring that trials are conducted smoothly and efficiently [16, 17].

In conclusion, AI is revolutionizing the CT landscape by addressing significant challenges related to data management, labor-intensive processes, remote patient monitoring, and trial complexities. By utilizing AI technologies, CTs can perform more efficiently, cost-effective, and patient-centric, ultimately accelerating the innovation in treatments and advancement of patient care outcomes.

21.3 AI-DRIVEN TOOLS IN CTS

AI-based tools can transform disease diagnosis, treatment, prognosis, and treatment planning [19]. AI-powered devices can come up with personalized recommendations based on the analysis of big data. Therefore, AI tools are thought to be more insightful than the foremost experts in any given field [20]. The balance between potential benefits and potential risks is important when using AI tools, as it can be inferred that there is a greater value in using the technology when the benefits outweigh the risks. Therefore, medical professionals and healthcare facilities that aim to increase the application of AI technologies in healthcare services are in a key position for decision makers involved in accepting or rejecting the use of these tools [19]. Additionally, according to a recently published study, healthcare professionals also have some basic concerns about the use of AI-supported tools [21]. To achieve this goal, AI researchers need to identify relevant current challenges more effectively, assess the requirements of healthcare systems, and create AI-powered tools to address these requirements [21]. Issues related to the use of AI tools can also be interpreted as a possible factor contributing to technostress, which happens when healthcare professionals cannot adapt to using technology. In the near future, patients may also have the right to choose recommendations from AI-powered tools instead of human care [22].

To maximize the benefits of AI in CTs, several key areas must be addressed:

21.3.1 Data Generation and Management

The production and coordination of medical data play an important role in the healthcare sector in better monitoring procedures and facilitating the management of healthcare services. Recently, the quantity and size of medical records obtained has steadily increased. For instance, healthcare providers and scientists can generate enormous amounts of data collected from several monitoring tools. This function is often performed through ML algorithms supported through data storage and computational capacity [23]. The combination of AI with healthcare can be called e-health [24]. E-health has brought about changes in the conventional practices in healthcare in many aspects, including patient monitoring, patient follow-up, diagnosis and treatment. AI-powered e-Health systems outperform traditional systems in the context of managing patient records. Therefore, the reduction in medical errors increases the overall effectiveness of the healthcare system. Complicated processes can be effectively digitized by reducing decision-making time and increasing the accuracy of decisions made. Fast data extraction, optimization of time requirements, efficient solutions, avoidance of unnecessary duplication, and improved speed when processing vast quantities of data are provided by AI [25].

AI-based tools in CTs rely heavily on vast quantities of data generated from different sources, such as EHRs, medical imaging, genetic data, and patient-reported outcomes [26]. Efficient data-management systems are crucial to handle this influx of information. These systems ensure data integrity, security, and accessibility, enabling seamless data sharing across different platforms. Proper data management not only facilitates real-time analysis but also supports the reproducibility of clinical studies, making it easier to validate results and draw reliable conclusions.

Effective data management is necessary for the successful completion of CTs, encompassing the acquisition, organization, and analysis of trial data to ensure its accuracy, integrity, and reliability. This meticulous process facilitates informed decision-making and valid conclusions, helping to detect trends, patterns, and possible risks within the trial. Additionally, exact and consistent data are crucial for adherence reports and safety tracking, ensuring adherence to regulatory standards and Good Clinical Practice guidelines.

Post-trial, data management remains essential by collecting real-life data from regular medical care, revealing the long-term advantages and risks associated with treatments in practical settings. However, the extensive process of gathering and organizing, and analyzing such large volumes of data can be resource-intensive and time-consuming, which is where AI can make a significant impact. AI has been increasingly integrated across various industries, including CTs. AI devices can enhance the achievements and effectiveness of clinical research by rapidly and accurately analyzing extensive patient data and health records. Specifically, AI can assist CTs in several ways: Research by analytics shows that approximately 80% of global CTs struggle to recruit and keep a sufficient number of patients within the deadline [27–29]. AI can streamline the identification of CT candidates by analyzing patient records and medical data, focusing the search on the best patient groups and speeding up enrollment. Additionally, AI can reduce the complexity of entry standards by analyzing patients' genetic composition, physiological information, and lifestyle habits.

Conventional CTs often face a 30% dropout rate, attributed to inconvenience, complicated procedures, and insufficient support. AI has the potential to improve patient experience and retention by facilitating remote health evaluations and real-time monitoring of medication adherence. AI-powered applications can send reminders and enable patients to monitor their own progress, thus enhancing their involvement in the study.

CTs generate enormous quantities of data that researchers need to sift through manually to extract valuable insights. AI can streamline this process by analyzing data more effectively, spotting trends that might elude human reviewers. AI-driven tools can forecast potential drug side effects, aid in choosing appropriate compounds, and even reveal insights for future studies. However, it is important to address inherent biases in data, and AI-based systems must safeguard the confidentiality and security of medical records to maintain confidentiality.

In CTs, free text entries from patients or staff can be ambiguous or incorrect, making manual review of this data from electronic Case Report Forms both laborious and time-consuming. To address this issue, Alcedis, a Huma company, has created Meteor, an AI-powered tool designed to simplify and expedite the process [30]. Meteor employs AI to automatically analyze text for mentions of adverse events, medications, treatments, and adherence to documentation standards. This AI-enhanced system not only reduces time and costs but also greatly improves the accuracy of CT data and enhances patient safety.

21.3.2 Patient-Centric Medical Record Tracking

Considered the most accurate health data in the medical system, the electronic medical record (EMR) is a system in which all medical processes related to the health history of patients, such as patient admissions, examinations, allergies, blood tests, treatment plans, medications, and surgeries, are stored. Many investigators have worked on executing several AI algorithms with EMR data (Figure 21.2). As medical AI platforms that support various algorithms have emerged, AI has become indispensable for medical quality improvement and healthcare management [31]. However, to enhance the clinical application of EMR data, data gathering phase alone is not sufficient; data quality should also be monitored at least as well as the data collection phase. Since EMR data are sensitive data including patient data, data security must be ensured, and an effective balance is mandatory during clinical use. Consequently, researchers do not need to overlook the elements that need to be considered for different AI algorithms using EMR data [32, 33].

EMR data can also facilitate AI-driven drug development. Specifically, it enables the prediction of certain diseases and the provision of personalized treatments based on individual patient characteristics. To enhance the effectiveness of AI research with EMR data, data standardization and curation are essential [31]. Therefore, to optimally utilize EMR data for clinical research, it is crucial to secure and verify the data beforehand to ensure its applicability for new medical studies. Moreover, researchers must approach the use of medical data with a strong sense of legal and ethical responsibility. When all these are handled in harmony, the use of EMR data can contribute to medical services and the improvement of patient health [34].

One of the significant advancements brought by AI in CTs is the ability to track patient medical records in a patient-centric manner. AI algorithms can aggregate data from multiple sources, providing a comprehensive and continuous view of a patient's health condition. This holistic approach ensures that all relevant information is considered during treatment planning and monitoring, improving patient outcomes. Additionally, AI-powered tools have the capability to uncover patterns and trends in patient data that might not be easily noticeable to human clinicians. This can facilitate the early identification of potential problems and enable the creation of more tailored care plans.

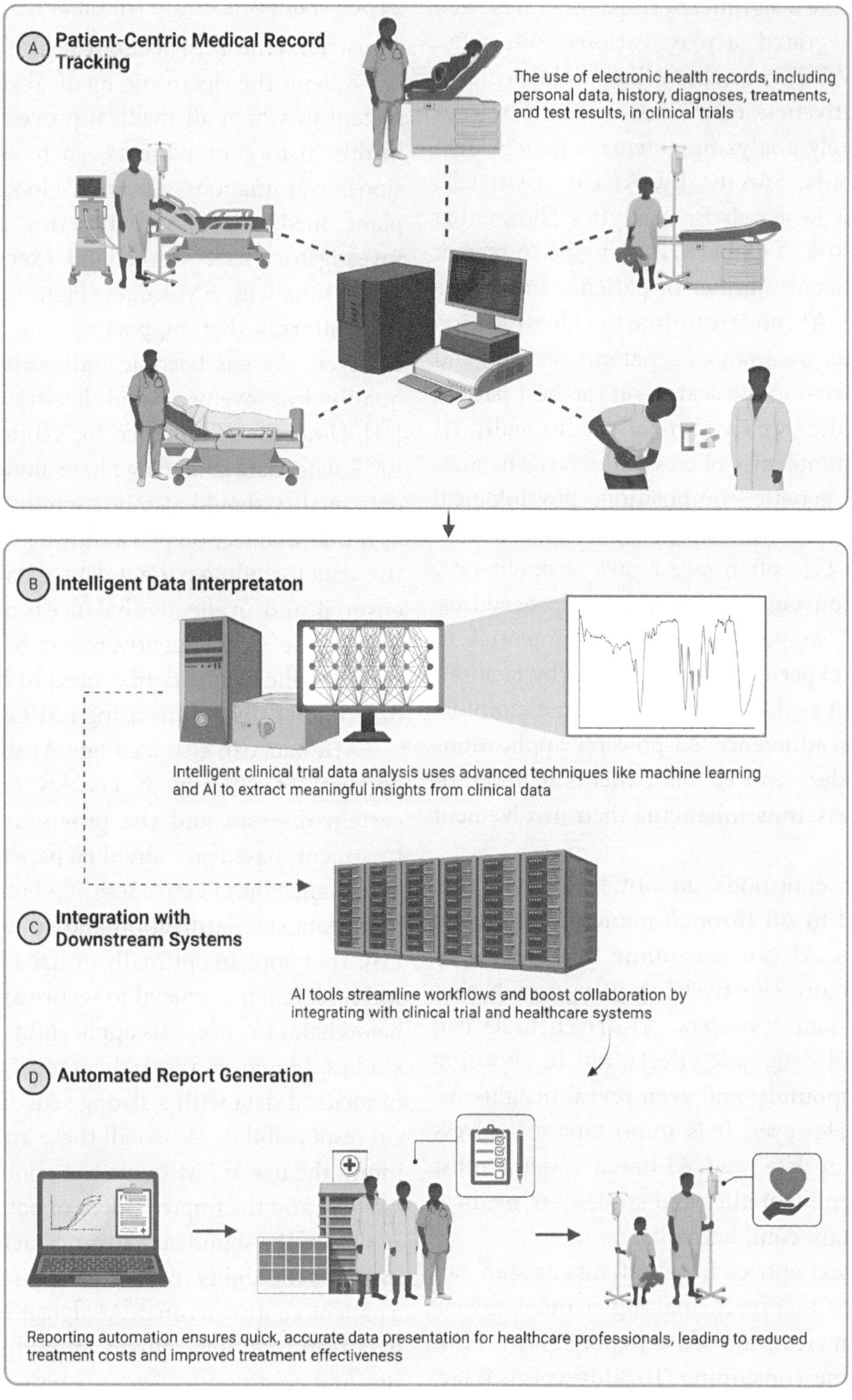

FIGURE 21.2 AI-driven tools in CTs. AI enhances CTs by automating the review of free-text data, improving the quality of EMRs, and enabling comprehensive, patient-centered monitoring that leads to better treatment planning and outcomes (A); AI-powered intelligent data interpretation enhances clinical studies by extracting valuable insights from complex datasets, optimizing treatment development, improving safety, and supporting better decision-making processes (B); Integrating AI into CDSSs enhances their efficiency and capabilities, while understanding doctors' viewpoints and establishing a smooth interface with existing systems are vital for successful acceptance and effective use in healthcare settings (C); Reporting automation using AI and ML streamlines data collection and analysis, increases accuracy, provides real-time insights, and enables customized reporting—saving time and improving decision-making in healthcare (D).

21.3.3 Intelligent Data Interpretation

Intelligent CT data analysis is the process of using advanced data analysis techniques, including ML and AI, to extract meaningful insights from CT data [35]. This can provide significant benefits for patients, healthcare providers, and pharmaceutical companies. For example, it could reduce the cost of CTs by identifying patients who are more likely to gain advantages from a specific treatment. Also, this could help in preventing serious side effects by identifying patients at risk of developing side effects from a particular treatment, thereby improving the safety of CTs [35, 36]. By identifying patients who are more expected to respond to a specific treatment, it can help develop new drugs and treatments more quickly. By providing insights into the risk and benefit profile of innovative drugs and interventions, it can help make better decisions about which drugs and treatments to develop further. This comprehensive analysis enables pharmaceutical companies to optimize trials, develop safer and more effective treatments, and ultimately benefit patients, healthcare professionals, and the entire industry [35, 36].

Control, assessment and tracking of therapeutic investigations (COATI) is a CT system for managing data derived from a general model of CT data [37]. The data model is embedded with a study plan representation that guides the statistical analysis of data, such as the specifics of the patient appointment schedule, the stages of the CT, and the patient data gathered at each visit. Data Analysis Package (PANDA) and Data Analysis and Reporting Tool (DART) are CT data assessment and reporting platforms that are set up with two distinct components integrated with COATI [35, 37, 38]. PANDA is a rule-based system that deals with the descriptive analysis of data. For example, the analysis includes the identification of acceptable patients and observations, and the closure of all outstanding time periods. DART is an algorithm-based system that participates in the evaluation of the structured research hypothesis. DART captures the key description of the plan for statistical analysis and includes a consulting system that helps the user make decisions about the analysis plan. The output and printing of the PANDA-DART systems is a full collection of tables mandated by ICH guidelines, with their generation being completely automated.

The expansive amount of data obtained in CTs can be overwhelming for human analysts. AI tools excel in intelligent data interpretation, extracting meaningful insights from complex datasets. ML algorithms can identify correlations, estimate consequences, and uncover hidden trends that could be overlooked by traditional analytical methods (Figure 21.2). These insights can inform decision processes, contributing to greater effective treatments and better patient outcomes. Additionally, AI can constantly evolve and adjust based on new information., improving its accuracy and predictive capabilities as time progresses.

21.3.4 Integration with Downstream Systems

Clinicians play an important role in incorporating new clinical innovations into care practices via user input. Understanding clinicians' roles and perceptions as users are able to give feedback on the functioning of investigated technologies for clinical validation, implementation, and downstream adoption [39]. Clinical decision support system (CDSS) is a commonly employed tool for enhancing healthcare system efficiency and improving clinical practice. CDSS is driven by EMRs and computerized workflows. For instance, treatment-related patient health outcomes shown in existing reports can improve the care delivery experience by providing physicians with insights into patient satisfaction and treatment outcomes [40].

The incorporation of AI into CDSS increases the efficiency of CDSS, expanding its capabilities to a broader area. Personalized medicine is an area of healthcare that leverages AI's high-level data examination for diagnostic purposes, outcome predictions, and individualized treatment methods [41]. AI in personalized medicine has a significant place in disease management, reducing adverse outcomes, and containing healthcare costs. Physician perspectives on new clinical tools are a key factor in both clinical integration and downstream adoption [39]. When looking at the research conducted to date, it is seen that physicians' perspectives on emerging technologies have not been investigated sufficiently. Therefore, it may be useful to investigate the suitability of new technologies for the provider in the future [42]. However, although physicians present numerous factors to consider, they are generally open to new technological ideas to advance their care practices. The necessary evidence generation, what is important for the safety of patients, how to make data applicable and usable, necessary understanding, clinical integration, and sustainable use are the key factors that complete technology adoption for physicians [43].

For AI-driven tools to be truly effective, they must seamlessly integrate with downstream systems used in CTs and healthcare settings. This integration ensures that insights generated by AI are easily accessible and actionable by healthcare professionals (Figure 21.3). AI tools can be connected with EHR systems, laboratory information management systems, and other CT management systems to streamline workflows and enhance collaboration among stakeholders [44, 45]. Effective integration reduces redundancy, minimizes errors, and accelerates the translation of AI-driven insights into clinical practice.

21.3.5 Automated Report Generation

Automated reporting involves the use of software and tools to streamline the process of gathering, analyzing, and producing reports without manual effort. This automation simplifies and speeds up the reporting workflow, ensuring that data are presented quickly, accurately, and efficiently. By eliminating manual steps, reports can be created much more quickly and facilitate rapid decision-making. Using technology in automated reporting minimizes human error by ensuring the reliability and accuracy of data with built-in consistency checks, validation procedures, and automated calculations. One of the key benefits of automated reporting is the ability to provide real-time insights. When new data are received, the

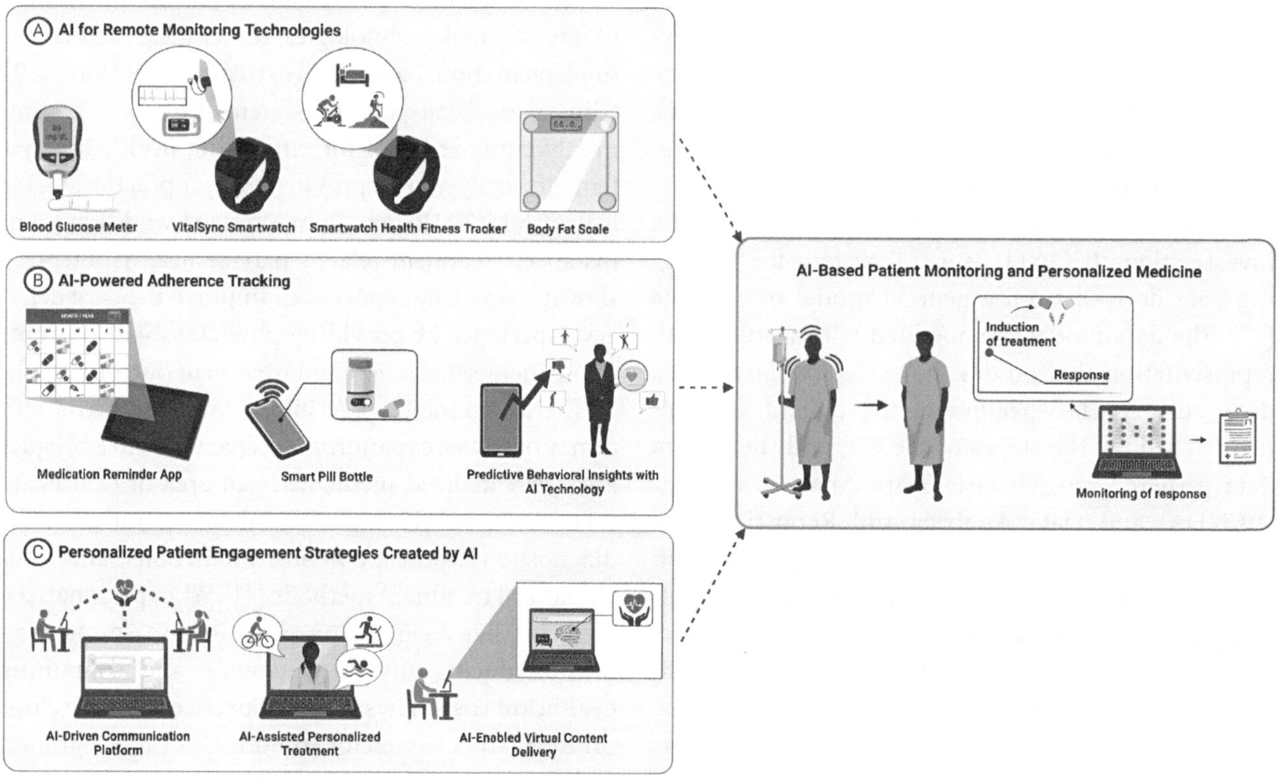

FIGURE 21.3 Improving patient monitoring and retention. AI-powered remote monitoring technologies, including wearable devices, smart home appliances, and telehealth platforms, are significantly improving patient management and outcomes by providing continuous and real-time health monitoring, personalized care, and timely interventions (A); AI-powered adherence monitoring tools, such as medication apps, smart medicine bottles, and predictive analytics, increase patient adherence to treatment plans by providing personalized reminders, tracking usage, and offering personalized interventions, leading to improved health outcomes (B); AI-powered personalized patient engagement strategies, such as personalized communication platforms, virtual health coaches, and dynamic content delivery, increase patient engagement and compliance by providing personalized support, feedback, and motivation, thereby improving health management and outcomes (C).

system can immediately analyze and refresh reports so users have instant access to the latest information. Contemporary automated reporting solutions offer a variety of customization features that allow you to customize the report format, integrate multiple data sources, and specify specific analysis criteria.

Today, ML and AI are transformative forces reshaping reporting automation. These advanced technologies improve the quality of reports and automate complex decision-making processes. As ML and AI technologies evolve, their incorporation into reporting tools will increasingly enhance their effectiveness and capabilities. We can talk about a future where reports not only provide data but also strategic insights and proactively guide decision-making [46].

These AI tools can automate the generation of reports, significantly reducing the time and effort required to compile and analyze data manually. Automated report generation ensures consistency and accuracy, providing healthcare professionals with reliable information to make informed decisions (Figure 21.2). These reports can include detailed analyses of patient outcomes, treatment efficacy, and adverse events, among other critical metrics [47]. By automating this process, AI liberates researchers and clinicians from routine tasks, enabling them to concentrate on higher-level, strategic responsibilities such as interpreting results and planning future studies.

21.4 ENHANCING PATIENT RECRUITMENT AND ENROLLMENT

21.4.1 AI Algorithms for Identifying Potential Participants

AI algorithms are transforming the landscape of CT recruitment by identifying potential participants with unprecedented accuracy and speed. These algorithms can analyze a multitude of information sources, including EHRs, genetic data, and social determinants of health, to find individuals who meet the specific inclusion and exclusion criteria for a study [16, 48]. For instance, IBM Watson for Clinical Trial Matching (CTM) uses NLP to read through medical records and CT databases, matching patients to trials in a way that requires only a small portion of the time compared to what a human would need [49]. IBM's Watson is an effective AI tool for CTM, having been tested across various states and tumor types. This platform interfaces with local EHR systems to categorize patients into 'Exclude' or 'Consider' categories for CTs [50, 51]. AI algorithms derive these classifications by analyzing trial protocols, patient data from EHRs or data archives, and attributes that are identified either entirely or partially through automated methods. Similarly, Deep 6 AI utilizes AI to parse through EHRs to find patients that match complex trial criteria, significantly increasing the recruitment pool while ensuring the selection of diverse and representative participants [52].

21.4.2 Streamlining Patient Matching

The process of patient matching can be labor-intensive and error-prone when done manually. AI-driven platforms are changing this by automating the matching process and ensuring real-time updates to patient databases. Tools like TrialMatch by the Alzheimer's Association use AI to continuously scan and match potential participants with ongoing CTs, significantly reducing the administrative burden on clinical staff [53]. Saama Technologies also provides advanced AI tools for CTs, specializing in analytics and AI solutions for the life sciences industry [54]. Their platform uses ML to optimize trial operations, from patient recruitment to data management and regulatory compliance [54]. Saama's AI-driven analytics streamline patient recruitment, enhance data integrity, and ensure regulatory adherence, helping researchers make informed decisions and accelerate treatment development. Clindata Insight is an advanced AI tool designed to enhance CT processes through automation of data cleaning, normalization, and analysis [55]. By leveraging cutting-edge algorithms, it alleviates the burden on research teams and accelerates trial timelines. The platform ensures data integrity and uncovers valuable insights within CT data, empowering researchers to make informed decisions for successful trial outcomes. Additionally, Clindata Insight offers real-time monitoring and reporting capabilities for proactive intervention, ensuring trials stay on track. Its intuitive interface and user-friendly features make it a valuable asset for trial sponsors, Contract Research Organizations, and research teams. These platforms can consider a wide range of variables, including medical history, genetic markers, and current health status, to accurately match patients with suitable trials. This automation not only enhances the precision of matching but also ensures that patients are promptly and appropriately assigned to trials, improving overall enrollment efficiency.

21.4.3 Reducing Recruitment Time and Costs

The recruitment phase of CTs is often lengthy and costly, posing a significant barrier to timely research progress. AI has the capability to simplify and improve efficiency in this process, shortening both time and expenses. By automating the identification and matching of potential participants, AI minimizes the delays caused by manual recruitment methods. For example, Antidote, an AI-powered recruitment platform, connects patients with CTs by leveraging ML to analyze patient data and identify suitable matches [56]. This approach not only accelerates the recruitment process but also decreases the costs associated with manual screening and outreach. Furthermore, AI can optimize recruitment strategies by analyzing data to identify the most effective outreach channels, such as social media, patient registries, or healthcare provider networks, ensuring that recruitment efforts are both targeted and efficient. This comprehensive approach leads to faster trial initiation, better resource allocation, and a quicker path to bringing new treatments to market. Examples of AI in Action:

1. PathAI—PathAI uses AI algorithms to analyze pathology slides for CT eligibility, ensuring that patients with specific biomarker profiles are identified quickly and accurately [57].
2. Flatiron Health—This company uses AI to aggregate and analyze oncology data from various sources, identifying potential candidates for cancer trials with high precision [58].
3. Deep Lens—Deep Lens leverages AI to connect community oncologists with CT opportunities, using real-time data from pathology and genetic testing to find eligible patients [59]. Deep Lens, acquired by Paradigm, was a pioneer in using AI to match cancer patients with CTs through their VIPER platform. VIPER analyzed clinical data and pathology reports for a comprehensive understanding of a patient's cancer, enabling optimal trial matching. By integrating with healthcare systems, VIPER scanned medical records in real-time, identifying eligible patients at diagnosis and accelerating enrollment in trials. This not only brought new cancer therapies to market faster but also improved patient access to potentially life-saving treatments, saving time and resources. Deep Lens' legacy of AI-powered patient-trial matching continues under Paradigm, shaping the future of cancer treatment.

These examples illustrate how AI is not only enhancing the efficiency and accuracy of patient recruitment and enrollment and is also reshaping the complete CT landscape to be more inclusive, faster, and cost-effective.

21.5 IMPROVING PATIENT MONITORING AND RETENTION

The advancements in AI have transformed various aspects of healthcare, particularly in patient monitoring and retention. This section delves into the significant role AI plays in remote monitoring technologies, adherence tracking, and personalized patient engagement strategies. These innovations not only enhance patient outcomes but also streamline CT processes, ensuring higher retention rates and more reliable data collection.

21.5.1 AI for Remote Monitoring Technologies

Remote monitoring technologies powered by AI have revolutionized how patients are observed outside traditional clinical settings. These technologies enable continuous, real-time tracking of patients' health statuses enabling prompt actions and individualized treatment (Figure 21.3).

21.5.1.1 Wearable Devices and Sensors

AI-driven wearable devices and sensors have become increasingly popular in remote patient monitoring. These devices are capable of monitoring different health indicators, including heartbeat, blood pressure, blood sugar levels, and activity levels. For example, the Apple Watch, equipped with an electrocardiogram app, uses AI algorithms to detect irregular heart rhythms, potentially identifying conditions like atrial fibrillation early. Similarly, continuous glucose monitors like the Dexcom G6 provide real-time glucose readings to diabetics, which are analyzed by AI to predict and manage blood sugar levels [60].

21.5.1.2 Smart Home Devices

AI-integrated smart home devices are also playing a crucial role in patient monitoring. Devices like smart scales, blood pressure monitors, and even smart mirrors can collect and analyze health data. For instance, the Withings Body Cardio scale measures weight, body composition, heart rate, and even vascular age [61]. This data is processed by AI to track trends and notify healthcare professionals of any major alterations that may require intervention.

21.5.1.3 Telehealth Platforms

Telehealth platforms have integrated AI to enhance remote consultations and monitoring. AI-driven chatbots and virtual health assistants like Sensely's Molly can conduct preliminary assessments, triage patients, and provide personalized health advice [62–64]. These platforms can analyze patient-reported symptoms and medical history, using NLP to provide accurate and timely responses.

21.5.1.4 Real-World Applications

In real-world applications, AI-powered remote monitoring has shown significant benefits. For example, the AI system developed by Current Health monitors patients' vital signs through wearable devices and uses ML to predict potential health deteriorations [64]. This has been particularly useful in managing chronic diseases like heart failure and chronic obstructive pulmonary disease, which help minimize hospital readmissions and emergency room visits.

21.5.2 AI-Powered Adherence Tracking

Patient adherence to treatment protocols is a critical factor in the success of CTs and overall patient health outcomes. AI-powered adherence tracking tools have been developed to monitor and encourage patients to follow their prescribed treatment plans (Figure 21.3).

21.5.2.1 Medication Adherence Apps

AI-powered medication adherence apps like Medisafe use ML to analyze patients' medication-taking patterns. These apps provide personalized reminders and educational content to help patients understand the importance of adherence. Medisafe's AI algorithms can predict when a patient is likely to miss a dose and send timely alerts to prevent non-adherence [65].

21.5.2.2 Smart Pill Bottles and Dispensers

Smart pill bottles and dispensers, such as the AdhereTech smart pill bottle, utilize AI to monitor medication usage [66]. These devices track when the bottle is opened and send adherence information to healthcare professionals. AI algorithms process and interpret this information to identify patterns of non-adherence and trigger interventions, such as reminders or caregiver notifications, to ensure patients take their medications as prescribed [66, 67].

21.5.2.3 Behavioral Insights and Predictive Analytics

AI-powered tools can provide behavioral insights and predictive analytics to improve adherence. For example, Proteus Discover, a digital medicine system, includes an ingestible sensor that tracks medication ingestion [68]. The data collected are analyzed by AI to provide insights into the patient's adherence behavior, allowing medical professionals to customize treatments according to each patient's specific requirements.

21.5.2.4 Case Study: Improving Adherence in Hypertension Patients

A notable case study involves AI-powered adherence tracking in hypertension patients [69]. Researchers used AI algorithms to analyze data from smart pill bottles and mobile apps. The study found that patients receiving AI-driven reminders and personalized feedback had significantly higher adherence rates compared to those receiving standard care. This improvement in adherence was linked to better blood pressure control and reduced incidence of hypertension-related complications.

21.5.3 Personalized Patient Engagement Strategies Created by AI

Personalized patient engagement strategies are crucial for maintaining patient interest and involvement in their health management and CTs. AI has enabled the development of highly tailored engagement strategies that resonate with individual patients (Figure 21.3).

21.5.3.1 AI-Driven Communication Platforms

AI-driven communication platforms like Conversa Health use NLP and ML to engage patients in meaningful conversations [64]. These platforms can send personalized messages, educational content, and motivational reminders based on the patient's health data and preferences. For instance, Conversa Health's platform can send customized follow-up messages after a medical appointment, addressing specific patient concerns and reinforcing care instructions [70].

21.5.3.2 Virtual Health Coaches

Virtual health coaches, powered by AI, provide personalized guidance and support to patients. Apps like Lark Health use AI to offer real-time health coaching based on patient data. Lark's AI analyzes users' activity levels, diet, and sleep patterns, providing personalized feedback and recommendations [71]. This continuous engagement helps patients make healthier lifestyle choices and adhere to treatment plans.

21.5.3.3 Dynamic Content Delivery

AI facilitates the dynamic distribution of content tailored to the patient's preferences and behavior. Platforms like HealthPrize use gamification and AI to deliver personalized content, quizzes, and rewards to motivate patients [72]. By analyzing engagement data, the AI can adjust the type and frequency of content to keep patients interested and involved in their health management.

21.5.3.4 Real-World Example: Improving Diabetes Management

In diabetes management, AI-powered personalized engagement strategies have shown promising results. The mySugr app uses AI to provide personalized diabetes coaching [73]. By analyzing blood sugar levels, food intake, and physical activity, mySugr's AI offers tailored advice and motivational messages. A study demonstrated that patients using mySugr had better glycemic control and higher engagement levels compared to those using traditional management methods [73, 74].

The role of AI in improving patient monitoring and retention is revolutionary, providing sophisticated tools and strategies that enhance patient care and streamline CT processes. From remote monitoring technologies and adherence tracking to personalized patient engagement strategies, AI is leading the shift toward healthcare that is more streamlined, impactful, and focused on patients. As these innovations progress, they hold great promise for enhancing patient care and advancing scientific research becomes increasingly apparent, heralding a new era of healthcare innovation.

21.6 CASE STUDIES AND REAL-WORLD APPLICATIONS

21.6.1 Successful Implementations of AI in CTS

Successful applications of AI in CTs are gaining momentum and are increasingly being integrated into real-world applications [4]. This section explores the successful use of AI used in CTs in medical diagnosis, medical treatment and new drug production in the context of real-world applications.

21.6.1.1 Medical Diagnosis

Successful applications of AI contribute to the effective support of patient diagnosis based on radiology, pathology and endoscopy with higher accuracy and lower workload [3].

21.6.1.1.1 AI in Radiology As indicated in Figure 21.4, radiology plays a role in diagnosing nearly every disease. There are many successful applications of AI in radiologic diagnosis [3]. For instance, Gong [75] analyzed 243 patients with pulmonary lesions to classify malignant and benign lung nodules and invented an AI-assisted diagnosis (CADx) system. The CADx system is applicable to accurately distinguish the nature of lung tumors derived from clinical computed tomography data and is a usable method for detecting lung cancer at an early stage [3, 75]. Similarly, during an AI research, Francesco discovered a highly effective algorithm for the early detection and swift diagnosis of retinal conditions [3, 76], and the usability of this algorithm has been confirmed by other studies [77, 78].

21.6.1.1.2 AI in Pathology The role of pathology in the diagnosis of tumors and lesions is indisputable. AI's use in pathology has demonstrated its potential for forecasting diagnoses of many high-accuracy results (Figure 21.4). For instance, Coudray confirmed that an AI-powered deep learning model has the capability to identify six genetic mutations linked to cancer. He also proved that doctors could detect the gene mutation with a remarkable accuracy rate of 97%, and its code was made available online [3, 79]. In addition, AI has been applied to diagnose epithelial tumors, basal

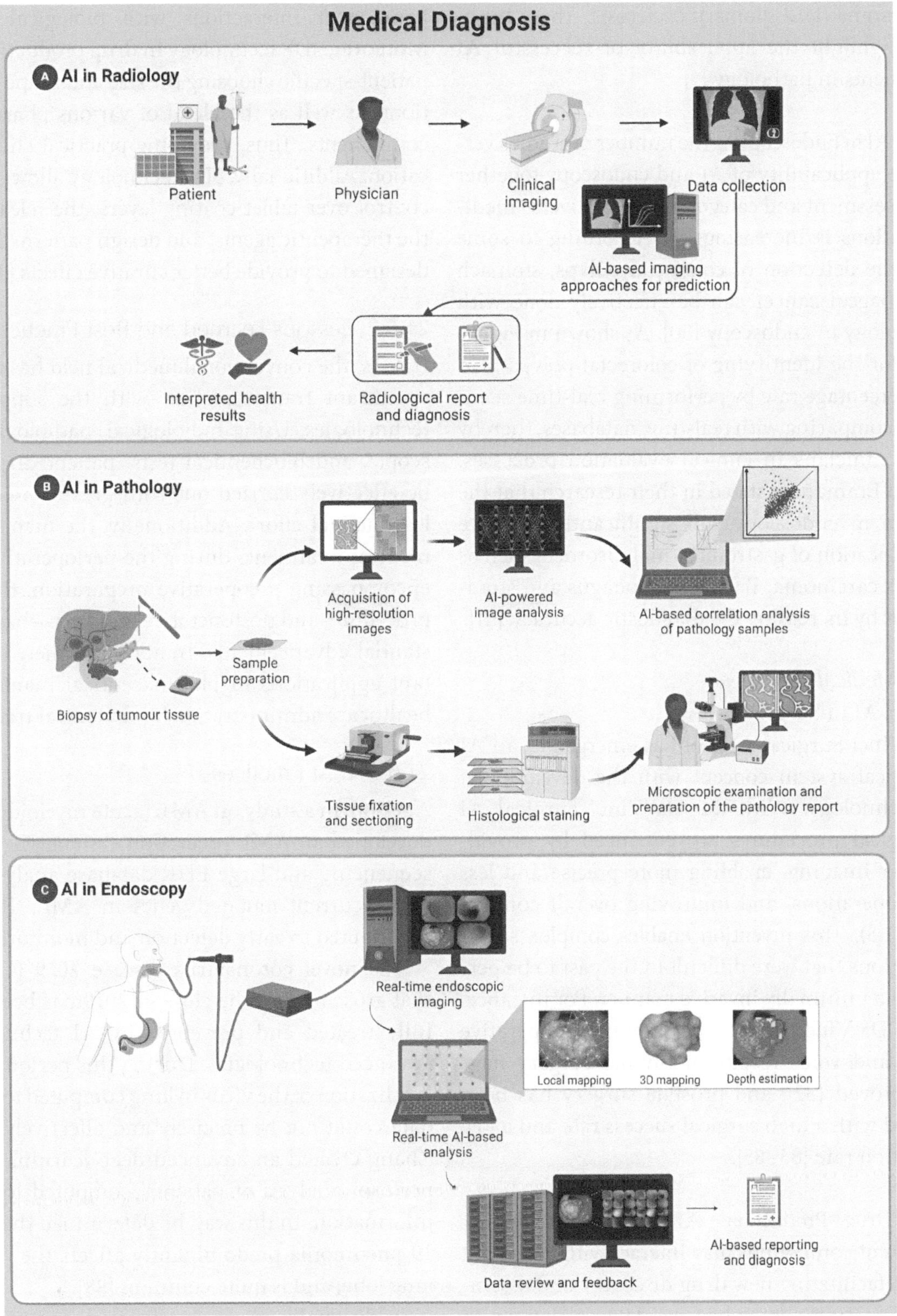

FIGURE 21.4 Case studies and real-world applications: Successful implementations of AI in CTs—medical diagnosis. AI has significantly advanced radiology by enhancing diagnostic accuracy and efficiency, demonstrated through innovative systems for early-stage identifying and classification of health conditions including lung cancer (A); AI has revolutionized pathology by achieving high accuracy in diagnosing tumors and genetic mutations, and proving valuable in identifying various tumors and lesions (B); AI has significantly enhanced endoscopy by improving the detection and classification of diseases such as colorectal polyps and stomach cancer, leading to more accurate diagnoses and better disease management (C).

cell carcinoma, and glomerulosclerosis. These innovations highlight the applicability of successful AI advancements in pathology [3].

21.6.1.1.3 AI in Endoscopy The number of studies verifying the applicability of AI and endoscopy together for the assessment and categorization of diverse medical conditions is increasing [3]. According to some experts, the detection of colorectal polyps, stomach and esophageal cancer can be effectively done with AI technology in endoscopy [80]. As shown in Figure 21.4, AI for the identifying of colorectal polyps provides a percentage rate by performing real-time scanning and comparing with real-time databases, thereby assisting clinicians in clinical evaluation processes. Gulat and Emmanuel stated in their research that the AI system in endoscopy will significantly facilitate the identification of gastrointestinal disorders such as squamous carcinoma, Barrett's esophagus and stomach tumor by increasing the diagnostic accuracy [81].

21.6.1.2 Medical Treatment

21.6.1.2.1 AI in Surgery

The Da Vinci Surgical AI tool has emerged as an AI plus surgical system concept with the development of AI technology. With the Da Vinci Surgical AI tool, surgical procedures are enhanced by providing clearer imaging, enabling more precise and less invasive operations, and improving overall comfort (Figure 21.5). This invention enables complex surgical operations that were difficult in the past to be performed in a minimally invasive manner. For instance, using the Da Vinci Surgical AI system, postoperative cosmetic and voice results in thyroid surgery have been improved [82], and prostate surgery has been performed with a high surgical success rate and a low complication rate [83–85].

21.6.1.2.2 Drug Production AI algorithms predict how different compounds may interact with biological pathways, facilitating new drug discovery and assembly by evaluating comprehensive data collections to uncover possible drug candidates. For instance, AI predictive models reduce research and development costs by accelerating CT processes [86]. As shown in Figure 21.5, AI facilitates the rapid identification and optimization of drug candidates, making the development process more efficient and cost-effective by predicting component interactions with biological pathways. Moreover, 3DP technology in drug production realizes patient-specific choosing the size and shape of medications, as well as the blend of various pharmaceutical components. Thus, providing practical clinical application. Additionally, 3DP technology allows for precise control over tablet coating layers, the release rates of the therapeutic agents, and design patterns can be predesigned to provide better curative effects [3].

21.6.2 Lessons Learned and Best Practices

In CTs, the conventional medical field has undergone significant transformation with the support of AI technologies. Using radiological, pathological, endoscopic, and biochemical tests, patient diagnosis can be effectively carried out with greater precision and less manual effort. Additionally, the management of medical treatments during the perioperative phase—encompassing preoperative preparation, the surgical procedure, and postoperative recovery—has seen substantial advancements. In addition, there are important applications in pharmaceutical manufacturing, healthcare administration, and medical training [3].

21.6.3 Best Practices

Abelson, in a study on AML (acute myeloid leukemia), developed an AML prediction system. He used deep sequencing and large EHR database analysis to analyze recurrent mutated genes in AML. This system contributed to early detection and monitoring [87].

The novel coronavirus disease 2019 (COVID-19) that arose toward the close of 2019 has been successfully treated and prevented by AI technology and advanced technologies. During this period, when the localization of the virus in lung computed tomography data could not be precisely and effectively analyzed, Zhang created an advanced deep learning system to perform analysis of patients' computed tomography information. In this way, he determined that COVID-19 pneumonia predominantly affects the right lower lung lobe, and is quite common [88].

Additionally, Sweta performed a smart screening to manage COVID-19 using a drug repurposing technique, identifying potentially useful drugs faster using AI and pharmacology-based methods [89]. This method was validated by other scientists, and eventually more than 80 drugs with great potential were found [90].

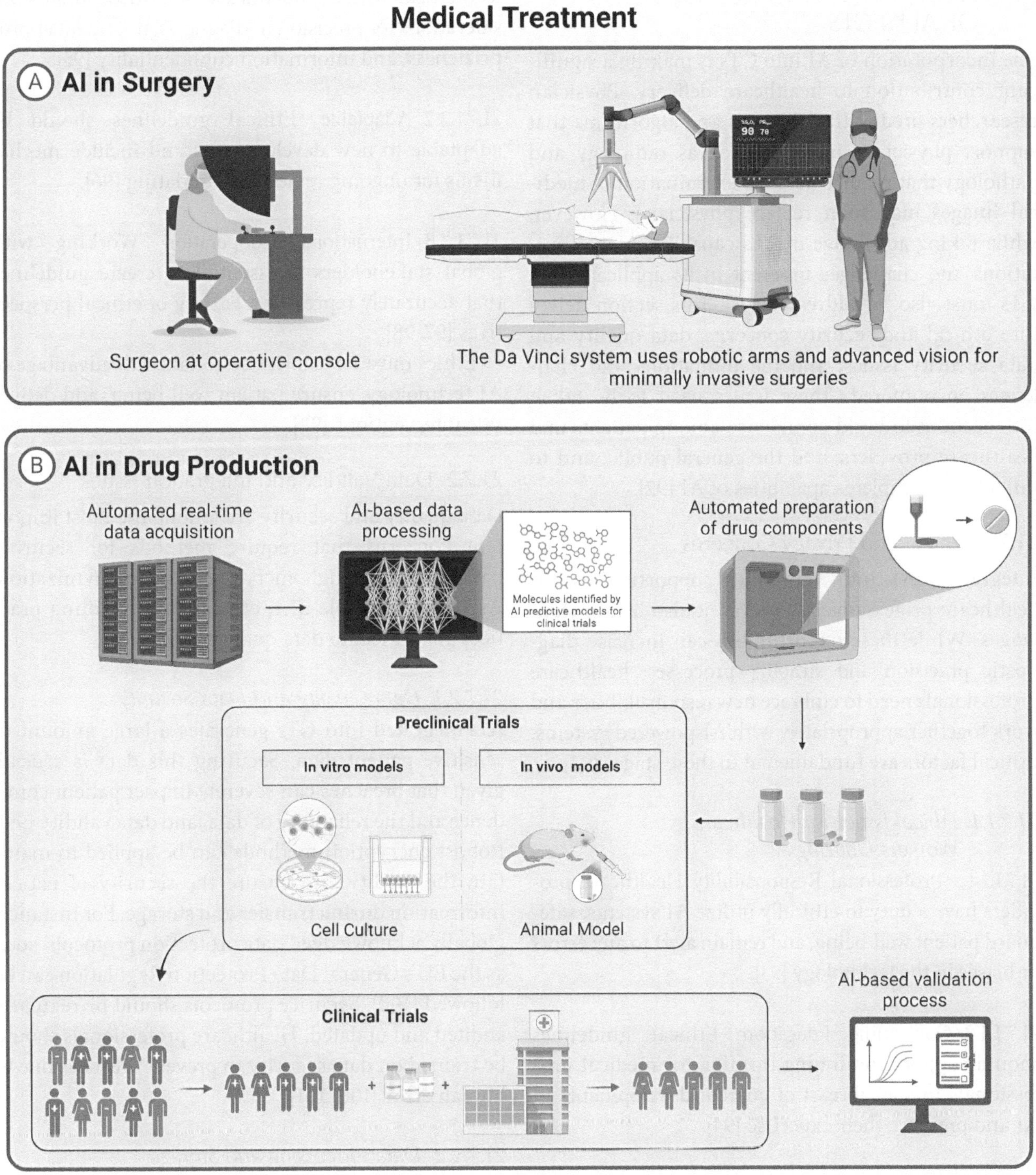

FIGURE 21.5 Case studies and real-world applications: Successful implementations of AI in CTs—medical treatment. The Da Vinci surgical AI tool enhances minimally invasive surgery by offering clearer imaging and more precise control, leading to improved outcomes, higher success rates, and fewer complications in procedures (A); AI accelerates drug production by predicting compound interactions with biological pathways, reducing research and development costs, and facilitating patient-specific drug design through technologies like 3D printing, which allows for tailored drug sizes, shapes, and release patterns (B).

21.7 LIMITATIONS AND RESTRICTIONS OF AI IN CTS

The incorporation of AI into CTs is making a significant contribution to healthcare delivery. Physician researchers predict that AI tools and algorithms that support physicians in fields such as radiology and pathology that require detailed examination of medical images may soon replace physicians. However, while taking advantage of AI's capabilities, the limitations and challenges inherent in its application in CTs must also be addressed [91]. This section delves into ethical and security concerns, data quality and data security issues, and the limitations and challenges encountered. These issues need to be taken into account to build confidence among patients and healthcare providers, and the general public, and to unlock the complete capabilities of AI [92].

21.7.1 Ethical and Privacy Concerns

Integrating AI into CTs offers opportunities for healthcare professionals; however, it also brings challenges. While these opportunities can increase diagnostic precision and simplify processes, healthcare professionals need to embrace new responsibilities and work together appropriately with AI-powered systems. Ethical factors are fundamental to these stages [92].

21.7.1.1 Ethical Issues for Healthcare Workers Utilizing AI

21.7.1.1.1 Professional Responsibility Healthcare providers have a duty to ethically utilize AI systems, safeguard patient well-being, and remain alert to any errors or biases in the technology [93].

21.7.1.1.2 Continuing Education Ethical guidelines should support continuing training for medical professionals to stay abreast of current developments in AI and preserve their expertise [94].

21.7.1.2 Need for Ethical Frameworks and Guidelines

In CTs, it is imperative to create suitable ethical standards and guidelines to ensure responsible ethical use. These frameworks protect ethical standards by drawing on established ethical principles [92].

21.7.1.2.1 CT-Based Guidelines Current ethical frameworks offer useful direction, but new guidelines are required to tackle the distinct ethical issues inherent to the field. These guidelines should incorporate considerations for precision in diagnosis, treatment appropriateness, and information confidentiality [95].

21.7.1.2.2 Adaptable Ethical guidelines should be adaptable to new developments and include mechanisms for ongoing review and updating [96].

21.7.1.2.3 International Cooperation Working with global stakeholders is essential to create guidelines that accurately represent a variety of ethical perspectives [97, 98].

Ethics must be prioritized to take full advantage of AI technology, ensure patient well-being, and deliver safe interventions [92].

21.7.2 Data Validity and Integration Issues

Data quality and security are among the most important concerns that require methods for securing patient data through encryption and anonymization. As shown in Table 21.1, ethical data handling practices are critical to data quality [92].

21.7.2.1 Data Quality and Data Security

AI integrated into CTs generates a large amount of sensitive patient data. Securing this data is crucial, given that breaches can severely impact patient confidence and the reliability of data, and data validity [99]. Robust encryption methods can be applied to maintain the quality and ensure the security of patient information during transfer and storage. For instance, globally acknowledged data protection protocols, such as the EU's General Data Protection Regulation can be followed [100]. Security protocols should be regularly audited and updated. Healthcare professionals should be trained on data security to prevent breaches due to human error [100, 101].

21.7.2.2 Data Processing and Storage

Healthcare organizations should adopt strong data anonymization methods, encryption, and secure data exchange protocols to collect, store, and use patient data. Patient data can be anonymized when individual privacy is desired. AI-specific data handling guidelines can be developed for CTs that outline data acquisition, preservation, and dissemination procedures. Data-management procedures should be reviewed regularly to guarantee adherence to ethical guidelines [102].

TABLE 21.1 A general framework for addressing data quality and integration issues related to the use of Artificial Intelligence (AI) Systems in Clinical Trials

Issue	Summary	Example
Unsafe failure mode	A system produces a prediction when it does not have enough information.	A clinician may not realize how unsafe that prediction is without any information about the confidence of the prediction.
Self-fulfilling prediction	A system indirectly reinforces the outcome it was designed to detect.	A system trained on outcome data predicts that certain cancer patients have a poor prognosis. This reinforces the learned behavior.
Reinforcement of outmoded practice	A system is trained on data that cannot adapt to new developments or sudden changes.	When a drug is withdrawn due to safety concerns, the AI decision support system cannot adapt because there is no data on an alternative.
Insensitivity to impact	In some cases, the system may make predictions that do not take into account the impact of false positive or false negative predictions in the clinical context.	An unsafe diagnostic system may occasionally miss malignant tumors while being trained to achieve maximum accuracy by correctly diagnosing benign tumors.
Unscalable oversight	A system requires excessive time-consuming monitoring.	A subcutaneous insulin pump requires the patient to provide detailed details of everything they eat so that the patient can adjust their insulin regimen.

With the integration of AI into clinical trials, research needs to be critically evaluated from a data quality perspective. Some clinical safety issues must be taken into account to achieve successful medical applications.

21.7.3 Regulatory and Compliance Challenges

AI integration in CTs involves numerous regulatory and compliance challenges that require rigorous evaluation. Therefore, adaptable and comprehensive regulatory frameworks are required. Addressing these adaptable regulations is important for ethical integration [92]. Ethical guidelines should highlight the need for interoperability standards to guarantee that AI is integrated in CTs in a compliant manner [103]. In the event of an AI-induced error, responsibility must be determined. Therefore, clear accountability frameworks must be established [104]. Moreover, the issue of informed consent comes to the fore when AI systems play a role in the decision-making process. Patients should be provided with the necessary information about the role of these systems in their care [105]. Overcoming the regulatory and compliance challenges of AI in CTs, balancing innovation with trust, and defining responsibility are key ethical requirements for AI to positively contribute to CTs [92].

21.7.4 Addressing AI Bias and Interpretability

Addressing AI bias and interpretability is important to prevent diagnostic and treatment disparities in CTs and to ensure fairness in healthcare decision-making [92]. AI algorithms may unintentionally reinforce biases present in historical data, which can lead to diagnostic and therapy disparities. It is necessary to develop algorithms that reduce to design algorithms that address bias and ensure equity in healthcare decision-making [106]. To strategically reduce bias, representative data can be used throughout the training of AI models, consistently monitored for bias, and data can be recalibrated if necessary [107].

21.8 FUTURE DIRECTIONS AND OPPORTUNITIES

21.8.1 Enhancing the Future of Clinical Research

The future of clinical development will be further augmented by combining AI through advanced innovations in personalized medicine and computer modeling. Virtual trials, made possible by modern digital technology, are emerging as a cost-effective alternative to traditional clinical studies [108, 109]. By leveraging AI, virtual trials can significantly lower the expenses and shorten the delays, and logistical challenges often experienced by patients. These trials enable remote participation, increasing patient retention and making it easier to recruit a diverse pool of participants. It's projected that as much as 50% of upcoming studies could be conducted using virtual trials, which would speed up the clinical development process and save human resources and optimize resource allocation.

Large pharmaceutical companies have already started to invest heavily in AI-driven research,

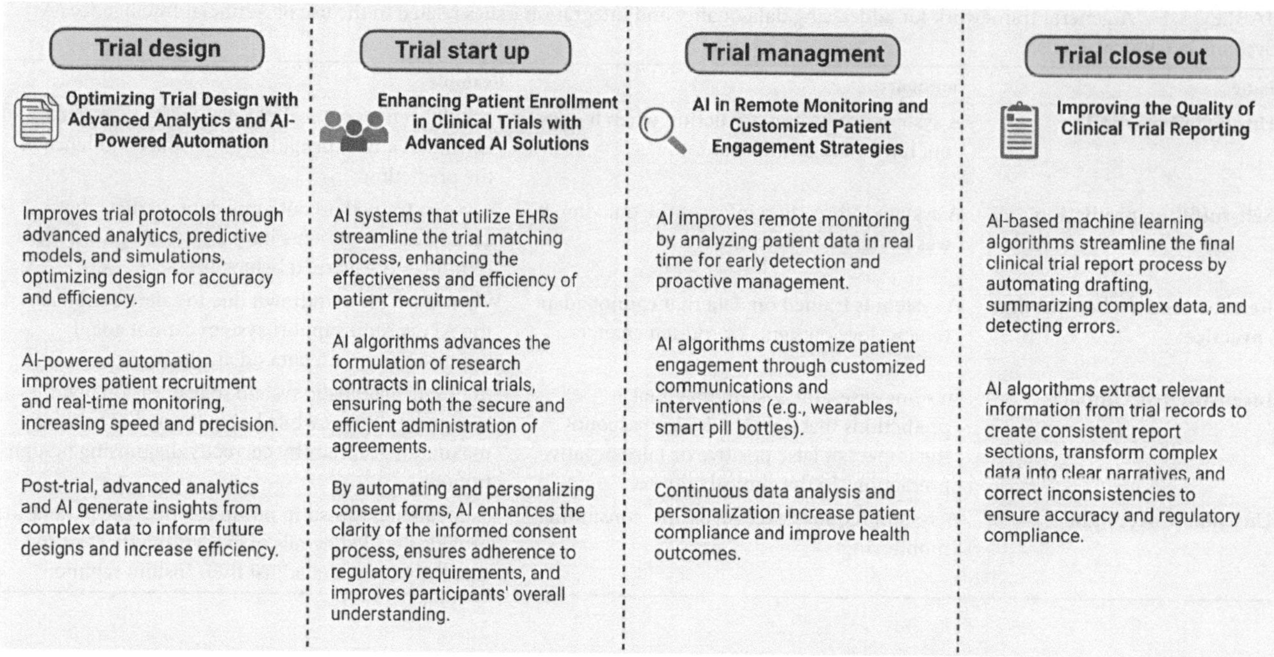

FIGURE 21.6 Future directions and opportunities.

recognizing its potential to revolutionize the clinical development landscape [108]. AI not only facilitates faster and more efficient trials but also enhances the safety and affordability of the research process. As shown in Figure 21.6, there is still considerable room for growth in the application of AI in clinical studies. By incorporating AI into every phase of clinical development, from initial study design to post-market surveillance, the industry can achieve breakthroughs that were previously unimaginable. The combination of AI, personalized healthcare, and advanced simulations will lead to a future where CTs are not only more effective but also more accessible to a broader range of patients.

The future of clinical development will be improved by combining AI with personalized healthcare and virtual trials, which provide a cost-effective alternative to traditional studies. AI-driven virtual trials can lower costs and logistical challenges while enhancing patient retention and recruitment. Major health companies are investing in AI to optimize all phases of clinical development, increasing both effectiveness and accessibility.

21.8.2 Potential for Personalized and Precision Medicine

AI's integration into CTs is not only transforming the research process but is also paving the path to achieving personalized and precision medicine. Personalized medicine, which customizes medical treatment based on each patient's unique characteristics, is becoming more attainable with the help of AI. Through the examination of a patient's genetic profile, lifestyle factors, and environmental exposures, AI can assist in discovering which treatments are most likely to be effective for specific individuals. This personalized approach ensures that patients get the most suitable therapies, reducing the trial-and-error aspect of therapy and minimizing the risk of harmful outcomes.

Moreover, AI's capacity to process extensive quantities of data enables the discovery of biomarkers—molecular signatures that can predict a patient's response to a particular treatment. Identifying these biomarkers is crucial for developing precision medicine strategies, as they allow for the stratification of patients into subgroups that are more likely to benefit from specific therapies. This not only improves treatment outcomes but also enhances the efficiency of CTs by focusing on the most responsive patient populations. AI is also facilitating the development of more sophisticated predictive models that can forecast disease progression and treatment outcomes. These models take into account a range of factors, such as genetic information, imaging results, and clinical history, to generate personalized risk assessments and treatment plans. As AI progresses, its role in enabling personalized and precision medicine will become even more pronounced, ultimately leading to more effective, patient-centered care.

In the future, the combination of AI with other cutting-edge technologies, including genomics, proteomics, and advanced imaging, will further accelerate the shift toward personalized medicine. By leveraging AI to interpret complex datasets and identify actionable insights, researchers and clinicians will have the capability to create more precise and individualized treatment regimens. This shift toward precision medicine will enhance patient outcomes and lower healthcare expenses by eliminating ineffective treatments and focusing resources on therapies that are more likely to succeed.

21.8.3 Collaborative Efforts between Industry and Regulatory Bodies

The complete realization of AI's capabilities in CTs and personalized medicine will depend on close collaboration between industry stakeholders, technology developers, and regulatory bodies. As AI technologies continue to evolve at a rapid pace, regulatory frameworks must keep up to ensure that these innovations are implemented safely, ethically, and effectively. This will require a concerted effort to establish clear guidelines, standards, and optimal approaches for the development, validation, and application of AI in clinical research.

A major challenge in this context is ensuring the transparency and explainability of AI algorithms. Regulatory agencies, including the FDA and EMA, are increasingly emphasizing the need for AI systems to be interpretable and auditable, particularly when they are used to make decisions that affect patient care. This necessitates the creation of AI models that are both precise and transparent in their decision-making processes. Collaborative efforts between AI developers, pharmaceutical companies, and regulators will be crucial in creating standards for algorithm transparency and ensuring that AI-driven decisions can be trusted by clinicians and patients alike.

Another important aspect of collaboration is addressing data confidentiality and protection concerns. Applying AI in CTs frequently requires handling substantial volumes of sensitive patient information, eliciting concerns about data protection and the potential for misuse. Industry stakeholders and regulatory bodies must work together to establish strong data-management frameworks that protect patient information while enabling the dissemination of data necessary for AI-driven research. This includes developing protocols for data anonymization, secure data storage, and controlled access to data, ensuring that patient privacy is maintained throughout the research process.

Furthermore, collaboration between industry and regulatory bodies will be essential in mitigating biases within AI models. Bias in AI can stem from various factors, such as skewed training data, and may result in disparate treatment outcomes across different patient populations. Regulatory agencies are increasingly focused on ensuring that AI technologies are designed and implemented in a manner that is fair and equitable. This requires a collaborative approach to identify potential sources of bias, validate AI models across diverse patient populations, and implement strategies to address and rectify biases in AI-driven decision-making.

In addition to tackling these issues, cooperation between industry and regulatory bodies can also foster innovation by fostering a conducive environment for AI research and development. Regulatory agencies can play a proactive role by providing clear guidance on the regulatory requirements for AI-based tools, facilitating early engagement with AI developers, and offering pathways for the expedited review and approval of AI-driven therapies and diagnostics. These collaborative efforts will be crucial to accelerating the implementation of AI in CTs and personalized medicine, ultimately benefiting patients by bringing new, effective therapies to market more quickly.

In conclusion, the future of CTs and personalized medicine will be shaped by the continued evolution of AI technologies and the collaborative efforts of industry and regulatory stakeholders. By working together, these entities can overcome the challenges associated with AI integration, guarantee the secure and ethical application of AI in clinical research, and unlock new opportunities for improving patient care. The capability of AI to transform CTs and personalized medicine is immense, and with the right partnerships and frameworks in place, this potential can be fully realized, marking the beginning of a new era of innovation in healthcare.

21.9 SUMMARY

In this chapter, we have explored the impact of AI on CTs and its capacity to transform the field of personalized and precision medicine. We have discussed how emerging AI technologies are enhancing various aspects of CTs, from patient recruitment and trial design to data analysis and monitoring. The integration of AI has been shown to enhance efficiency, cut expenses, and boost the accuracy of outcomes, thereby accelerating the development of new therapies. We also highlighted the potential for AI to enable personalized medicine, allowing treatments to be tailored to individual patient profiles, improving efficacy, and minimizing adverse effects. Additionally, we examined the importance of collaborative efforts between industry stakeholders and regulatory bodies in guaranteeing the secure, ethical, and efficient implementation of AI in clinical research.

As we look to the future, the path forward for AI in CTs is marked by both opportunities and challenges. The continued advancement of AI technologies will bring about even more sophisticated tools that can further optimize every stage of the CT process. Virtual trials, supported by AI, are poised to become a mainstream approach, reducing barriers to participation and making trials more inclusive and efficient. However, the successful integration of AI will require ongoing collaboration between pharmaceutical companies, technology developers, and regulatory agencies. It will be essential to develop robust frameworks that tackle concerns like data privacy, transparency of algorithms, and reducing bias. By fostering innovation while maintaining rigorous standards, the industry can ensure that AI-driven CTs not only meet regulatory requirements but also achieve their full potential in improving patient care.

The future of CTs is bright, with AI at the forefront of innovation. As AI technologies continue to evolve, they will become increasingly pivotal in influencing the way we conduct clinical research and develop new therapies. The integration of AI promises to make CTs more streamlined, economical, and focused on patient needs, ultimately leading to better healthcare outcomes. However, realizing this potential will require a coordinated effort from everyone involved. By embracing AI and working together to overcome the challenges associated with its implementation, we can usher in a new era of clinical research that is more adaptive, individualized, and responsive to the needs of patients. The journey ahead is both exciting and challenging, but with the right approach, AI has the power to transform CTs and revolutionize medicine as we know it.

REFERENCES

1. Bhat, M., Rabindranath, M., Chara, B. S., & Simonetto, D. A. (2023). Artificial intelligence, machine learning, and deep learning in liver transplantation. *Journal of Hepatology, 78*(6), 1216–1233.
2. Secinaro, S., Calandra, D., Secinaro, A., Muthurangu, V., & Biancone, P. (2021). The role of artificial intelligence in healthcare: A structured literature review. *BMC Medical Informatics and Decision Making, 21*(1), 125.
3. Liu, P. R., Lu, L., Zhang, J. Y., Huo, T. T., Liu, S. X., & Ye, Z. W. (2021). Application of artificial intelligence in medicine: An overview. *Current Medical Science, 41*(6), 1105–1115.
4. FDA (2024, May 30). *The Role of Artificial Intelligence in Clinical Trial Design and Research with Dr. ElZarrad.* Retrieved August 4, 2024, from https://www.fda.gov/drugs/news-events-human-drugs/role-artificial-intelligence-clinical-trial-design-and-research-dr-elzarrad#:~:text=AI%2Denabled%20algorithms%20have%20the,events%20in%20clinical%20trial%20participants
5. Panwar, H., Gupta, P. K., Siddiqui, M. K., Morales-Menendez, R., & Singh, V. (2020). Application of deep learning for fast detection of COVID-19 in X-rays using nCOVnet. *Chaos, Solitons, and Fractals, 138*, 109944.
6. Davenport, T., & Kalakota, R. (2019). The potential for artificial intelligence in healthcare. *Future Healthcare Journal, 6*(2), 94–98.
7. Mayorga-Ruiz, I., Jiménez-Pastor, A., Fos-Guarinos, B., López-González, R., Garcia-Castro, F., & Alberich-Bayarri, Á. (2019). The role of artificial intelligence in clinical research. In E. Ranschaert, S. Morozov, & P. Algra (Eds.), *Artificial Intelligence in Medical Imaging* (pp. 189–199). Cham, Switzerland: Springer.
8. Panch, T., Szolovits, P., & Atun, R. (2018). Artificial intelligence, machine learning and health systems. *Journal of Global Health, 8*(2), 020303.
9. de Jong, A. J., van Rijssel, T. I., Zuidgeest, M. G. P., van Thiel, G. J. M. W., Askin, S., Fons-Martínez, J., De Smedt, T., de Boer, A., Santa-Ana-Tellez, Y., & Gardarsdottir, H. Trials@Home Consortium (2022). Opportunities and challenges for decentralized clinical trials: European Regulators' perspective. *Clinical Pharmacology and Therapeutics, 112*(2), 344–352.
10. Petrini, C., Mannelli, C., Riva, L., Gainotti, S., & Gussoni, G. (2022). Decentralized clinical trials (DCTs): A few ethical considerations. *Frontiers in Public Health, 10*, 1081150.

11. Piekarz, D., & Sorokin, A. (2024, April 16). *AI in Clinical Trials: Increase Speed and Efficiency While Cutting Cost and Error Rates*. DataArt. Retrieved August 3, 2024, from https://www.dataart.com/blog/ai-in-clinical-trials-increase-speed-and-efficiency-while-cutting-cost-and-error-rates#:~:text=Clinical%20trials%20have%20traditionally%20been,lab%20to%20the%20pharmacy%20shelf
12. Juneja, A., Chaturvedi, S., Shrivastav, S., Dhar, H., & Beri, A. (2022). Artificial intelligence—Revolutionizing clinical trial process. *IJBPAS*, *11*(12), 5625–5642.
13. Sarma, J. (2021). Role of artificial intelligence in medicine and clinical research. *International Journal of Research*, *9*(8), 1512–1518.
14. Venigandla, K., & Tatikonda, V. M. (2023). Optimizing clinical trial data management through RPA: A strategy for accelerating medical research. *International Journal of Enhanced Research in Science, Technology & Engineering*, *12*(8), 1–9. ISSN 2319-7463
15. Boikanyo, K., Zungeru, A. M., Sigweni, B., Yahya, A., & Lebekwe, C. (2023). Remote patient monitoring systems: Applications, architecture, and challenges. *Scientific African*, *20*, e01638.
16. Askin, S., Burkhalter, D., Calado, G., & El Dakrouni, S. (2023). Artificial intelligence applied to clinical trials: Opportunities and challenges. *Health and Technology*, *13*(2), 203–213.
17. Krishnan, G., Singh, S., Pathania, M., Gosavi, S., Abhishek, S., Parchani, A., & Dhar, M. (2023). Artificial intelligence in clinical medicine: Catalyzing a sustainable global healthcare paradigm. *Frontiers in Artificial Intelligence*, *6*, 1227091.
18. Maleki Varnosfaderani, S., & Forouzanfar, M. (2024). The role of AI in hospitals and clinics: Transforming healthcare in the 21st century. *Bioengineering (Basel, Switzerland)*, *11*(4), 337.
19. Esmaeilzadeh, P. (2020). Use of AI-based tools for healthcare purposes: A survey study from consumers' perspectives. *BMC Medical Informatics and Decision Making*, *20*(1), 170.
20. Jarrahi, M. H. (2018). Artificial intelligence and the future of work: Human-AI symbiosis in organizational decision making. *Business Horizons*, *61*(4), 577–586.
21. Laï, M. C., Brian, M., & Mamzer, M. F. (2020). Perceptions of artificial intelligence in healthcare: Findings from a qualitative survey study among actors in France. *Journal of Translational Medicine*, *18*(1), 14.
22. Zhao, X., Xia, Q., & Wayne, W. H. (2020). Impact of technostress on productivity from the theoretical perspective of appraisal and coping processes. *Information & Management*, 103265.
23. Kitsios, F., Kamariotou, M., Syngelakis, A. I., & Talias, M. A. (2023). Recent advances of artificial intelligence in healthcare: A systematic literature review. *Applied Sciences*, *13*(14), 7479.
24. Chen, J., & See, K. C. (2020). Artificial intelligence for COVID-19: Rapid review. *Journal of Medical Internet Research*, *22*(10), e21476.
25. Katiyar, S., & Farhana, A. (2022). Artificial intelligence in e-health: A review of current status in healthcare and future possible scope of research. *Journal of Computer Science*, *18*(10), 928–939. https://doi.org/10.3844/jcssp.2022.928.939
26. Shlobin, N. A., Baig, A. A., Waqas, M., Patel, T. R., Dossani, R. H., Wilson, M., Cappuzzo, J. M., Siddiqui, A. H., Tutino, V. M., & Levy, E. I. (2022). Artificial intelligence for large-vessel occlusion stroke: A systematic review. *World Neurosurgery*, *159*, 207–220.e1.
27. Alcedis (a Huma company) (2024, February 12). *How AI Data Management can Transform Your Clinical Trial*. Clinical Trials Arena. Retrieved August 7, 2024, from https://www.clinicaltrialsarena.com/sponsored/how-ai-data-management-can-transform-your-clinical-trial/
28. Devlin, H. (2023, February 27). *Patients Losing Out amid Slump in NHS Clinical Trials, Warn Top Clinicians*. Retrieved July 10, 2024, from https://www.theguardian.com/science/2023/feb/27/patients-losing-out-amid-slump-in-nhs-clinical-trials-warn-top-clinicians?trk=article-ssr-frontend-pulse_little-text-block
29. MAC Clinical Research (2024, June 26). *Why Participant Recruitment Remains the Biggest Challenge in Clinical Trials*. LinkedIn. Retrieved July 4, 2024, from https://www.linkedin.com/pulse/why-participant-recruitment-remains-biggest-challenge-arb0e
30. Alcedis (a Huma Company) (2024, November 12). *Meteor*. Alcedis. Retrieved August 4, 2024, from https://www.alcedis.de/en/meteor
31. Lee, S., & Kim, H. S. (2021). Prospect Of artificial intelligence based on electronic medical record. *Journal of Lipid and Atherosclerosis*, *10*(3), 282–290.
32. Jeun, Y. J. (2013). EMR system and patient medical information protection. *Korean J Health Serv Manag*, *7*, 213–224.
33. Kim, H. S., Kim, D. J., & Yoon, K. H. (2019). Medical big data is not yet available: Why we need realism rather than exaggeration. *Endocrinology and Metabolism (Seoul, Korea)*, *34*(4), 349–354.
34. Korea Legislation Research Institute (2020, February 4). *Personal Information Protection Act*. Retrieved July 4, 2024, from https://elaw.klri.re.kr/eng_service/lawView.do?hseq=53044&lang=ENG
35. Oliveira, A., & Salgado, N. (2005). Intelligent data analysis of clinical trials. *AMIA Annual Symposium Proceedings*, *2005*, 1064.
36. Aimlprogramming (n.d.). *Intelligent Clinical Trial Data Analysis*. Retrieved August 7, 2024, from https://aimlprogramming.com/download/pdf/intelligent-clinical-trial-data-analysis-1711129082.pdf

37. DATAMEDICA (n.d.). *COATI*. Retrieved August 5, 2024, from https://www.datamedica.pt/download/popup/coatiuk.htm
38. Gouveia-Oliveira, A., & Salgado, N. C. (1999). A database system for integrated clinical trial management, control, statistical analysis and ICH-compliant reporting. *Proceedings. AMIA Symposium*, 72–76.
39. Vijayakumar, S., Lee, V. V., Leong, Q. Y., Hong, S. J., Blasiak, A., & Ho, D. (2023). Physicians' perspectives on AI in clinical decision support systems: Interview study of the CURATE.AI personalized dose optimization platform. *JMIR Human Factors, 10*, e48476.
40. Sutton, R. T., Pincock, D., Baumgart, D. C., Sadowski, D. C., Fedorak, R. N., & Kroeker, K. I. (2020). An overview of clinical decision support systems: Benefits, risks, and strategies for success. *NPJ Digital Medicine*, *3*, 17.
41. de Jong, J., Cutcutache, I., Page, M., Elmoufti, S., Dilley, C., Fröhlich, H., & Armstrong, M. (2021). Towards realizing the vision of precision medicine: AI based prediction of clinical drug response. *Brain: a Journal of Neurology*, *144*(6), 1738–1750.
42. Ganapathi, S., & Duggal, S. (2023). Exploring the experiences and views of doctors working with artificial intelligence in English healthcare: A qualitative study. *PLoS One*, *18*(3), e0282415.
43. Blasiak, A., Khong, J., & Kee, T. (2020). CURATE. AI: Optimizing personalized medicine with artificial intelligence. *SLAS Technology*, *25*(2), 95–105.
44. Fegan, G. W., & Lang, T. A. (2008). Could an open-source clinical trial data-management system be what we have all been looking for? *PLoS Medicine*, *5*(3), e6.
45. Park, Y. R., Yoon, Y. J., Koo, H., Yoo, S., Choi, C. M., Beck, S. H., & Kim, T. W. (2018). Utilization of a clinical trial management system for the whole clinical trial process as an integrated database: system development. *Journal of Medical Internet Research*, *20*(4), e103.
46. Datrics (n.d.). *What Is Automated Reporting? A Key to Data Intelligence*. Retrieved July 4, 2024, from https://www.datrics.ai/articles/what-is-automated-reporting-a-key-to-data-intelligence
47. Narrativa (2023, May 19). *Automation of Clinical Study Reports (CSRs)*. Retrieved August 5, 2024, from https://www.narrativa.com/automation-of-clinical-study-reports/#:~:text=Automation%20of%20Clinical%20Report%20Writing,also%20the%20number%20of%20reviews
48. Atieh, D., & Domanska, O. (2024, March 26). *Finding the Right Patients for the Right Treatment with AI*. Retrieved August 6, 2024, from https://www.avenga.com/magazine/how-ai-advances-patient-recruitment-in-clinical-trials/#:~:text=Its%20algorithms%20can%20identify%20potential,%2Dworld%20data%20(RWD)
49. Wang, K., Cui, H., Zhu, Y., Hu, X., Hong, C., Guo, Y., An, L., Zhang, Q., & Liu, L. (2024). Evaluation of an artificial intelligence-based clinical trial matching system in Chinese patients with hepatocellular carcinoma: A retrospective study. *BMC Cancer, 24*(1), 246.
50. Alexander, M., Solomon, B., Ball, D. L., Sheerin, M., Dankwa-Mullan, I., Preininger, A. M., … Herath, D. M. (2020). Evaluation of an artificial intelligence clinical trial matching system in Australian lung cancer patients. *JAMIA Open*, *3*(2), 209–215.
51. Beck, J. T., Rammage, M., Jackson, G. P., Preininger, A. M., Dankwa-Mullan, I., Roebuck, M. C., Torres, A., Holtzen, H., Coverdill, S. E., Williamson, M. P., Chau, Q., Rhee, K., & Vinegra, M. (2020). Artificial intelligence tool for optimizing eligibility screening for clinical trials in a large community cancer center. *JCO Clinical Cancer Informatics*, *4*, 50–59.
52. Deep 6 AI (n.d.). *Patient Recruitment Software with AI - Deep 6 AI*. Retrieved July 4, 2024, from https://deep6.ai/
53. Alzheimer's Association (n.d.). *TrialMatch: Find Clinical Trials for Alzheimer's and Other Dementia*. Retrieved August 4, 2024, from https://www.alz.org/alzheimers-dementia/research_progress/clinical-trials/trialmatch
54. Saama (n.d.). *Saama: #1 in AI Clinical Analytics*. Retrieved July 3, 2024, from https://www.saama.com/
55. Clindata Insight (n.d.). *Clindata Insight: Biometrics Consulting*. Retrieved August 3, 2024, from https://www.clindatainsight.com/
56. Antidote (2024, March 8). *Clinical Trial Patient Recruitment | Antidote*. Retrieved August 5, 2024, from https://www.antidote.me/
57. PathAI (2022, February 1). *PathAI*. Retrieved July 7, 2024, from https://www.pathai.com
58. Flatiron Health (n.d.). *OncoTrials*. Retrieved August 6, 2024, from https://flatiron.com/clinical-research-solutions/oncotrials
59. Deep Lens (2020, July 24). *Deep Lens*. Rev1 Ventures. Retrieved August 6, 2024, from https://www.rev1ventures.com/blog/video-founders-story-with-deep-lens-ceo/
60. Dexcom (n.d.). *Dexcom CGM Clinical Trials on Glycemic Control*. Retrieved August 5, 2024, from https://provider.dexcom.com/education-research/clinical-evidence/clinical-studies
61. Collier, S. R., McCraw, C., Campany, M., Lubkeman, A., StClair, P., Ji, H., Sandberg, K., Morgan, J. W., & Smith, C. J. (2020). Withings body cardio versus gold standards of pulse-wave velocity and body composition. *Journal of Personalized Medicine*, *10*(1), 17.
62. Ohannessian, R., Duong, T. A., & Odone, A. (2020). Global telemedicine implementation and integration within health systems to fight the COVID-19 pandemic: A call to action. *JMIR Public Health and Surveillance*, *6*(2), e18810.
63. Sensely (n.d.). *Sensely: Conversational AI to Improve Health and Drive Member Engagement*. Retrieved July 3, 2024, from https://sensely.com/

64. Sun, G., & Zhou, Y. H. (2023). AI in healthcare: Navigating opportunities and challenges in digital communication. *Frontiers in Digital Health*, *5*, 1291132.
65. Medisafe: Digital Health, Platform, Adherence, Persistence. https://www.medisafe.com/ (accessed on 4 July 2024).
66. AdhereTech (2024, July 23). *AdhereTech: Medication Adherence Solutions for Better Health*. Retrieved August 6, 2024, from https://adheretech.com/
67. Proteus Digital Health (n.d.). *Proteus Digital Health*. Retrieved July 6, 2024, from https://www.lottiecrumbleholme.com/proteus-discover
68. Bohlmann, A., Mostafa, J., & Kumar, M. (2021). Machine learning and medication adherence: Scoping review. *JMIRx Med*, *2*(4), e26993.
69. Cho, J. S., & Park, J. H. (2024). Application of artificial intelligence in hypertension. *Clinical Hypertension*, *30*(1), 11.
70. Plaster, L. (2020, April 6). *Conversa Is Using Its Chatbot to Help Top Hospitals Increase Capacity During the COVID-19 Pandemic*. Medium. Retrieved August 6, 2024, from https://healthtransformer.co/conversa-is-using-its-chatbot-to-help-top-hospitals-increase-capacity-during-the-covid-19-pandemic-d4bedd421532
71. Lark Health (n.d.). *Lark Health*. Retrieved August 5, 2024, from https://www.lark.com/
72. HealthPrize (n.d.). *HealthPrize*. Retrieved August 5, 2024, from https://healthprize.com
73. Debong, F., Mayer, H., & Kober, J. (2019). Real-world assessments of mySugr Mobile health app. *Diabetes Technology & Therapeutics*, *21*(S2), S235–S240.
74. Aguillard, K., & Garson, A. Jr (2020). Patient use of cardiovascular devices and apps: Are we getting our money's worth. *Methodist DeBakey Cardiovascular Journal*, *16*(4), 291.
75. Gong, J., Liu, J. Y., Sun, X. W., Zheng, B., & Nie, S. D. (2018). Computer-aided diagnosis of lung cancer: The effect of training data sets on classification accuracy of lung nodules. *Physics in Medicine and Biology*, *63*(3), 035036.
76. Sorrentino, F. S., Jurman, G., De Nadai, K., Campa, C., Furlanello, C., & Parmeggiani, F. (2020). Application of artificial intelligence in targeting retinal diseases. *Current Drug Targets*, *21*(12), 1208–1215.
77. Heydon, P., Egan, C., Bolter, L., Chambers, R., Anderson, J., Aldington, S., Stratton, I. M., Scanlon, P. H., Webster, L., Mann, S., du Chemin, A., Owen, C. G., Tufail, A., & Rudnicka, A. R. (2021). Prospective evaluation of an artificial intelligence-enabled algorithm for automated diabetic retinopathy screening of 30000 patients. *The British Journal of Ophthalmology*, *105*(5), 723–728.
78. Xie, Q., Liu, Y., Huang, H., Hong, B., Wang, J., Han, H., & Liu, Y. (2020). An innovative method for screening and evaluating the degree of diabetic retinopathy and drug treatment based on artificial intelligence algorithms. *Pharmacological Research*, *159*, 104986.
79. Coudray, N., Ocampo, P. S., Sakellaropoulos, T., Narula, N., Snuderl, M., Fenyö, D., Moreira, A. L., Razavian, N., & Tsirigos, A. (2018). Classification and mutation prediction from non-small cell lung cancer histopathology images using deep learning. *Nature Medicine*, *24*(10), 1559–1567.
80. Namikawa, K., Hirasawa, T., Yoshio, T., Fujisaki, J., Ozawa, T., Ishihara, S., Aoki, T., Yamada, A., Koike, K., Suzuki, H., & Tada, T. (2020). Utilizing artificial intelligence in endoscopy: A clinician's guide. *Expert Review of Gastroenterology & Hepatology*, *14*(8), 689–706.
81. Gulati, S., Emmanuel, A., Patel, M., Williams, S., Haji, A., Hayee, B., & Neumann, H. (2020). Artificial intelligence in luminal endoscopy. *Therapeutic Advances in Gastrointestinal Endoscopy*, *13*, 2631774520935220.
82. Tae, K. (2021). Robotic thyroid surgery. *Auris, Nasus, Larynx*, *48*(3), 331–338.
83. Jones, R., Dobbs, R. W., Halgrimson, W. R., Vigneswaran, H. T., Madueke, I., Wilson, J., Abern, M. R., & Crivellaro, S. (2020). Single port robotic radical prostatectomy with the Da Vinci SP platform: A step by step approach. *The Canadian Journal of Urology*, *27*(3), 10263–10269.
84. Lenfant, L., Wilson, C. A., Sawczyn, G., Aminsharifi, A., Kim, S., & Kaouk, J. (2020). single-Port robot-assisted dismembered pyeloplasty with mini-Pfannenstiel or peri-umbilical access: Initial experience in a single center. *Urology*, *143*, 147–152.
85. Winder, A., Strauss, D. C., Jones, R. L., Benson, C., Messiou, C., Chaudry, M. A., & Smith, M. J. (2020). Robotic surgery for gastric gastrointestinal stromal tumors: A single center case series. *Journal of Surgical Oncology*, *122*(4), 691–698.
86. Rahmaniar, W., Maarif, A., Haq, Q. M. U., & Iskandar, M. E. (2023). AI in industry: Real-world applications and case studies. *TechRxiv*.
87. Abelson, S., Collord, G., Ng, S. W., Weissbrod, O., Mendelson Cohen, N., Niemeyer, E., & Shlush, L. I. (2018). Prediction of acute myeloid leukaemia risk in healthy individuals. *Nature*, *559*(7714), 400–404.
88. Zhang, H. T., Zhang, J. S., Zhang, H. H., Nan, Y. D., Zhao, Y., Fu, E. Q., Xie, Y. H., Liu, W., Li, W. P., Zhang, H. J., Jiang, H., Li, C. M., Li, Y. Y., Ma, R. N., Dang, S. K., Gao, B. B., Zhang, X. J., & Zhang, T. (2020). Automated detection and quantification of COVID-19 pneumonia: CT imaging analysis by a deep learning-based software. *European Journal of Nuclear Medicine and Molecular Imaging*, *47*(11), 2525–2532.
89. Mohanty, S., Harun Ai Rashid, M., Mridul, M., Mohanty, C., & Swayamsiddha, S. (2020). Application of artificial intelligence in COVID-19 drug repurposing. *Diabetes & Metabolic Syndrome*, *14*(5), 1027–1031.
90. Ke, Y. Y., Peng, T. T., Yeh, T. K., Huang, W. Z., Chang, S. E., Wu, S. H., Hung, H. C., Hsu, T. A., Lee, S. J., Song, J. S., Lin, W. H., Chiang, T. J., Lin, J. H., Sytwu, H. K., & Chen, C. T. (2020). Artificial intelligence approach fighting COVID-19 with repurposing drugs. *Biomedical Journal*, *43*(4), 355–362.

91. Char, D. S., Shah, N. H., & Magnus, D. (2018). Implementing machine learning in health care – Addressing ethical challenges. *The New England Journal of Medicine, 378*(11), 981–983.
92. Elendu, C., Amaechi, D. C., Elendu, T. C., Jingwa, K. A., Okoye, O. K., John Okah, M., Ladele, J. A., Farah, A. H., & Alimi, H. A. (2023). Ethical implications of AI and robotics in healthcare: A review. *Medicine, 102*(50), e36671.
93. Akinci D'Antonoli, T. (2020). Ethical considerations for artificial intelligence: An overview of the current radiology landscape. *Diagnostic and Interventional Radiology, 26*(5), 504.
94. Price, W. N. 2nd, & Cohen, I. G. (2019). Privacy in the age of medical big data. *Nature Medicine, 25*(1), 37–43.
95. Calo, R. (2017). Artificial intelligence policy: A primer and roadmap. *University of Bologna Law Review, 3*, 180–218.
96. Ruger, J. P. (2004). Ethics of the social determinants of health. *Lancet (London, England), 364*(9439), 1092–1097.
97. Denecke, K., Bamidis, P., Bond, C., Gabarron, E., Househ, M., Lau, A. Y., Mayer, M. A., Merolli, M., & Hansen, M. (2015). Ethical issues of social media usage in healthcare. *Yearbook of Medical Informatics, 10*(1), 137–147.
98. Sivarajah, U., Wang, Y., Olya, H., & Mathew, S. (2023). Responsible artificial intelligence (AI) for digital health and medical analytics. *Information Systems Frontiers: A Journal of Research and Innovation*, 1–6. Advance online publication.
99. Emam, K. E., & Arbuckle, L. (Eds.) (2013). *Anonymizing Health Data: Case Studies and Methods to Get You Started*. Sebastopol, CA: O'Reilly Media.
100. Suresh, H., & Guttag, J. (2021). A framework for understanding sources of harm throughout the machine learning life cycle. In *Proceedings of EAAMO '21: Equity and Access in Algorithms, Mechanisms, and Optimization (EAAMO '21)* (pp. 1–9). ACM.
101. Calders, T., & Verwer, S. (2010). Three naive Bayes approaches for discrimination-free classification. *Data Mining and Knowledge Discovery, 21*, 277–292.
102. Buolamwini, J., & Gebru, T. (2018). Gender shades: Intersectional accuracy disparities in commercial gender classification. *Proceedings of the 1st Conference on Fairness, Accountability and Transparency, 81*, 77–91.
103. Naik, N., Hameed, B. M. Z., Shetty, D. K., Swain, D., Shah, M., Paul, R., Aggarwal, K., Ibrahim, S., Patil, V., Smriti, K., Shetty, S., Rai, B. P., Chlosta, P., & Somani, B. K. (2022). Legal and ethical consideration in artificial intelligence in healthcare: Who takes responsibility? *Frontiers in Surgery, 9*, 862322.
104. Brady, A. P., & Neri, E. (2020). Artificial intelligence in radiology-ethical considerations. *Diagnostics (Basel, Switzerland), 10*(4), 231.
105. Wheeler, T. (2023, January 15). *The Three Challenges of AI Regulation*. Brookings. Retrieved July 4, 2024, from https://www.brookings.edu/articles/the-three-challenges-of-ai-regulation/
106. The European Data Protection Board (2020, January 30). *Guidelines 3/2019 on Processing of Personal Data through Video Devices*. European Data Protection Board. Retrieved August 3, 2024, from https://edpb.europa.eu/our-work-tools/our-documents/guidelines/guidelines-32019-processing-personal-data-through-video_en
107. Corbett-Davies, S., Gaebler, J. D., Nilforoshan, H., Shroff, R., & Goel, S. (2023). The measure and mismeasure of fairness. *The Journal of Machine Learning Research, 24*(1), 14730–14846.
108. Soni, N., Sharma, E. K., Singh, N., & Kapoor, A. (2020). Artificial intelligence in business: From research and innovation to market deployment. *Procedia Computer Science, 167*, 2200–2210.
109. Wang, Z., Gao, C., Glass, L., & Sun, J. (2022). Artificial intelligence for in silico clinical trials: A review. *ArXiv*. https://arxiv.org/abs/2209.09023

CHAPTER 22

Augmented Reality and Virtual Reality in Medical Training with AI

Muhammad Sulaman Akhtar, Ayesha Khalid, Muneeb Faiz, Amatul Hadi Hamdana, Ali Abbas Mankani, Mohammed Hisham Alamin, and Muhammad Yaqoob Akhtar

22.1 OVERVIEW

Augmented Reality (AR) and Virtual Reality (VR) has transformed medical teaching by providing immersive, realistic settings in which healthcare workers may practice without compromising patient safety. AR augments the surroundings with graphical overlays such as 3D anatomical models, allowing students to perceive and engage with medical information in ways that textbooks and lectures cannot. On the other hand, VR offers entirely immersive, computer-generated environments that enable medical trainees to practice procedures, experience various scenarios, and enhance their abilities in a controlled environment [1]. These technologies are especially effective when coupled with artificial intelligence (AI), which improves the training experience by providing real-time feedback, personalized learning, and adaptive simulations suited to each learner's progress. The combination of AR, VR, and AI is making medical education more effective and tailored. AI-powered simulations create scenarios for students to practice complicated operations and build expertise and confidence before dealing with actual patients. This promotes faster learning and skill improvement [2]. Similarly, AR is revolutionizing the teaching of anatomy. AR enhances knowledge and retention over traditional textbook learning by allowing students to interact with 3D representations of the human body [3].

The integration of AI into AR and VR systems enables the creation of flexible scenarios that adjust immediately in accordance with learner's performance. Consequently, we provide tailored supervision to medical trainees, ensuring they receive appropriate challenges and support throughout their training [4]. This adaptability is essential for giving a more effective learning experience. Despite the obvious benefits, there are still hurdles. High price and technical requirements of these technologies limit their use, particularly in underfunded medical facilities. Furthermore, ensuring that these sophisticated training tools are available in a wider range of healthcare settings is also an ongoing issue [5].

22.2 APPLICATIONS IN MEDICAL TRAINING

Early research into AR and VR technologies in medicine was mostly experimental, with an emphasis on the tools' ability to improve visualizations and simulations. The early uses of VR in thoracic surgery training demonstrated its potential to give thorough, realistic simulations of complicated operations, allowing trainees to practice without endangering patients [6]. These early attempts paved the way for wider adoption by demonstrating the practical benefits of immersive learning environments in medical education. The previous five years have brought remarkable breakthroughs in AR and VR technology, due to considerable improvements in hardware and software [7]. The healthcare sector has progressively used AR and VR, especially in the field of diagnostic imaging. These technologies are going from being simple

DOI: 10.1201/9781003533351-22

learning tools to important parts of planning and carrying out surgeries, which shows how important they are in clinical practice [1]. These technologies have progressed from just experimental tools to must use instructional materials. These technologies have proven important in training patient-facing communication skills, providing realistic settings in which medical students may communicate with virtual patients [8]. AR and VR have been widely accepted as essential components of medical education. These technologies are used in Pediatrics for accessibility of medical procedures and making them engaging for the children [9]. These tools are beneficial for different specialties and ages.

22.3 INTEGRATION OF AI IN AR AND VR

AI has changed the medical education, improving capabilities of AR and VR technologies. The incorporation of AI has allowed these technologies to become more adaptable, personalized, and sensitive to the specific requirements of medical trainees. AI-powered improvements include real-time feedback, individualized learning experiences, and complex scenario development, all of which lead to a more effective and engaging educational experience [10].

The AI enables AR and VR systems to assess data from user interactions tailoring them for the. For example, AI can monitor the precision and accuracy of a surgical simulation, offering quick feedback and suggestions for improvement. This dynamic integration establishes a loop of continuous learning in which the system adjusts to the learner's demands, improving the educational result [10]. AI systems are able to instantly analyze data by processing a learner's activities in a simulation. For example, in a simulated operation, the system may evaluate incision accuracy, tool management, and adherence to procedural procedures. This rapid feedback loop enables trainees to learn from errors in controlled setting [1]. By monitoring a trainee's performance over time, AI may discover areas of strength and weakness, adjusting the intensity and emphasis of future training sessions accordingly. For example, if a learner repeatedly has difficulties with a certain diagnostic process, the system might provide more practice situations and resources for strengthening that ability. This tailored strategy not only speeds up learning but also improves memory and understanding of difficult medical procedures [2, 1]. One of the most important advances in AI integration with AR and VR is the capacity to create intricate medical scenarios. AI-powered simulations may generate from common operations to unusual and difficult cases, which enable trainees to practice in a safe setting, preparing them for real-world medical difficulties.

For example, AI may create a simulated patient who is suffering an uncommon problem following surgery. Trainees may practice reacting to this scenario, obtaining significant experience that is difficult to obtain in a standard training environment [2].

22.4 APPLICATIONS OF AR AND VR IN MEDICAL TRAINING

AR and VR have changed medical training. These allow medical professionals to engage in realistic simulations of surgical procedures, anatomical studies, and collaborative learning environments without risking patient safety.

22.4.1 Surgical Training

VR is most widely used in surgical simulation and is therefore the most utilized in surgical training platforms. AR and Mixed Reality are useful in real surgical settings, specifically in preoperative planning and intraoperative guidance [11]. These technologies offer high efficiency in surgical simulations and can replace traditional training procedures.

VR surgical simulations have 3D images that allow for detailed anatomic representations along with the extent of tissue deformation that help in a better understanding of any surgical case. A study demonstrated that VR models significantly increased surgical success rates among trainees, achieving an 86% success rate compared to 74% with traditional methods [12]. AR is revolutionizing surgery by creating an immersive, interactive and real learning space, significantly improving skills and confidence of surgical trainees. It addresses the bias in traditional surgical assessment methods by providing objective performance metrics [13]. Trainees can now perform difficult surgical operations in a risk-free and secure setting by utilizing AR. This helps develop skills necessary for real-life surgeries [14].

22.4.2 Immersive Anatomy Learning

AR allows visualization of human anatomy in 3D rather than usual 2D visualization. This innovative approach allows students and trainees to learn Anatomy in a dynamic environment, addressing the limitations of traditional learning methods. Hence, makes learning easier along with a better understanding of complex organs.

Several studies have explored the applications of AR in anatomy learning. AR tools such as Microsoft HoloLens II help students to see complex anatomic structures in 3D enhancing comprehension and pictorial retention [15]. The miracle system uses a magic mirror concept to augment CT datasets onto the user's body, allowing for interactive slice-based visualization [16].

An Android app has been developed to display 3D anatomical models in real space when scanning images [17]. Screen-based AR and tablet applications have been demonstrated to provide better learning experiences compared to traditional textbooks, making anatomy education easily available [3].

22.4.3 Interventional and Diagnostic Training

Procedural and diagnostic training in healthcare is revolutionized by AR and VR technologies, which provide immersive and interactive learning. These technologies improve skill acquisition as well as cognitive challenges faced by trainees. AR can overlay diagnostic data (CT scans, MRI, ultrasound images) directly onto patients' bodies thus allowing the healthcare providers to see patients' internal structures in real-time and correct anatomical context [18]. AR improves CT scan interpretation by providing 3D reconstructions, aiding radiologists in visualizing lesions and anatomical details more efficiently [1]. It enhances the accuracy of procedures in different interventions, such as needle placements, by offering real-time guidance and reducing radiation exposure [19]. VR could be used for central venous catheter placement, indicating its potential to improve the learning experience for novice trainees [20].

AR can assist with ultrasound-guided needle biopsies. In a trial, medical team inserted needle for biopsy into a synthetic tumor and stated that process was easy, simple, and reliable [21].

AR and VR are revolutionizing interventional radiology (IR) by enhancing visualization, procedural guidance, and training. AR technology enhances the accuracy of needle placement by providing 3D or 2D anatomical structures along with the target area. This visual aid helps healthcare providers achieve more precise results compared to conventional methods [19]. So, all procedure carried on with IR can be performed more efficiently with AR and VR assistance.

22.4.4 Collaborative Learning

Multiplayer VR systems allow learners in different roles (doctors, nurses, patients, etc.) to interact within the same virtual environment and deal with the same clinical scenario: this enhances teamwork, communication, and a better understanding of inter-professional roles. Multiplayer VR systems are anticipated to become progressively accessible, enabling learners to visualize and engage with each other and the simulated patient within a virtual context. Medical students can participate in virtual rounds and discuss the case findings together. This innovation will enhance interpersonal and professional competencies by prioritizing humanistic care and collaboration, in addition to simulating real-life clinical experiences. Distance is no longer a barrier to medical education. AR/VR enables learners to engage in the same training process from anywhere worldwide [22].

The outbreak of Coronavirus disease increases the reliance on digital tools such as VR in medical education, which forced a shift significant and accelerated change in the remote learning environment [23].

22.5 BENEFITS OF AR AND VR WITH AI

AR\VR transforms medical concepts into 3D models and simulations. AI algorithms create a personalized learning path for every student by analyzing his performance and providing feedback and guidance through the entire learning process [24]. VR also can present complex clinical scenarios understandably [25].

The interactive nature of VR increases student involvement compared to traditional learning methods which enhances knowledge retention and result in greater self-efficacy regarding the ability to replicate the procedural steps [22]. VR/AR technologies significantly improve spatial awareness compared to textbooks. Realistic simulations of complex surgeries and emergencies allow students to enhance their critical thinking in a controlled space to practice skills. These situations can be programmed with different disease cases, anatomical variations, investigations, and complications offering to broad range of situations [3].

Students can practice procedure as many times as they want without harming the patient and can become expert before applying the skills on real patients. VR-based endoscopy provides a standardized training experience by tracking performance metrics such as procedure time, and movement errors [26].

22.6 LIMITATIONS AND CHALLENGES

One of the biggest challenges that have limited the use of AR/VR in all institutions is the hefty payment for these systems. AR/VR for healthcare are still in their infancy and are often considered a luxury rather than a necessity by most of the education sector. This coupled with the lack of tools needed to establish and increase access to interactive healthcare education handicaps the use of these systems in medical training. Developing AR- and VR-based curriculums requires top-quality hardware, high-speed computers and graphics, high-resolution displays, and highly specialized tools, all of which are extremely costly [27].

The high cost of these systems also raises concerns about accessibility. Although these technologies do offer innovative and immersive learning opportunities, their high costs and the need for specialized equipment can create barriers for training institutions, particularly those in underfunded or rural areas. This disparity can lead to unequal access to advanced training resources, widening the gap between well-resourced and underserved educational environments.

Another critical challenge is its potential impact on face-to-face communications and how it may hamper the relationship between an instructor and a student. Ultimately, over-reliance on AR and VR simulations may lead to reduced opportunities for students to develop empathy and learn effective communication skills and bedside techniques—key elements in patient care and medical education. This digital shift could create a generation of healthcare professionals who are technically proficient but potentially lacking in the interpersonal skills necessary to establish trust and a strong interpersonal relationship with patients. Consequently, this reduced human interaction may undermine AR and VR's holistic approach to patient care and the nurturing of supportive and dynamic learning environments.

Introducing any new technology requires a lot of training of its own—for both trainers and trainees. Unfortunately, while many people may have heard of AR/VR for gaming and other forms of digital media, they are completely unaware of utilization of AR/VR in medicine. This lack of knowledge combined with the gap in training leads to a lack of confidence among instructors and students alike, hindering the development of innovative teaching methods that could enhance learning outcomes. Furthermore, without adequate awareness, educators may be reluctant to shift from traditional teaching methods. Consequently, this gap in education and training can prevent the medical community from fully leveraging AR, VR, and AI technologies, which could otherwise revolutionize medical education and improve patient care outcomes. In addition, the lack of standardized protocols for integrating AR and VR into curricula may lead to inconsistent educational experiences across different regions and institutions. Thus, the development of research-based curricula and stringent guidelines that define the best use of AR and VR is the need of the hour.

There are a number of ethical and safety considerations that must also be taken into account when it comes to AI-driven medical training. Data privacy is a key issue. Large volumes of patient data are analyzed by AI systems, which prompts how these private data are kept secure from intrusions. Additionally, bias in AI algorithms can perpetuate existing inequalities in healthcare if the training data are not diverse, leading to a lack of proper training on treating underrepresented groups. Additionally, if a mishap occurs due to an AI suggestion, it's often unclear who is responsible: the developer, the educator, or the trainee.

VR can also interfere with mental functions, affecting a person's ability to process experiences, recall information, make decisions, and differentiate between their identity and the external world [28]. Therefore, studying and managing the safety and health implications is extremely important. Nausea, dizziness, disorientation, postural instability, and fatigue have been reported with utilization of VR [22]. Often referred to as cyber sickness, the exact mechanism of action of these symptoms are not known, but they are thought to be a consequence of a disconnect between the digital world and reality. These ethical and safety concerns must be addressed to ensure AI-driven medical training enhances, rather than undermines, the quality and fairness of healthcare education.

22.7 EXAMPLES AND CASE STUDIES

Notable implementations of AR/VR with AI in medical education have demonstrated significant benefits, including enhanced expertise [29, 30], improved safety, repeatability, and efficiency in surgical training [31]. These implementations have transformed healthcare delivery, improved AR display accuracy, and enhanced COVID-19 diagnostics using XR and deep neural models [30]. Success stories from leading medical institutions and organizations include Lausanne University Hospital's use of the metaverse for virtual surgical operations, providing invaluable hands-on experience, VR learning environments—for counseling, enhancing comfort, smart glasses for medical professionals, and AR technology for dental implant placement and urology surgical education [30]. Studies have shown the success of using AR/VR with AI in medical education, such as improved learning curves for hip arthroscopy trainees, enhanced understanding of orthopedic surgical processes (Cecil et al., 2018), reduced learning time and errors in orthognathic surgery (Medellin-Castillo et al., 2020), and improved anatomical information visualization and decision-making accuracy in neurosurgical training (Si et al., 2019). Additionally, AR has been used to provide real-time feedback during CPR, showing improved performance, and its application in supporting clinical care delivery and education [9]. Evaluation of outcomes and impact on medical professionals' training reveals mixed results, with benefits like improved learning curves, enhanced understanding of surgical processes, and reduced learning time and errors [29, 31]. However, limitations exist, such as high production costs and device immaturity [31]. Despite this, VR/AR-based training methods have unlocked benefits like safety, repeatability, and efficiency, particularly during the COVID-19 pandemic [29]. Some studies have shown no clear advantage while others have demonstrated improved CPR performance [9].

22.8 EMERGING TRENDS

As AR, VR, and AI technologies continue to evolve, their applications in medical education are poised to expand significantly. The integration and development of these technologies are revolutionizing how medical personnel are trained, offering immersive, interactive, and highly personalized learning experiences. But it doesn't just stop there. New evolutions and discoveries every day, the sky is the limit. In the near future, we can expect AR and VR technologies to become more sophisticated with improved graphics, higher resolution, and more realistic simulations. These advancements will enable more detailed and accurate anatomical representations, for exploring complex anatomical structures and procedures. New surgical training VR programs are being developed that allow students to practice a number of surgical operations with zero risk, enhancing their skills and building their confidence. Enhanced haptic feedback systems are also being created, allowing users to feel textures and resistance, which is crucial for practicing surgical techniques [17]. AI technologies are also advancing, with machine learning algorithms becoming more adept at understanding complex datasets. Real-time feedback is being increasingly integrated into AR and VR training modules. These AI-powered systems optimize learning outcomes.

Predictive analytics, powered by AI, is important in medical training. Predictive models can anticipate which areas a trainee is likely to struggle with and provide targeted training suggestions before those challenges become even more difficult. This proactive approach can help reduce training time and costs while improving competency and confidence among medical professionals. The use of AI-driven predictive analytics can also help identify trends and gaps in training programs, enabling educators to refine curricula and ensure that they are aligned with the latest advancements in medical practice.

VR and AR can also play a critical role in emergency medicine training. Different VR and AR tools provide users with a wide variety of immersion experiences that allow users to create simulations mimicking real-life scenarios that can help medical professionals learn how to make quick decisions in high-stress situations [32].

AI is also paving the way for future applications in remote surgery and global medical education. These technologies can be used for remote collaboration between surgeons, allowing experts to participate in complex surgeries. AI can enhance these remote surgeries giving decision support, ensuring precision and reducing the likelihood of errors.

Several technological breakthroughs are anticipated in the coming years that will have profound implications for the use of AR, VR, and AI in medical training. One such breakthrough is the development of more advanced neural networks that can better simulate human thought processes and decision-making, enhancing the realism and effectiveness of AR and VR training environments. Additionally, improvements can make interactions within virtual environments more intuitive [33].

Quantum computing could change data processing capabilities, allowing for more sophisticated simulations and real-time analysis. This could lead to the development of highly detailed and tailored training modules that adjust instantly to a trainee's actions, providing a more immersive and responsive learning experience. Furthermore, advancements in 5G and beyond will facilitate faster data transmission, reducing latency in AR and VR applications, which is critical for real-time training and remote surgery [17, 32, 23]. As these technologies continue to grow and develop, they offer the opportunity to transform medical education, making it more immersive and adaptive.

22.9 ETHICAL CONCERNS

The use of AI presents substantial challenges related to ethics, regulation, and privacy [34]. Ethical considerations in AI-driven training simulations include ensuring prevention from harm, data privacy, and adherence to current healthcare regulations [35], compliance with principles like respect for individual autonomy, harm prevention and transparency, with a focus on vulnerable groups and thorough risk assessment [34], and addressing potential biases in data distribution [36]. Moreover, the need for large volumes of data increases security risks, and the potential for bias in data distribution raises ethical concerns [36], Regulatory standards and compliance for medical AR/VR technologies are crucial [35], and require validation of AI algorithms, consideration of accuracy, safety, and equity, and transparency of algorithms [34]. Furthermore, an emphasis on patient agency and consent, including requirements for recurrent informed consent and the right to withdraw data, is necessary [37]. Data privacy issues and patient confidentiality in AI and VR environments are critical, with concerns about near-constant surveillance, excessive data collection, and potential privacy attacks that can compromise patient confidentiality. Additionally, AI applications often involve protected health information and unprotected data, which can be compromised through triangulation with other identifiable datasets, leading to tangible adverse effects, such as discrimination and inflated insurance premiums, as well as subjective, immeasurable effects, like loss of control and mental trauma [36]. Ensuring privacy preservation is a tedious task, and AI ethics advocates for fair, unbiased, and transparent algorithms, with a trade-off between accuracy and ethical correctness [20].

22.10 SUMMARY

This chapter discusses the advantages and limitation of these technologies in medical education.

Key applications include:

- Surgical Training: Simulating surgeries in virtual environments for hands-on practice without risks.
- Anatomy Learning: 3D visualization of human anatomy.

Procedural and Diagnostic Training: Mastering diagnostic tools through AR/VR simulations.

- Collaborative Learning: Allows for collaborative procedures and diagnostics in multi-user settings.

The benefits of integrating AR/VR and AI into medical education are:

- Improved memory and comprehension of complicated medical topics.
- Increased learner engagement and motivation.
- Provide realistic and immersive practice chances.
- Provide safe, consistent conditions for skill mastery.

The challenges and limitations are outlined below:

- The high price and technical restrictions of AR/VR systems.
- Accessibility and equality issues.
- Ethical concerns for AI-driven training.

The chapter continues by exploring future directions and new trends, such as advances in AI-powered adaptive learning and predictive analytics. It also examines the ethical, regulatory, and privacy issues surrounding these technologies.

REFERENCE

1. Lastrucci, A., Wandael, Y., Barra, A., Ricci, R., Maccioni, G., Pirrera, A., & Giansanti, D. (2023). Augmented Reality in diagnostic imaging: A review. *Journal of Clinical Medicine*, 14(13), 1333.
2. Koshak, B. O., Smiyan, S. I., & Biluha, A. V. (2024). Virtual reality in medical education: A systematic review. *Medical Journal*, 28(4), 20–30.
3. Noël, S. L., & Smuts, H. (2024). Key factors affecting the implementation of immersive technologies in ophthalmology education. *ResearchGate*, 1–15. https://www.researchgate.net/publication/382423491_Key_Factors_Affecting_the_Implementation_of_Immersive_Technologies_in_Ophthalmology_Education (accessed on 16 July 2024).
4. Sánchez-Jáuregui Descalzo, T., Gutiérrez-Pérez, N., Abad Balboa, T., Chías, P., Giordano, A., Russo, M., & Spallone, R. (Eds.) (2024). Immersion through extended reality as a tool applied to wayfinding inside hospitals. In *Advances in Representation. Digital Innovations in Architecture, Engineering and Construction* (pp. 711–722). Cham: Springer.
5. Nykyporets, S. S., Kot, S. O., Boiko, Y. V., Melnyk, M. B., & Chopliak, V. V. (2024). Advanced integration of virtual information environments (VIEs) in contemporary educational methodologies. *Society and National Security*, 4(4), 139–154.
6. Yang, J., Liu, Y., Li, J., Cheng, W., Li, B., & Jiang, P. (2024). Application of virtual reality technology in enhancing thoracic surgery skills training. *Journal of Medicine and Physical Education*, 1(2), 1–7.
7. Blumstein, G., Zukotynski, B., Cevallos, N. et al. (2020). Randomized trial of a virtual reality tool to teach surgical technique for tibial shaft fracture intramedullary nailing. *Journal of Surgical Education*, 77(4), 969–977. https://doi.org/10.1016/j.jsurg.2020.01.002.
8. Khodabakhshian, N., Gaeul Lee, K., Marawi, T., Sorkhou, M., Vyravanathan, S., & Harnett, N. (2024). Virtual reality for developing patient-facing communication skills in medical and graduate education: Protocol for a scoping review. *JMIR Research Protocols*, 13, e53901. https://www.researchprotocols.org/2024/1/e53901; https://doi.org/10.2196/53901
9. Cheng, A., Fijacko, N., Lockey, A., Greif, R., Abelairas-Gomez, C., Gosak, L., & Lin, Y. (2024). Use of augmented and virtual reality in resuscitation training: A systematic review. *Resuscitation Plus*, 18, 100643. https://doi.org/10.1016/j.resplu.2024.100643.
10. Bashirynejad, M., Soleymani, F., Nikfar, S., Zackery, A., Kebriaeezadeh, A., Majdzadeh, R., Fatemi, B., & Zare, N. (2025). Trends analysis and future study of medical and pharmacy education: A scoping review. *BMC Medical Education*, 25, Article 1527.
11. Lungu, A. J., Swinkels, W., Claesen, L., Tu, P., Egger, J., & Chen, X. (2021). A review on the applications of virtual reality, augmented reality, and mixed reality in surgical simulation: An extension to different kinds of surgery. *Expert Review of Medical Devices*, 18(1), 47–62. https://doi.org/10.1080/17434440.2021.1860750 (accessed on 1 August 2024).
12. Peng, M. J., Chen, H. Y., Chen, P. et al. (2024). Virtual reality-based surgical planning simulator for tumorous resection in FreeForm modeling: An illustrative case of clinical teaching. *Quantitative Imaging in Medicine and Surgery*, 14(2), 2060068–2062068. https://doi.org/10.21037/qims-23-1151.
13. Barrios, C. J. C., Hereñú, M. P., & Francisco, S. M. (2023). Augmented reality for surgical skills training: Update on the topic. *Gamification and Augmented Reality*, 1(1), 8–8. https://doi.org/10.56294/gr20238
14. Whig, P., Kouser, S., Sharma, A., Bhatia, A. B., & Nadikattu, R. R. (2023). AR/VR revolutions in future healthcare. In *Immersive Virtual and Augmented Reality in Healthcare* (pp. 1–30). CRC Press.
15. ISJEM Journal. (2024). Revolutionizing Healthcare Through Augmented Reality Integration in Medical Training (pp. 1–15). https://isjem.com/download/revolutionizing-healthcare-through-augmented-reality-integration-in-medical-training/ (accessed on 2 August 2024).
16. Blum, T., Kleeberger, V., Bichlmeier, C., & Navab, N. (2012). Mirracle: Augmented Reality in-situ visualization of human anatomy using a magic mirror. In *2012 IEEE Virtual Reality (VR)* (pp. 169–170). https://doi.org/10.1109/VR.2012.6180934
17. Khang, A., & Muthmainnah, M. (2023). AI-aided teaching model for the education 5.0 ecosystem. In *AI-Based Technologies and Applications in the Era of the Metaverse* (pp. 1–20). IGI Global Press. https://doi.org/10.4018/978-1-6684-8851-5.ch004
18. IIP Series. (2024). *Conferences & Edited Books* (pp. 1–20). Selfypage Developers Pvt. Ltd. (India). Retrieved August 5, 2024, from https://iipseries.org/
19. Dessai, A. B. G., Shetty, S. K., Jijo, D. N., & Gaonkar, O. U. (2024). Augmented Reality system guidance for computed tomography-based needle insertion: A narrative review. *JCDR*, 1–15. https://doi.org/10.7860/JCDR/2024/70665.19631
20. Savir, S., Khan, A. A., Yunus, R. A. et al. (2024). Virtual reality training for Central venous catheter placement: An interventional feasibility study incorporating virtual reality into a standard training curriculum of novice trainees. *Journal of Cardiothoracic and Vascular Anesthesia*, 1–10. https://doi.org/10.1053/j.jvca.2024.07.002.

21. Mistry, D., Brock, C. A., & Lindsey, T. (2023). The present and future of virtual reality in medical education: A narrative review. *Cureus*, 15(12), e51124. https://doi.org/10.7759/cureus.51124
22. Wolf, M. A., Mergen, M., Winter, P., Landgraeber, S., & Orth, P. (2024). Revolutionizing medical education: Surgery takes the lead in virtual reality research. *Surgery Open Science*, 20, 151–155.
23. Paulsen, L., Dau, S., & Davidsen, J. (2024). Designing for collaborative learning in immersive virtual reality: A systematic literature review. Virtual Reality, 28(1), 63–79. https://doi.org/10.1007/s10055-024-00975-4
24. Kassutto, S. M., Baston, C., & Clancy, C. (2021). Virtual, augmented, and alternate reality in medical education: Socially distanced but fully immersed. *ATS Scholar*, 2(5), 1–8. https://doi.org/10.34197/ats-scholar.2021-0002RE
25. Lundin, R. M., Yeap, Y., & Menkes, D. B. (2023). Adverse effects of virtual and augmented reality interventions in psychiatry: Systematic review. *JMIR Mental Health*, 10, e43240.
26. Mahmood, T., Scaffidi, M. A., Khan, R., & Grover, S. C. (2018). Virtual reality simulation in endoscopy training: Current evidence and future directions. *World Journal of Gastroenterology*, 24(48), 5439–5445. https://doi.org/10.3748/wjg.v24.i48.5439
27. Rizzo, A. A., Strickland, D., & Bouchard, S. (2004). The challenge of using virtual reality in telerehabilitation. *Telemedicine Journal and e-Health*, 10(2), 184–195.
28. Baniasadi, T., Ayyoubzadeh, S. M., & Mohammadzadeh, N. (2020). Challenges and practical considerations in applying virtual reality in medical education and treatment. *Oman Medical Journal*, 35(3), e125.
29. Clay, C. J., Budde, J. R., Hoang, A. Q., & Gushchyan, A. (2024). An evaluation of the effectiveness of immersive virtual reality training in non-specialized medical procedures for caregivers and students: A brief literature review. *Frontiers in Virtual Reality*, 5, 1–12. https://doi.org/10.3389/frvir.2024.1402093.
30. Saudagar, A. K., Kumar, A., & Khan, M. B. (2024). Mediverse beyond boundaries: A comprehensive analysis of AR and VR integration in medical education for diverse abilities. *Journal of Disability Research*, 3(1), 1–15. https://doi.org/10.57197/jdr-2023-0066.
31. Motaharifar, M., Norouzzadeh, A., Abdi, P., Iranfar, A., Lotfi, F., Moshiri, B., & Taghirad, H. D. (2021). Applications of haptic technology, virtual reality, and artificial intelligence in medical training during the COVID-19 pandemic. *Frontiers in Robotics and AI*, 8, 1–18. https://doi.org/10.3389/frobt.2021.612949.
32. Vashishth, T. K., Sharma, V., Sharma, K. K., Kumar, B., Chaudhary, S., & Panwar, R. (2023). Virtual reality (VR) and augmented reality (AR) transforming medical applications. In A. Khang (Ed.), *AI and IoT-Based Technologies for Precision Medicine* (pp. 324–348). Hershey, Pennsylvania, USA: IGI Global.
33. Xu, Y., Liu, X., Cao, X., Huang, C., Liu, E., Qian, S., Liu, X., Wu, Y., Dong, F., Qiu, C. W., Qiu, J., Hua, K., Su, W., Wu, J., Xu, H., Han, Y., Fu, C., Yin, Z., Liu, M., Roepman, R., Dietmann, S., Virta, M., Kengara, F., Zhang, Z., Zhang, L., Zhao, T., Dai, J., Yang, J., Lan, L., Luo, M., Liu, Z., An, T., Zhang, B., He, X., Cong, S., Liu, X., Zhang, W., Lewis, J. P., Tiedje, J. M., Wang, Q., An, Z., Wang, F., Zhang, L., Huang, T., Lu, C., Cai, Z., Wang, F., & Zhang, J. (2021). Artificial intelligence: A powerful paradigm for scientific research. *Innovation (Cambridge)*, 2(4), 100179.
34. Karimian, G., Petelos, E., & Evers, S. M. (2022). The ethical issues of the application of artificial intelligence in healthcare: A systematic scoping review. *AI and Ethics*, 2(4), 539–551. https://doi.org/10.1007/s43681-021-00131-7.
35. Mennella, C., Maniscalco, U., De Pietro, G., & Esposito, M. (2024). Ethical and regulatory challenges of AI technologies in healthcare: A narrative review. *Heliyon*, 10(4), 1–10. https://doi.org/10.1016/j.heliyon.2024.e26297.
36. Yadav, N., Pandey, S., Gupta, A., Dudani, P., Gupta, S., & Rangarajan, K. (2023). Data privacy in healthcare: In the era of artificial intelligence. *Indian Dermatology Online Journal*, 14(6), 788–792. https://doi.org/10.4103/idoj.idoj_543_23.
37. Murdoch, B. (2021). Privacy and artificial intelligence: Challenges for protecting health information in a new era. *BMC Medical Ethics*, 22(1), 1–12. https://doi.org/10.1186/s12910-021-00687-3.

CHAPTER 23

AI in Healthcare Workforce

Opportunities and Challenges

Mariam Chokor, Hiba Zreik, Mohammad H. Awada, Abbas Al Bazzal, Zahraa Takesh, and Zeinab Hazime

23.1 INTRODUCTION

Artificial intelligence (AI) refers to computing technologies that mimic human intelligence-assisted procedures including reasoning, deep learning, engagement, flexibility, and sensory perception [1, 2]. Certain devices may perform tasks that usually need human decision-making and interpretation [3, 4]. These approaches may be used in a variety of fields, including health and medicine, because of their multidisciplinary approach [5]. Interest and progress in medical AI applications have increased significantly in recent years because of the much-expanded capacity of current computers and the large quantity of digital information available [6]. The use of modern technology and AI for medical purposes offers the potential to solve some of the supply-and-demand issues. The growing availability of multimodal data (demographic, economic, genomics, clinical, and phenotypic) [7] has the potential to improve medical care by improving patient safety, health outcomes, and the cost and accessibility of high-quality treatment [8].

There have already been several studies showing that AI may perform equally well or more effectively than humans in essential healthcare activities like illness diagnosis [9]. For example, the importance of AI in radiography has expanded tremendously in recent years [10]. Deep learning algorithms in radiology tend to increase the clinical finding detection abilities of radiologists [11], particularly younger clinicians [12] and have enabled decreases in mean interpretation time [11, 13]. Another advantage is that AI can speed up and improve the efficiency of time-consuming, repetitive processes [14]. Furthermore, the application of AI allows healthcare personnel to manage treatment for a greater number of patients. In nursing, for instance, it has been observed that using AI-powered technologies increases productivity by 30–50% [15].

In contrast, AI has several ethical issues concerning autonomy, privacy and data protection, informed consent, and others [16]. In addition, one disadvantage of AI is the lack of an emotional connection between the health provider and the patient. The harmonious doctor-patient connection plays a critical role in the accomplishment of therapy and healing and, hence is regarded as a cornerstone of care [17]. In other words, medical personnel have a special ability to behave toward and diagnose patients with empathy that no machine or robot can match [18].

This chapter offers a thorough examination of the potential benefits and drawbacks of AI implementation within the healthcare field.

23.2 OPPORTUNITIES

23.2.1 Enhanced Decision-Making and Diagnostics

AI is increasingly being used in a wide range of healthcare applications, such as medical diagnostics and imaging, virtual assistants, wearable technology, risk management, and drug learning [19]. As for diagnosis, accurate identification of diseases is still seen as a difficult process globally, despite all the advancements in medical science. Because of the different disease processes and sometimes vague and complicated signs and symptoms, developing techniques that allow early definitive diagnosis is a continuous challenge. AI has

DOI: 10.1201/9781003533351-23

the potential to substantially improve the accuracy and effectiveness of disease diagnosis in healthcare [20].

AI technologies are becoming the preferred approach for image analysis in radiology [21]. This covers a variety of imaging techniques, as well as a range of applications like disease prediction, image segmentation, and tumor detection [22]. AI, in particular algorithms based on deep learning, has shown to be remarkably effective at extracting useful details from medical pictures [23]. By training these deep learning models on extensive datasets, they can detect complex structures and features that may be imperceptible to the human eye [24]. It has been shown that with deep learning algorithms, pneumonia from lung imaging may be identified with a specificity and sensitivity of 64% and 96%, respectively, as opposed to radiologists' 73% and 50%, respectively [25]. Furthermore, compared to radiologists, the AI-powered diagnosis had a 90% sensitivity rate for detecting breast cancer with mass against a 78% rate for radiologists. Moreover, AI detected early breast cancer 91% of the time, compared to radiologists (74%) [26]. In addition, early AI-based identification of diseases is essential because it can result in immediate measures that might save lives and enhance treatment success [24]. A study involving 625 cases was conducted to identify acute appendicitis early in order to anticipate the need for appendectomy. The findings indicated that AI performed best, correctly identifying appendicitis in 83.75% of cases with a precision of 84.11%, specificity of 81.01%, and sensitivity of 81.08% [27]. Equally important, AI serves as a promising tool in the field of genetics. It offers considerable potential for studies on gene therapy and genetic engineering. Moreover, a gene model's prior data can aid in the identification of genetic illnesses and heredity [28].

23.2.2 Disease Prevention

AI applications are attracting considerable interest in the area of disease prevention. An example of such applications is "polygenic risk scores," which allow the identification of patients with a genetically higher likelihood of a certain medical condition. This allows the physician to monitor the patient closely and detect early disease manifestations. Such precautionary measures can prevent the disease from progressing into more severe forms that require aggressive therapy [29].

In addition, AI can be very effective in monitoring patients remotely. It can identify unusual behaviors or irregular occurrences and categorize physical activities. This can protect remotely monitored patients and ensure their compliance with the prescribed treatment [30]. In the field of anesthesiology, for instance, the integration of AI technology would be revolutionary. Remote monitoring systems in the intensive care unit (ICU) and post-anesthesia care unit (PACU) will allow the detection of patients who are more prone to complications, enabling swift intervention and protective measures. Additionally, using wearable devices that can monitor the recovery process and track vital signs can allow earlier hospital discharge and facilitate home care [31].

23.2.3 Automation of Administrative Tasks

AI can automate administrative tasks like scheduling appointments, transcribing notes, and managing data. By automating these tasks, AI allows healthcare workers to focus more on patient care and making medical decisions [32]. For instance, tree-based machine learning (ML) models can utilize scheduling data to predict the time of a patient's operation, thereby indicating when the patient should begin fasting an essential requirement for procedures involving anesthesia [33]. Another example is how AI software programs can predict the likelihood of patients missing their ophthalmological appointment. Predicting non-attendance optimizes scheduling by allowing healthcare providers to reallocate the time slots of missed appointments by accommodating other patients. Improved appointment scheduling not only minimizes wasted time but also reduces patient wait times [34].

Additionally, AI can play a pivotal role in the field of organ transplantation. The liver transplantation process often involves a significant number of patients on the waiting list, necessitating a robust system for prioritization through risk stratification. The traditional MELD (Model for End-Stage Liver Disease) score, however, fails to account for variations in mortality risk among patients with hepatocellular carcinoma (HCC) based on differing tumor characteristics. Contemporary ML models can now predict the risk of waitlist dropout for HCC patients across 3-, 6-, and 12-month intervals. Patients identified as having a higher risk may benefit more from expedited services and could be prioritized accordingly [35]. AI applications in liver transplantation extend beyond risk prediction. For instance, AI models have demonstrated their efficacy in managing waiting lists. An Artificial Neural Network (ANN)–based model enhances donor-recipient matching by predicting the safest grafting outcomes. While the MELD score automates the classification of recipients based on the urgency of their condition, the ANN model further refines the process by considering potential interactions between the donor and recipient, hence ensuring compatibility [36].

Other administrative tasks that can be improved by AI are medical billing and the process of submitting claims to insurance companies for payment. Billing procedures rely on assigning a symbol for each medical procedure or evaluation by a code. Executing such tasks manually not only consumes a lot of time, but is prone to mistakes. The first model that allows prediction of billing code was developed in 2019. It's a system that automatically assigns codes to magnetic resonance imaging (MRI) procedures based on the information recorded during the scan [37]. ML can probabilistically match data. Insurance companies must ensure that payments are made correctly. The process of reliable identification, analysis, and correction of coding errors and inaccurate claims would be less costly and laborious than the manual alternative [9].

23.2.4 Burnout Prevention and Augmenting Quality of Care

Burnout of healthcare providers characterized by emotional exhaustion, depersonalization, and diminished ability to cope with and balance both work and life responsibilities has been reported in one in three physicians in the United States [38, 39]. The effect of burnout on physicians raises concerns regarding patients' welfare due to the increased probability of medical errors which doubles the risk of compromising patients' safety [38, 40]. Burnout also affects patient-doctor relationships because affected physicians are less likely to empathize with their patients, thus contributing to their dissatisfaction [39].

An inventive way to lessen the clinical documentation burden, which is one of the main causes of burnout, is ambient AI. By using cutting-edge speech recognition and natural language processing (NLP) technology, ambient AI systems can reduce the time clinicians spend on documentation duties by transcribing clinician-patient conversations and formatting them into the proper clinical documentation format [41, 42]. This change may give healthcare professionals more time for one-on-one caregiving and interpersonal interactions, frequently mentioned as more satisfying facets of their work [43, 44]. Incorporating AI into electronic health records (EHRs) is one of the most important achievements that aim to reduce physicians' excess workload.

Chatbots are software applications in the healthcare industry that use NLP and ML to understand and reply to human questions. Chatbots increase productivity for both sides by gathering vital patient data, including demographics, medical history, and current symptoms. They also act by handling duties like appointment booking and patient triage [45].

Healthcare chatbots enable users to receive prompt responses to frequently requested questions (such as those regarding lifestyle modifications) without consulting a doctor. By integrating chatbots into an EHR system, doctors may quickly access patient records, such as lab results, medical histories, and treatment plans, and spend less time responding to repetitive and non-emergency inquiries [38]. AI, specifically NLP, had an important role during the COVID-19 pandemic, as its integration into EHR allowed physicians to identify urgent and high-priority cases while demonstrating high accuracy by correctly identifying 96% of positive COVID-19 cases and all negative cases [46]. Besides chatbots, healthcare organizations have incorporated a clinical digital assistant (CDA) such as one named Suki which generates structured medical notes for the EHR automatically using speech recognition technology during physician-patient conversations. Following a 62% reduction in documentation time, burnout rates decreased by 17% within 30 days of Suki intervention [38]. Medical image processing, such as MRI, is one of the most time-consuming processes for doctors, often taking up to two hours. An ML algorithm developed by the Massachusetts Institute of Technology can analyze 3D images 1,000 times quicker than other techniques; a crucial characteristic in emergency cases, which not only resolves physicians' time limits in analyzing images but also develops diagnostic and treatment approaches [38, 47].

23.2.5 Overcoming Language Obstacles

Language obstacles made physician-patient interactions more difficult, leading to patients expressing lower levels of satisfaction with the communication they received from medical professionals [48]. Moreover, patients with language challenges may be less likely to acquire informed consent, be scheduled for appropriate follow-up sessions, or comprehend medical conditions [49]. Patients who experience language challenges during healthcare contacts are entitled to a professional interpreter, these services can be rendered in-person remotely, or virtually [50]. However, the shortage of interpreters and lack of awareness by clinicians of the legal requirement for using interpreters have been a barrier to the utilization of these interpretation specialists [51]. Besides, challenges with easily arranging and coordinating convenient times for in-person interpreters have been another barrier [52]. This discourages a lot of physicians from utilizing in-person interpreters and can occasionally lead to communication delays, avoidance, or poor communication when in-person interpreters are not used. AI can overcome these language barriers by improving communication, and possibly lessening inequalities in care [53].

Moreover, AI can be used to prioritize patients who are most in need throughout the entire institution, a task that human interpretation coordinators are unable to accomplish [53]. Additionally, digital advances played a pivotal role in the COVID-19 pandemic, where healthcare organizations that previously relied on in-person interpreters transitioned almost entirely to virtual video interpretation [54]. Furthermore, certain healthcare specialties can especially benefit from AI's potential to break down complex medical procedures for patients. In the field of dentistry, for instance, platforms driven by AI can provide thorough explanations of dental operations in a way that suits patients' cognitive levels and learning preferences. It can translate technical dentistry jargon into simple phrases that patients may comprehend [55]. AI also helps to overcome clinician bias, improve the standard of care, and enhance interpreter preparation and input [53].

23.3 CHALLENGES

23.3.1 Ethical Concerns and Data Privacy

It should be highlighted that introducing AI into the healthcare sector raises several difficulties, including privacy, data protection, empathy, and ethical issues [16]. Clinical medicine is a field in which ethics are fundamental to its practice [56]. Thus, experts and practitioners should take into account the four ethical principles—autonomy, nonmaleficence, beneficence, and justice in all facets of healthcare—before incorporating AI into the healthcare system [56, 57]. Concerning privacy and data collection, AI applications require the use of considerably greater amounts of data than telemedicine does [19], which renders data security even more important [58]. The maintenance and distribution of massive data repositories is due to ML's requirement for vast amounts of data. This can have significant implications for individual patients if their sensitive health information is connected to them and shared without their explicit permission and consent [58]. In addition, robot-collected clinical data can be hacked and misused, compromising privacy and security [16]. Another concern for powerful AI is that clinicians may tend to excessively rely on it [59], which, unconsciously or consciously, can be harmful to the patient and cause negative consequences [60]. One added challenge is the potential biases that AI may perpetuate. A scoping review of AI clinical research from PubMed articles published in 2019 showed that the vast majority of clinical AI datasets were of US (40.8%) or Chinese (13.7%) origin. Such homogeneity in data collection could lead to healthcare disparities when it comes to the development of AI algorithms and training of AI models [61]. Hence, justice is jeopardized when equality for all patients is not met [62].

A 2024 study exploring the different ethical issues arising with AI implementation in healthcare established 3 key themes based on the input of 41 AI experts. The first theme argues that AI integration in healthcare should have a well-defined goal and true beneficence. The second one highlights the experts' growing concern regarding the conflict of financial interest that may arise with AI introduction and emphasizes that the welfare of doctors and patients should come above all. The last theme draws attention to the importance of designing "context-sensitive AI" guided by its core theoretical framework [63]. All of this shows that AI integration will require first addressing any ethical issues that could compromise the potential benefits of AI adoption.

23.3.2 Regulatory Measures and Legal Liability

The present regulatory framework for AI in healthcare in most countries is that of Software as a Medical Device (SaMD), meaning the same traditional framework used to asses software programs for other electronic healthcare devices. However, those regulatory measures lack considerably as they fail to encompass several AI implementations such as those that assist in administrative tasks or those that aid the physician in clinical decision-making. Moreover, these regulations adopt mostly a "soft-law" approach worldwide. The drawback of such an approach is that these interventions are not mandatory, meaning that institutions can choose not to abide by them [64].

A paper written by Van Laere and colleagues concluded that both American and European officials have failed to address all the concerns regarding the regulation of Clinical Decision Support (CDS) systems (i.e., any SaMD device that supports physicians with clinical decision-making) [65]. That can be explained given that regulators are often faced with five key challenges [66].

1. **Ensuring Accurate AI Performance:** Data entry devices must be properly calibrated, and the data itself should follow a standardized format to ensure reliability. For instance, users should not input low-quality images into an AI system trained exclusively on high-quality datasets.
2. **Establishing a Dynamic Regulatory Framework:** Developing a responsive and adaptive framework capable of keeping pace with AI's continuously evolving nature.
3. **Addressing Potential Biases in AI Algorithms:** Ensuring fairness and transparency in AI decision-making.
4. **Reconciling AI's Data Demands with Data Protection Laws:** Balancing innovation with compliance with privacy and data protection regulations.
5. **Safeguarding Patient Privacy:** Implementing stringent measures to protect personal health information amid large-scale data collection.

Another growing concern is the legal liability of AI technology. Some have proposed associating AI with legal personhood, but that is problematic because holding physicians accountable could decrease their willingness to utilize AI devices. In a similar sense, establishing liability through product laws presents difficulties due to the fluid nature of AI tools that differs from that of traditional medical devices. Others have suggested focusing on compensation for damages as an alternative to culpability. Nevertheless, such measures could present economic and reputational hurdles. This shows that further research is needed to address these complex issues and perhaps even to adopt a more reconciliatory approach. This would require a body of independent and reliable specialists capable of authentically assessing the technology [67].

The importance of efficiently regulating AI applications lies in guaranteeing the safety of patients, holding AI stakeholders accountable, and providing real tangible benefits. It also serves to facilitate widespread adoption of AI and reinforce physician trust in the technology [68]. Hence, policymakers must develop further regulatory measures that address the shortcomings of the current regulations, provide more thorough governance, and build strategies capable of rapidly adapting to the ever-developing field of AI technology [64, 68].

23.3.3 Job Displacement

Job displacement, defined as unemployment or shift in job plans for healthcare professionals, has become a growing concern among clinicians [69]. This is illustrated by the fact that AI can affect employment through substitution and creation effects. Substitution results from the replacement of human labor with novel technical tools that have much lower cost, since the salary and resources needed for human employees will cost more than the actual machinery and equipment. All of this could potentially cause healthcare workers to lose their jobs [70]. One healthcare profession expected to be affected remarkably by AI is radiology, where as far as of early 2023, the Food and Drug Administration (FDA) has approved more than 200 AI-driven software radiology programs [10]. A

2019 international survey involving 1,041 radiologists showed that 38% expressed genuine fear about being replaced by AI [71]. The effect of AI on the job market is further echoed by medical students' future career choices. A study done in 2022 found that 50% of students who considered radiology as one of their top three potential specialties expressed mounting apprehension toward AI implementation and the future of radiology [72].

Moreover, this escalating concern regarding job displacement is shared by workers in different domains other than the medical field. Thus, this phenomenon has been labeled "techno-insecurity" [73]. Technostress is a psychophysiological condition marked by cognitive symptoms like inability to focus, irritability, and memory problems as well as elevated levels of hormones sensitive to stress [74]. It was seen to lower overall job satisfaction and organizational engagement [73]. It theoretically affects the progress of healthcare practitioners' work, suggesting that future practices and actions should be taken to manage this issue [69].

23.3.4 Training and New Skillsets Required

As AI enters the medical field, doctors must be competent enough to welcome it. A realistic and lasting implementation of AI needs training and education for both healthcare personnel and surgeons. Surgeons are still unprepared to incorporate AI into their work [75]. The future of physicians in the AI era emphasizes this idea. Even those who agree that AI won't substitute doctors admit that only physicians who collaborate with AI tools will prosper. Implementing "Human-in-the-Loop," a view concerning the cooperation between AI and skilled physicians, needs readily well-trained medical personnel. Human expertise is necessary to manage AI applications, lead their workflow, analyze their results, and comment on their shortcomings. All of this could serve to enhance its power, precision, and future development. Integrating AI in the healthcare domain requires building pedagogical strategies and courses to teach physicians AI principles, proper applications, and AI-assisted medical service provision. Training doctors shouldn't only be at the personal level, the healthcare institution as a whole should be ready and prepared to implement AI technology [76].

Besides healthcare organizations, medical school is where training must commence. AI education early on can help graduate physicians comprehend different algorithms, and hence boost AI utility. However, numerous obstacles impede the integration of AI into the medical curriculum, some of which include the absence of solid evidence of its success as well as technical difficulties related to building AI software applications [77].

Doctors should master different skill sets to adapt to AI integration into the medical practice. To begin with, doctors must learn about AI-enabled missions, the domains in which AI might be introduced, and how it affects traditional practice. Moreover, they should be mindful and cautious while utilizing AI technology, and hence capable of criticizing its different applications. That can be met by assessing the precision of the results and evaluating whether or not the device was able to meet its end target. Furthermore, physicians should be able to assess the likelihood of bias and inequality. Finally, they must be trained on how to utilize AI devices and what measures to take in case misdiagnosis occurs [78].

23.3.5 Empathy

AI lacks an emotional connection with the sick, which is essential for successful therapy. Maintaining a healthy doctor-patient relationship is an important responsibility for physicians. Doctors elaborate on the potential treatments, ensure patients participate in decision-making, and respect their privacy. Even though some AI tools are programmed to boost empathy, nothing compares to the humanistic physical contact and vocal tone. Also, physicians can notice the subtle facial expressions and body language that tell a lot about the patient's circumstances, thanks to their humanistic nature. Moreover, they can be aware of the patient's background, which goes beyond their medical record. This allows physicians to make a more accurate diagnosis, readily adapt treatment plans, and provide more sustainable and effective precautionary measures compared to smart machines.

Physicians can empathize with patients by appreciating their souls and humanity, but AI cannot [17]. This is because these innovative technologies are unable to interact with patients at the same level as the physician, where doctors are capable of comforting and empathizing with the individuals under their care. Such limitations render AI devices unfit to replace physicians [79]. A study held in Bahrain examined the doctors' insights toward the future of AI in medicine. The authors did a cross-sectional survey with 114 doctors. Despite their optimistic views, most physicians worried about the incapacity of AI to empathize with patients as humans do [80].

23.4 THE ATTITUDES, PERCEPTIONS, AND ACCEPTANCE OF AI AMONG HEALTHCARE WORKERS

The use of AI in the upcoming years is expected to increase significantly [81]. Even though the employment of physicians may be threatened by the increasing use of advanced technological devices, the work performed by doctors would nevertheless be at lower risk for automation compared to other professions [82]. According to some AI specialists, doctors' roles typically involve working with machines as a team [81]. This led to mixed opinions concerning the future of AI in the medical field.

Many factors were attributed to hindering AI acceptance among physicians, some of which include: assumed job burden, perceived reliability, and readiness to undergo AI training. Others include the absence of accountability and associated risks. Such hurdles lead some physicians to regularly express their lack of confidence in AI technology, especially regarding medical decision-making [83].

Studies done on physicians show that 17.6% of respondents' pathologists are concerned about their future job security, while most accept AI [84]. Another study involving radiologists showed that 77% expressed support for the use of AI, and 89% said they were not worried about losing their jobs [85]. Ninety percent of German doctors from various specialties who were contacted predicted that AI and human intelligence would be combined to practice medicine in the future [86]. Conversely, a UK study found that general practitioners (GPs) were skeptical about AI's capacity to deliver compassionate treatment [87]. Similarly, there are notable differences in expectations and worries about the legal elements of AI applications according to an Asian study assessing physicians' attitudes toward AI [80]. A survey assessing physician attitudes toward AI in healthcare found that 36.3% of respondents viewed AI as crucial for reducing misdiagnosis, while 37.6% expressed skepticism [88].

Furthermore, a study showed that most clinicians and medical school undergraduates had a positive outlook on AI use and strongly believed that cooperation between AI and healthcare providers should be encouraged. However, they did admit to a reserved stance. This can be explained by the fact that the majority had limited real-world experience and sufficient knowledge regarding AI technology. This in return can justify any present anxiety surrounding AI implementation, hence highlighting the need for healthcare practitioners to undergo additional education on this topic [89].

23.5 SUMMARY

This chapter provides a comprehensive overview of how AI is reshaping the healthcare sector. It highlights several key opportunities, including the enhancement of diagnostic accuracy, improved disease prevention strategies, and automation of administrative tasks, which collectively aim to reduce burnout among healthcare professionals and overcome language barriers in physician-patient interactions. The chapter also addresses significant challenges associated with AI integration. These include ethical considerations around patient data and potential healthcare disparities, legal liabilities, and the need for new regulations to govern AI use. Additionally, there are concerns about the impact of AI on empathy in patient care, potential job displacement within the workforce, and the necessity for healthcare workers to acquire new skills and competencies. Attitude and acceptance of AI among healthcare professionals also pose a challenge, as they must adapt to technological changes while maintaining a focus on patient-centered care.

REFERENCES

1. Tagliaferri, S. D., Angelova, M., Zhao, X., Owen, P. J., Miller, C. T., Wilkin, T., et al (2020). Artificial intelligence to improve back pain outcomes and lessons learnt from clinical classification approaches: Three systematic reviews. *NPJ Digital Medicine*, *3*(1), 93.
2. Tran, B. X., Vu, G. T., Ha, G. H., Vuong, Q. H., Ho, M. T., Vuong, T. T., et al (2019). Global evolution of research in artificial intelligence in health and medicine: A bibliometric study. *Journal of Clinical Medicine*, *8*(3), 360.
3. Hamid, S. (2016). The opportunities and risks of artificial intelligence in medicine and healthcare. *Cambridge University Repository*. (Accessed: 20 August 2024). https://www.repository.cam.ac.uk/items/4494a9ef-7256-4bd3-9d642ec9ab04

4. Panch, T., Szolovits, P., & Atun, R. (2018). Artificial intelligence, machine learning and health systems. *Journal of Global Health*, *8*(2), 020303.
5. Secinaro, S., Calandra, D., Secinaro, A., Muthurangu, V., & Biancone, P. (2021). The role of artificial intelligence in healthcare: A structured literature review. *BMC Medical Informatics and Decision Making*, *21*(1), 125.
6. Meskó, B., Drobni, Z., Bényei, É, Gergely, B., & Győrffy, Z. (2017). Digital health is a cultural transformation of traditional healthcare. *mHealth*, *3*, 38.
7. Bajwa, J., Munir, U., Nori, A., & Williams, B. (2021). Artificial intelligence in healthcare: Transforming the practice of medicine. *Future Healthcare Journal*, *8*(2), e188–194.
8. Silcox, C., Zimlichmann, E., Huber, K., Rowen, N., Saunders, R., McClellan, M., Kahn, C. N., 3rd, Salzberg, C. A., & Bates, D. W. (2024). The potential for artificial intelligence to transform healthcare: Perspectives from international health leaders. *NPJ Digital Medicine*, *7*(1), 88.
9. Davenport, T., & Kalakota, R. (2019). The potential for artificial intelligence in healthcare. *Future Healthcare Journal*, *6*(2), 94–98.
10. Liu, D. S., Abu-Shaban, K., Halabi, S. S., & Cook, T. S. (2023). Changes in radiology due to artificial intelligence that can attract medical students to the specialty. *JMIR Medical Education*, *9*, e43415.
11. Seah, J. C. Y., Tang, C. H. M., Buchlak, Q. D., Holt, X. G., Wardman, J. B., Aimoldin, A., et al (2021). Effect of a comprehensive deep-learning model on the accuracy of chest x-ray interpretation by radiologists: A retrospective, multireader multicase study. *Lancet Digital Health*, *3*(8), e496–506.
12. Hwang, E. J., Nam, J. G., Lim, W. H., Park, S. J., Jeong, Y. S., Kang, J. H., et al (2019). Deep learning for chest radiograph diagnosis in the emergency department. *Radiology*, *293*(3), 573–580.
13. Kim, J. H., Kim, J. Y., Kim, G. H., Kang, D., Kim, I. J., Seo, J., Andrews, J. R., & Park, C. M. (2020). Clinical validation of a deep learning algorithm for detection of pneumonia on chest radiographs in emergency department patients with acute febrile respiratory illness. *Journal of Clinical Medicine*, *9*(6), 1981.
14. Hazarika, I. (2020). Artificial intelligence: Opportunities and implications for the health workforce. *International Health*, *12*(4), 241–245.
15. Manyika, J. (2017). AI, automation, employment, and productivity. *McKinsey & Company*. https://www.mckinsey.com/~/media/mckinsey/featured%20insights/Digital%20Disruption/Harnessing%20automation%20for%20a%20future%20that%20works/MGI-A-future-that-works-Executive-summary.ashx. (Accessed: 20 August 2024).
16. Farhud, D. D., & Zokaei, S. (2021). Ethical issues of artificial intelligence in medicine and healthcare. *Iranian Journal of Public Health*, *50*(11), i–v.
17. Cordero, D. (2024). The downsides of artificial intelligence in healthcare. *The Korean Journal of Pain*, *37*(1), 87–88.
18. Rahman, M. A., Victoros, E., Ernest, J., Davis, R., Shanjana, Y., & Islam, M. R. (2024). Impact of artificial intelligence (AI) technology in healthcare sector: A critical evaluation of both sides of the coin. *Clinical Pathology (Thousand Oaks, Ventura County, California)*, *17*. https://doi.org/10.1177/2632010X241226887
19. Hosny, A., Parmar, C., Quackenbush, J., Schwartz, L. H., & Aerts, H. J. W. L. (2018). Artificial intelligence in radiology. *Nature Reviews. Cancer*, *18*(8), 500–510.
20. Alowais, S. A., Alghamdi, S. S., Alsuhebany, N., Alqahtani, T., Alshaya, A. I., Almohareb, S. N., Aldairem, A., Alrashed, M., Bin Saleh, K., Badreldin, H. A., Al Yami, M. S., Al Harbi, S., & Albekairy, A. M. (2023). Revolutionizing healthcare: The role of artificial intelligence in clinical practice. *BMC Medical Education*, *23*(1), 689.
21. Mylonas, A., Keall, P. J., Booth, J. T., Shieh, C. C., Eade, T., Poulsen, P. R., & Nguyen, D. T. (2019). A deep learning framework for automatic detection of arbitrarily shaped fiducial markers in intrafraction fluoroscopic images. *Medical Physics*, *46*(5), 2286–2297.
22. Tang, X. (2019). The role of artificial intelligence in medical imaging research. *BJR Open*, *2*(1), 20190031.
23. Ghaffar Nia, N., Kaplanoglu, E., & Nasab, A. (2023). Evaluation of artificial intelligence techniques in disease diagnosis and prediction. *Discover Artificial Intelligence*, *3*(1), 5.
24. Pinto-Coelho, L. (2023). How artificial intelligence is shaping medical imaging technology: A survey of innovations and applications. *Bioengineering (Basel, Switzerland)*, *10*(12), 1435.
25. Becker, J., Decker, J. A., Römmele, C., Kahn, M., Messmann, H., Wehler, M., Schwarz, F., Kroencke, T., & Scheurig-Muenkler, C. (2022). Artificial intelligence-based detection of pneumonia in chest radiographs. *Diagnostics (Basel, Switzerland)*, *12*(6), 1465.
26. Kim, H. E., Kim, H. H., Han, B. K., Kim, K. H., Han, K., Nam, H., Lee, E. H., & Kim, E. K. (2020). Changes in cancer detection and false-positive recall in mammography using artificial intelligence: A retrospective, multireader study. *The Lancet. Digital Health*, *2*(3), e138–e148.
27. Mijwil, M. M., & Aggarwal, K. (2022). A diagnostic testing for people with appendicitis using machine learning techniques. *Multimedia Tools and Applications*, *81*(5), 7011–7023.
28. Vilhekar, R. S., & Rawekar, A. (2024). Artificial intelligence in genetics. *Cureus*, *16*(1), e52035.
29. Schork, N. J. (2019). Artificial intelligence and personalized medicine. *Cancer Treatment and Research*, *178*, 265–283.

30. Khosravi, M., Zare, Z., Mojtabaeian, S. M., & Izadi, R. (2024). Artificial intelligence and decision-making in healthcare: A thematic analysis of a systematic review of reviews. *Health Services Research and Managerial Epidemiology*, *11*, 1–11.
31. Feinstein, M., Katz, D., Demaria, S., & Hofer, I. S. (2024). Remote monitoring and artificial intelligence: Outlook for 2050. *Anesthesia and Analgesia*, *138*(2), 350–357.
32. Zheng, L., Beck, J. C., Mafeld, S., Parotto, M., Matthews, A., Alexandre, S., & Conway, A. (2024). Determining pre-procedure fasting alert time using procedural and scheduling data. *Health Informatics Journal*, *30*(2), 1–21.
33. Breeze, F., Hossain, R. R., Mayo, M., & McKelvie, J. (2023). Predicting ophthalmic clinic non-attendance using machine learning: Development and validation of models using nationwide data. *Clinical & Experimental Ophthalmology*, *51*(8), 764–774.
34. Rothenberg, S., Bame, B., & Herskovitz, E. (2022). Prospective evaluation of a machine-learning prediction model for missed radiology appointments. *Journal of Digital Imaging*, *35*(6), 1690–1693.
35. Kwong, A., Hameed, B., Syed, S., Ho, R., Mard, H., Arshad, S., Ho, I., Suleman, T., Yao, F., & Mehta, N. (2022). Machine learning to predict waitlist dropout among liver transplant candidates with hepatocellular carcinoma. *Cancer Medicine*, *11*(6), 1535–1541.
36. Seyahi, N., & Ozcan, S. G. (2021). Artificial intelligence and kidney transplantation. *World Journal of Transplantation*, *11*(7), 277–289.
37. Denck, J., Landschütz, W., Nairz, K., Heverhagen, J. T., Maier, A., & Rothgang, E. (2019). Automated billing code retrieval from MRI scanner log data. *Journal of Digital Imaging*, *32*(6), 1103–1111.
38. Fawzy Eid, M. (2024). Using artificial intelligence in electronic health record systems to mitigate physician burnout: A roadmap. *Journal of Healthcare Management/American College of Healthcare Executives*, *69*(4), 244–254.
39. Nan, J., Herbert, M. S., Purpura, S., Henneken, A. N., Ramanathan, D., & Mishra, J. (2024). Personalized machine learning-based prediction of wellbeing and empathy in healthcare professionals. *Sensors (Basel, Switzerland)*, *24*(8), 2640.
40. Misurac, J., Knake, L. A., & Blum, J. M. (2024). Impact of ambient artificial intelligence notes on provider burnout. *medRxiv*. https://doi.org/10.1101/2024.07.18.24310656. (Accessed: 24 August 2024).
41. Coiera, E., Kocaballi, B., Halamka, J., & Laranjo, L. (2018). The digital scribe. *npj Digital Medicine*, *1*, 58.
42. Ghatnekar, S., Faletsky, A., & Nambudiri, V. E. (2021). Digital scribe utility and barriers to implementation in clinical practice: A scoping review. *Health and Technology*, *11*(4), 803–809.
43. Friedberg, M. W., Chen, P. G., Van Busum, K. R., Aunon, F., Pham, C., Caloyeras, J., Mattke, S., Pitchforth, E., Quigley, D. D., Brook, R. H., Crosson, F. J., & Tutty, M. (2014). Factors affecting physician professional satisfaction and their implications for patient care, health systems, and health policy. *Rand Health Quarterly*, *3*(4), 1.
44. West, C. P., Dyrbye, L. N., & Shanafelt, T. D. (2018). Physician burnout: Contributors, consequences and solutions. *Journal of Internal Medicine*, *283*(6), 516–529.
45. Grassini, E., Buzzi, M., Leporini, B., & Vozna, A. (2024). A systematic review of chatbots in inclusive healthcare: Insights from the last 5 years. *Universal Access in the Information Society*. https://doi.org/10.1007/s10209-024-01118-x. (Accessed: 24 August 2024).
46. Mermin-Bunnell, K., Zhu, Y., Hornback, A., Damhorst, G., Walker, T., Robichaux, C., Mathew, L., Jaquemet, N., Peters, K., Johnson, T. M., 2nd, Wang, M. D., & Anderson, B. (2023). Use of natural language processing of patient-initiated electronic health record messages to identify patients with COVID-19 infection. *JAMA Network Open*, *6*(7), e2322299.
47. Mohamed, K., & Albadawy, M. (2024). AI in diagnostic imaging: Revolutionising accuracy and efficiency. *ResearchGate*. https://www.researchgate.net/publication/378756461_AI_in_Diagnostic_Imaging_Revolutionising_Accuracy_and_Efficiency. (Accessed: 24 August 2024).
48. Schenker, Y., Lo, B., Ettinger, K. M., & Fernandez, A. (2008). Navigating language barriers under difficult circumstances. *Annals of Internal Medicine*, *149*(4), 264–269.
49. Schenker, Y., Karter, A. J., Schillinger, D., Warton, E. M., Adler, N. E., Moffet, H. H., Ahmed, A. T., & Fernandez, A. (2010). The impact of limited English proficiency and physician language concordance on reports of clinical interactions among patients with diabetes: The DISTANCE study. *Patient Education and Counseling*, *81*(2), 222–228.
50. Jacobs, E. A., Press, V. G., & Vela, M. B. (2015). Use of interpreters by physicians. *Journal of General Internal Medicine*, *30*(11), 1589.
51. Kale, E., & Syed, H. R. (2010). Language barriers and the use of interpreters in the public health services. A questionnaire-based survey. *Patient Education and Counseling*, *81*(2), 187–191.
52. Espinoza Suarez, N. R., Urtecho, M., Nyquist, C. A., Jaramillo, C., Yeow, M. E., Thorsteinsdottir, B., Wilson, M. E., & Barwise, A. K. (2021). Consequences of suboptimal communication for patients with limited English proficiency in the intensive care unit and suggestions for a way forward: A qualitative study of healthcare team perceptions. *Journal of Critical Care*, *61*, 247–251.
53. Barwise, A. K., Curtis, S., Diedrich, D. A., & Pickering, B. W. (2024). Using artificial intelligence to promote equitable care for inpatients with

language barriers and complex medical needs: Clinical stakeholder perspectives. *Journal of the American Medical Informatics Association, 31*(3), 611–621.

54. Yang, C., Prokop, L., & Barwise, A. (2023). Strategies used by healthcare systems to communicate with hospitalized patients and families with limited English proficiency during the COVID-19 pandemic: A narrative review. *Journal of Immigrant and Minority Health, 25*(6), 1393–1401.
55. Thorat, V., Rao, P., Joshi, N., Talreja, P., & Shetty, A. R. (2024). Role of artificial intelligence (AI) in patient education and communication in dentistry. *Cureus, 16*(5), e59799.
56. Singer, P. A., Pellegrino, E. D., & Siegler, M. (2001). Clinical ethics revisited. *BMC Medical Ethics, 2*, E1.
57. Jahn, W. (2011). The 4 basic ethical principles that apply to forensic activities are respect for autonomy, beneficence, nonmaleficence, and justice. *Journal of Chiropractic Medicine, 10*(3), 225–226.
58. Yadav, N., Pandey, S., Gupta, A., Dudani, P., Gupta, S., & Rangarajan, K. (2023). Data privacy in healthcare: In the era of artificial intelligence. *Indian Dermatology Online Journal, 14*(6), 788–792.
59. Goddard, K., Roudsari, A., & Wyatt, J. C. (2012). Automation bias: A systematic review of frequency, effect mediators, and mitigators. *Journal of the American Medical Informatics Association, 19*(1), 121–127.
60. Carter, S. M., Degeling, C., Doust, J., & Barratt, A. (2016). A definition and ethical evaluation of overdiagnosis. *Journal of Medical Ethics, 42*(11), 705–714.
61. Celi, L. A., Cellini, J., Charpignon, M. L., Dee, E. C., Dernoncourt, F., Eber, R., Mitchell, W. G., Moukheiber, L., Schirmer, J., Situ, J., Paguio, J., Park, J., Wawira, J. G., & Yao, S., for MIT Critical Data (2022). Sources of bias in artificial intelligence that perpetuate healthcare disparities-a global review. *PLOS Digital Health, 1*(3), e0000022.
62. Layman, E. J. (2008). Ethical issues and the electronic health record. *The Health Care Manager, 27*(2), 165–176.
63. Arbelaez Ossa, L., Lorenzini, G., Milford, S. R., Shaw, D., Elger, B. S., & Rost, M. (2024). Integrating ethics in AI development: A qualitative study. *BMC Medical Ethics, 25*(1), 10.
64. Palaniappan, K., Lin, E. Y. T., & Vogel, S. (2024). Global regulatory frameworks for the use of artificial intelligence (AI) in the healthcare services sector. *Healthcare (Basel, Switzerland), 12*(5), 562.
65. Van Laere, S., Muylle, K. M., & Cornu, P. (2022). Clinical decision support and new regulatory frameworks for medical devices: Are we ready for it? - A viewpoint paper. *International Journal of Health Policy and Management, 11*(12), 3159–3163.
66. McKee, M., & Wouters, O. J. (2023). The challenges of regulating artificial intelligence in healthcare comment on "Clinical Decision Support and New Regulatory Frameworks for Medical Devices: Are We Ready for It? - A viewpoint paper. *International Journal of Health Policy and Management, 12*, 7261.
67. Bottomley, D., & Thaldar, D. (2023). Liability for harm caused by AI in healthcare: An overview of the core legal concepts. *Frontiers in Pharmacology, 14*, 1297353.
68. Mennella, C., Maniscalco, U., De Pietro, G., & Esposito, M. (2024). Ethical and regulatory challenges of AI technologies in healthcare: A narrative review. *Heliyon, 10*(4), e26297.
69. Ahmed, M. I., Spooner, B., Isherwood, J., Lane, M., Orrock, E., & Dennison, A. (2023). A systematic review of the barriers to the implementation of artificial intelligence in healthcare. *Cureus, 15*(10), e46454. https://doi.org/10.7759/cureus.46454
70. Wang, X., Chen, M., & Chen, N. (2024). How artificial intelligence affects the labour force employment structure from the perspective of industrial structure optimisation. *Heliyon, 10*(5), e26686. https://doi.org/10.1016/j.heliyon.2024.e26686
71. Huisman, M., Ranschaert, E., Parker, W., Mastrodicasa, D., Koci, M., Pinto de Santos, D., Coppola, F., Morozov, S., Zins, M., Bohyn, C., Koç, U., Wu, J., Veean, S., Fleischmann, D., Leiner, T., & Willemink, M. J. (2021). An international survey on AI in radiology in 1,041 radiologists and radiology residents part 1: Fear of replacement, knowledge, and attitude. *European Radiology, 31*(9), 7058–7066.
72. Reeder, K., & Lee, H. (2022). Impact of artificial intelligence on US medical students' choice of radiology. *Clinical Imaging, 81*, 67–71.
73. Nitiéma, P. (2023). Artificial intelligence in medicine: Text mining of health care workers' opinions. *Journal of Medical Internet Research, 25*, e41138. https://doi.org/10.2196/41138. (Accessed: 23 August 2024).
74. Bahamondes-Rosado, M. E., Cerdá-Suárez, L. M., Dodero Ortiz de Zevallos, G. F., & Espinosa-Cristia, J. F. (2023). Technostress at work during the COVID-19 lockdown phase (2020-2021): A systematic review of the literature. *Frontiers in Psychology, 14*, 1173425. https://doi.org/10.3389/fpsyg.2023.1173425
75. Amin, A., Cardoso, S. A., Suyambu, J., Abdus Saboor, H., Cardoso, R. P., Husnain, A., Isaac, N. V., Backing, H., Mehmood, D., Mehmood, M., & Maslamani, A. N. J. (2024). Future of artificial intelligence in surgery: A narrative review. *Cureus, 16*(1), e51631. https://doi.org/10.7759/cureus.51631
76. Sezgin, E. (2023). Artificial intelligence in healthcare: Complementing, not replacing, doctors and healthcare providers. *Digital Health, 9*, 20552076231186520. https://doi.org/10.1177/20552076231186520

77. Chan, K. S., & Zary, N. (2019). Applications and challenges of implementing artificial intelligence in medical education: Integrative review. *JMIR Medical Education*, *5*(1), e13930. https://doi.org/10.2196/13930
78. Liaw, W., Kueper, J. K., Lin, S., Bazemore, A., & Kakadiaris, I. (2022). Competencies for the use of artificial intelligence in primary care. *Annals of Family Medicine*, *20*(6), 559–563.
79. Ahuja, A. S. (2019). The impact of artificial intelligence in medicine on the future role of the physician. *PeerJ*, *7*, e7702. https://doi.org/10.7717/peerj.7702
80. Al-Medfa, M. K., Al-Ansari, A. M. S., Darwish, A. H., Qreeballa, T. A., & Jahrami, H. (2023). Physicians' attitudes and knowledge toward artificial intelligence in medicine: Benefits and drawbacks. *Heliyon*, *9*(4), e14744. https://doi.org/10.1016/j.heliyon.2023.e14744
81. Verghese, A., Shah, N. H., & Harrington, R. A. (2018). What this computer needs is a physician: Humanism and artificial intelligence. *JAMA*, *319*(1), 19–20.
82. Krittanawong, C. (2018). The rise of artificial intelligence and the uncertain future for physicians. *European Journal of Internal Medicine*, *48*, e13–e14.
83. Choudhury, A., & Asan, O. (2022). Impact of accountability, training, and human factors on the use of artificial intelligence in healthcare: Exploring the perceptions of healthcare practitioners in the US. *Human Factors in Healthcare*, *2*,10002. https://doi.org/10.1016/j.hfh.2022.100021
84. Shebib, R., Bailey, J. F., Smittenaar, P., Perez, D. A., Mecklenburg, G., &Hunter, A. (2019). Randomized controlled trial of a 12-week digital care program in improving low back pain. *npj Digital Medicine*, *2*, 1. https://www.nature.com/articles/s41591-019-0325-3. (Accessed: 23 August 2024).
85. Coppola, F., Faggioni, L., Regge, D., Giovagnoni, A., Golfieri, R., Bibbolino, C., Miele, V., Neri, E., & Grassi, R. (2021). Artificial intelligence: Radiologists' expectations and opinions gleaned from a nationwide online survey. *La Radiologia medica*, *126*(1), 63–71.
86. Maassen, O., Fritsch, S., Palm, J., Deffge, S., Kunze, J., Marx, G., Riedel, M., Schuppert, A., & Bickenbach, J. (2021). Future medical artificial intelligence application requirements and expectations of physicians in German University Hospitals: Web-based survey. *Journal of Medical Internet Research*, *23*(3), e26646.
87. Blease, C., Kaptchuk, T. J., Bernstein, M. H., Mandl, K. D., Halamka, J. D., & DesRoches, C. M. (2019). Artificial intelligence and the future of primary care: Exploratory Qualitative Study of UK general practitioners' views. *Journal of Medical Internet Research*, *21*(3), e12802.
88. Ahmed, Z., Bhinder, K. K., Tariq, A., Tahir, M. J., Mehmood, Q., Tabassum, M. S., Malik, M., Aslam, S., Asghar, M. S., & Yousaf, Z. (2022). Knowledge, attitude, and practice of artificial intelligence among doctors and medical students in Pakistan: A cross-sectional online survey. *Annals of Medicine and Surgery (2012)*, *76*, 103493.
89. Chen, M., Zhang, B., Cai, Z., Seery, S., Gonzalez, M. J., Ali, N. M., Ren, R., Qiao, Y., Xue, P., & Jiang, Y. (2022). Acceptance of clinical artificial intelligence among physicians and medical students: A systematic review with cross-sectional survey. *Frontiers in Medicine*, *9*, 990604.

CHAPTER 24

Future Directions

Emerging Technologies and Innovations in AI for Healthcare

Manisha Rani

24.1 INTRODUCTION

Healthcare arrangements across the globe face significant encounters in achieving the 'quadruple aim'—enhancing public health, improving patient experiences, ensuring healthcare provider well-being, and controlling rising costs. The growing burden of chronic diseases, aging populations, and escalating healthcare expenses have prompted governments, policymakers, and healthcare providers to explore innovative solutions and reshape service distribution models. Integrating technology and artificial intelligence (AI) into healthcare has the potential to address these challenges effectively [1].

AI is transforming healthcare by enhancing diagnostics, improving treatment precision, and optimizing patient overhaul. AI's potential to process huge quantities of data, recognize patterns, and make conclusions has catalyzed numerous innovations across the healthcare industry. This review explores the latest emerging technologies and innovations in AI, providing a comprehensive overview of how these advancements are transforming healthcare [2, 3].

24.2 EMERGING TECHNOLOGIES IN AI FOR HEALTHCARE

As AI continues to advance, cutting-edge technologies are redefining the possibilities within healthcare. These innovations have the power to transform diagnostics, treatment approaches, patient care, and overall system efficiency. Below are some of the most impactful emerging AI technologies driving the future of healthcare.

24.2.1 Machine Learning (ML) and Deep Learning (DL)

ML and DL are pivotal in modern healthcare. They have found applications in diagnostic imaging, predictive analytics, and tailored medicine. DL models, predominantly Convolutional Neural Networks, have been employed in image-based diagnosis for circumstances like as cancer, cardiovascular diseases, and retinal disorders. Predictive analytics (i.e. a technique in which data is used to see future outcomes), of ML algorithms, are used for risk stratification, disease progression forecasting, and patient outcome predictions [4].

Recent innovations include the development of transformer models and reinforcement learning (RL). Transformers, initially designed for natural language processing (NLP), have been adapted for analyzing electronic health records (EHRs) and genomic data. Reinforcement learning, which enables AI models to learn from interactions with their environment, is being explored for personalized treatment planning and adaptive radiotherapy [5].

24.2.2 NLP

NLP plays a vital role in managing amorphous data in healthcare, particularly in EHRs. It enables automated extraction of relevant clinical information, supports clinical decision-making, and facilitates patient interaction through chatbots and virtual assistants. NLP is also used in processing and synthesizing vast amounts of medical prose, making it easier for healthcare experts to stay modernized with the newest research [6].

DOI: 10.1201/9781003533351-24

Recent advances include the adaptation of great language models such as GPT-4 (Generative Pre-trained Modernizers) and BERT [7] (Bidirectional Encoder Symbols from Transformers) to the healthcare domain. These models are modified for specific errands like summarizing patient annals, predicting patient outcomes based on clinical notes, and supporting decision-making in complex cases. Innovations in multilingual NLP models are also making healthcare more accessible in non-English-speaking regions [4].

24.2.3 Robotics and Automation

Robotics and mechanization have significant applications in surgery, rehabilitation, and hospital operations. Robotic systems, like as the da Vinci Surgical System, have been used for minimally destructive surgeries, offering greater exactness and reduced recapture times. In rehabilitation, robotic exoskeletons assist patients with mobility impairments. Automation in administrative tasks, such as scheduling and inventory management, enhances operational efficiency in healthcare facilities.

Emerging innovations include the development of autonomous surgical robots that can perform complex procedures with minimal human intervention. AI-powered prosthetics that adapt to users' movements and needs are also advancing rapidly. Additionally, Robotic Process Automation is being increasingly adopted in administrative workflows, dipping the burden on healthcare staff and diminishing errors [8].

24.2.4 Computer Vision

Computer vision is widely used in imaging diagnostics, pathology, and surgical assistance. AI-driven image analysis has been instrumental in detecting anomalies in radiographs, MRIs, and CT scans. In pathology, computer vision assists in analyzing tissue samples, improving the accuracy of cancer diagnoses. During surgery, AI-powered vision systems provide real-time guidance to surgeons, enhancing precision and outcomes.

Recent innovations include the integration of AI with progressive imaging procedures like 3D imaging and holography. AI models are also being trained to perform real-time analysis during endoscopic procedures, identifying polyps or lesions with high accuracy. Another innovation is the development of AI systems that can analyze and interpret images from different modalities, enabling comprehensive diagnostics [9].

24.2.5 AI-Driven Wearables and IoT

AI-driven wearables and the Internet of Things (IoT) are transforming remote monitoring, chronic disease management, and patient assignation. Vesture devices furnished with AI systems can screen dynamic signs, sense irregularities, and alert healthcare earners in real-time. IoT devices enable seamless data exchange between patients and healthcare systems, facilitating continuous care [10].

Innovations include the development of smart wearables that not solitary monitor health parameters but likewise provide personalized health endorsements. These devices are increasingly integrating AI for predictive analytics, helping in early recognition of illnesses like diabetes and cardiovascular conditions. The incorporation of IoT with AI-driven healthcare arrangements is enabling the creation of smart hospitals, where patient data are continuously monitored and analyzed for optimal care [11].

24.3 INNOVATIVE AI APPLICATIONS IN HEALTHCARE

AI is revolutionizing the healthcare commerce, enhancing competence, correctness, and accessibility. From early diagnosis to personalized treatments, AI applications are addressing some of healthcare's most pressing challenges. Below are important areas where AI is creation significant treads:

24.3.1 Telemedicine and Virtual Health Assistants

AI is enhancing telemedicine through the use of computer-generated health assistants and AI-driven chatbots. These technologies offer patients with 24/7 access to healthcare information, care triage, and facilitate virtual consultations. AI-powered platforms can analyze patient symptoms, suggest possible diagnoses, and recommend treatment options, dropping the need for in-person visits.

The impact of AI in telemedicine is profound, particularly in improving accessibility to healthcare services. Virtual health assistants reduce the burden on healthcare providers, allowing them to focus on more complex cases. The cost-effectiveness of telemedicine, combined with AI-driven personalization, has made healthcare more accessible to underserved populations [12].

24.3.2 Predictive Analytics for Disease Outbreaks

AI-driven prognostic analytics are playing a vital role in epidemiology and public well-being. By analyzing vast datasets, counting social media, travel designs, and healthcare annals, AI models can predict disease outbreaks with high accuracy. ML models are being used to forecast the spread of diseases like COVID-19, influenza, and malaria.

The ability to predict disease outbreaks has significant implications for public health planning and response. Early detection allows for timely interventions, such as vaccination campaigns and travel restrictions, potentially saving lives and reducing the economic impact of pandemics. AI models are also helping in real-time tracking of disease spread, providing crucial information to policymakers.

24.3.3 Drug Discovery and Development

AI is converting drug detection and expansion by accelerating the documentation of potential drug contenders and enhancing clinical trials. AI models can analyze vast chemical libraries to predict the efficacy of compounds against specific targets. ML is also used to identify biomarkers for patient stratification in clinical trials, increasing the chances of success.

The impact of AI on drug finding is significant, with several AI-discovered drugs entering clinical trials at unprecedented speeds. AI decreases the period and cost associated with drug development, enabling faster delivery of new therapies to patients. Personalized medicine, where AI helps in designing treatments tailored to individual genetic profiles, is another major advancement [13].

24.3.4 Personalized Medicine

Modified medicine, driven by AI, involves dressmaking medical behaviors to individual patient role based on their genetic makeup, existence, and atmosphere. AI models analyze genomic data, recognize biomarkers, and predict specific responses to treatments. This approach is particularly beneficial in oncology, where precision medicine is becoming the standard of care.

The shift towards personalized medicine is transforming patient care by improving treatment outcomes and reducing adverse effects. AI enables healthcare providers to design treatment plans that are highly specific to the patient's unique profile, increasing the effectiveness of therapies. In addition, AI is facilitating the expansion of targeted therapies, which are more active and have less side belongings associated to old-style treatments.

24.3.5 AI in Mental Health

Emerging AI technologies are being industrialized to diagnose and treat mental health conditions through the analysis of speech patterns, facial expressions, and behavior [14].

AI is transforming mental healthcare by enhancing early diagnosis, providing personalized interventions, and making mental health services more accessible. From virtual mental health assistants to predictive analytics and emotion recognition technologies, AI is authorizing persons to take control of their mental well-being while supporting healthcare providers with valuable insights. However, addressing principled challenges related to privacy, bias, and oversight will be crucial to ensuring that AI is used responsibly and equitably in mental healthcare. As AI endures to progress, it holds the promise of reshaping the future of mental health, offering advanced solutions to approximately of the most persistent challenges in the arena.

24.3.6 AI-Driven Clinical Trials

AI can enhance clinical trial processes by improving study design, streamlining patient recruitment, and analyzing data more efficiently. This leads to faster and more cost-effective trials, ultimately accelerating the development of new treatments.

AI is revolutionizing clinical trials by improving patient enrolment, optimizing trial design, and enabling real-time data analysis and monitoring. These innovations are accelerating the development of new therapies, dropping costs, and increasing the overall effectiveness of clinical research. As AI continues to evolve, it promises to make clinical trials more personalized, accessible, and adaptive, benefiting patients, researchers, and the healthcare industry as a whole. However, ethical considerations around privacy, transparency, and fairness will need to be cautiously addressed to ensure accountable and equitable practice of AI in clinical trials.

24.4 CHALLENGES AND ETHICAL CONSIDERATIONS

While AI holds great promise in transforming healthcare, its integration also offerings several contests and ethical considerations. These issues must be addressed to ensure that AI submissions are safe, effective, and reasonable.

24.4.1 Data Privacy and Security

The integration of AI in healthcare raises significant concerns about data privacy and safety. Patient data, which is highly sensitive, needs to be protected against unauthorized access and breaches. Adhering to regulatory frameworks like the General Data Protection Guideline and the Health Guard Portability and Accountability Act is vital to ensuring data privacy and safety in healthcare [15].

24.4.2 Bias in AI Models

Bias in AI mockups is a critical issue, predominantly in healthcare, where biased processes can lead to unequal treatment and outcomes. AI models trained on non-representative data may perpetuate existing health disparities, particularly across racial, gender, and socio-economic lines.

Mitigating bias requires the use of diverse and representative datasets in training AI models. Performances such as bias improvement algorithms and the enclosure of impartiality metrics in model evaluation are also important. Transparency in AI development and deployment, along with regular audits, can help ensure fairness in AI-driven healthcare applications.

24.4.3 Regulatory Challenges

The rapid advancement of AI technologies in healthcare offerings challenges for regulatory bodies. Current regulatory frameworks may not be equipped to evaluate and approve AI-driven medical devices and applications quickly enough. The dynamic nature of AI, where models continuously learn and evolve, complicates the regulatory approval process [16].

24.4.4 Ethical AI Usage

The principled practice of AI in healthcare requires lecturing issues associated to transparency, accountability, and patient consent. AI decisions, especially those impacting patient care, need to be explainable and understandable by healthcare providers and patients alike. Confirming that patient role are informed about the practice of AI in their care and gaining their permission is vital for ethical AI application.

24.5 FUTURE DIRECTIONS AND TRENDS

The incorporation of AI in healthcare is still in its primary phases, but ongoing advancements and emerging trends indicate a transformative future for the industry. Below are some key future directions and trends that will shape the upcoming of AI in healthcare:

24.5.1 Integration of AI with Emerging Fields

The integration of AI with other emerging fields like quantum computing and synthetic biology holds great promise for the future of healthcare. Quantum computing could exponentially increase the processing power available for AI, enabling the analysis of enormously complex living systems. AI combined with synthetic biology could lead to the development of innovative therapeutics and bioengineered tissues [17].

24.5.2 Collaborative AI Models

Collaborative AI representations, such as federated learning, are becoming increasingly popular. These models allow for the training of AI algorithms across multiple institutions without sharing sensitive data, preserving patient privacy. Multi-institutional collaborations are also helping to build more robust and generalizable AI models that can be applied across different healthcare settings.

24.5.3 AI in Global Health

AI has the latent to address healthcare disparities globally, predominantly in low-resource settings. AI-driven solutions can help in the early uncovering of diseases, optimizing resource allocation, and provided that remote care to underserved populations. The development of AI models that are adaptable to different healthcare environments is crucial for global health equity [18].

24.5.4 Sustainability of AI in Healthcare

As AI develops more integrated into healthcare, considerations around the sustainability of these technologies are becoming more important. Energy-efficient AI models, the economic impact of widespread AI adoption, and the long-standing viability of AI-driven healthcare arrangements are dangerous areas of focus. Ensuring that AI innovations are both environmentally and economically sustainable will be key to their long-term success.

24.6 SUMMARY

The landscape of healthcare is rapidly evolving with the advent of AI technologies. From diagnostics to treatment, AI is playing an gradually central character in improving patient outcomes besides healthcare delivery. However, as these technologies continue to develop, it is essential to discourse the challenges and ethical deliberations that come with them. The upcoming of AI in healthcare looks promising, with the potential to make healthcare more personalized, accessible, and efficient.

REFERENCES

1. Bajwa, J., et al. (2021). Artificial intelligence in healthcare: Transforming the practice of medicine. *Future Healthcare Journal, 8*(2), 188–194.
2. Ali, O., Abdelbaki, W., Shrestha, A., Elbasi, E., Alryalat, M. A. A., & Dwivedi, Y. K. (2023). A systematic literature review of artificial intelligence in the healthcare sector: Benefits, challenges, methodologies, and functionalities. *Journal of Innovation & Knowledge, 8*, Article 100333.
3. Paranjape, K. et al. (2019). Introducing artificial intelligence training in medical education. *JMIR Medical Education, 5*, Article 16048.
4. Ahmed, M. et al. (2021). A deep-learning-based smart healthcare system for patient's discomfort detection at the edge of Internet of Things. *IEEE Internet of Things Journal, 8*, 10318–10326.
5. Davenport, T., & Kalakota, R. (2019). The potential for artificial intelligence in healthcare. *Future Healthcare Journal, 6*, 94–98.
6. Iroju, O. G., & Olaleke, J. O. (2015). A systematic review of natural language processing in healthcare. *International Journal of Information Technology and Computer Science, 7*, 44–50.
7. Li, Y., Rao, S., Solares, J., Hassaine, R. A., Ramakrishnan, A., Canoy, R., Zhu, D., Rahimi, Y., & Salimi-Khorshidi, K. (2020). BEHRT: Transformer for electronic health records. *Scientific Reports, 10*, Article 7155.
8. Jiang, F., Jiang, Y., Zhi, H., Dong, Y., Li, H., Ma, S., Wang, Y., Dong, Q., Shen, H., & Wang, Y. (2017). Artificial intelligence in healthcare: Past, present and future. *Stroke and Vascular Neurology, 2*, 230–243.
9. Gazzah, S., & Bencharef, A. (2020). A survey on how computer vision can respond to urgent needs to contribute in COVID-19 pandemics. In *The International Conference on Intelligent Systems and Computer Vision*, IEEE2020 (pp. 1–5).
10. Huang, J., Wu, X., Huang, W., Wu, X., & Wang, S. (2021). Internet of Things in health management systems: A review. *International Journal of Communication Systems, 34*(4), 4683.
11. Elbasi, A. et al. (2021). A survey on machine learning and Internet of Things for COVID-19. *IEEE World AI IoT Congress*, 0115–0120.
12. Pacis, D. M. M., Subido, E. D., & Bugtai, N. T. (2018). Trends in telemedicine utilizing artificial intelligence. *AIP Conference Proceedings, 1993*, Article 040009.
13. Paul, D., Sanap, G., Shenoy, S., Kalyane, D., Kalia, K., & Tekade, R. K. (2021). Artificial intelligence in drug discovery and development. *Drug Discovery Today, 26*, 80–93.
14. Graham, S., Depp, C., Lee, E. E., Nebeker, C., Tu, X., Kim, H.-C., & Jeste, D. V. (2019). Artificial intelligence for mental health and mental illnesses: An overview. *Current Psychiatry Reports, 21*, Article 116.
15. Aldwean, A., & Tenney, D. (2024). Artificial intelligence in healthcare sector: A literature review of the adoption challenges. *Open Journal of Business and Management, 12*, 129–147.
16. Gerke, S., Minssen, T., & Cohen, G. (2020). Ethical and legal challenges of artificial intelligence-driven healthcare. In A. Bohr, & K. Memarzadeh (Eds.), *Artificial Intelligence in Healthcare* (pp. 295–336). Academic Press. doi: 10.1016/B978-0-12-818438-7.00012-5. Epub 2020 Jun 26. PMCID: PMC7332220.
17. Mishra, H. (2023). Future directions of artificial intelligence and machine learning in healthcare: A systematic analysis and mapping study. In *Proceedings of the 6th International Conference on Information Systems and Computer Networks (ISCON) GLA University, Mathura, India*.
18. Tran, B. X., et al. (2019). Global evolution of research in artificial intelligence in health and medicine: A bibliometric study. *Journal of Clinical Medicine, 8*, Article 360.

CHAPTER 25

Interoperability Challenges and Solutions in AI-Driven Healthcare Systems

Ahmed Tharwat Al-Namrouti, Mohamed N. Elshabrawi, Reem Sayad, Ihda Mahmoud Bani Khalaf, Haytham Jaoun, Esraa Marwan Morshedy, Karim Samir Shata, and Gamila Ali Attaelmanan

25.1 INTRODUCTION

25.1.1 Understanding Interoperability in Healthcare

Healthcare interoperability plays a significant role in contemporary healthcare systems by facilitating efficient communication and data exchange across disparate information systems. As healthcare technology continues to grow, the requirement for interoperability becomes progressively pressing to ensure that patient information can be accessed and utilized efficiently across diverse treatment settings [1]. This capability not only increases the continuity of treatment but also facilitates informed choices by healthcare personnel, thereby improving the results for patients [2].

The interoperability concept can be classified into many levels, each displaying different degrees of integration and consistency between systems:

25.1.1.1 Level 0 or No Interoperability

At this level, systems function independently without any ability to exchange data or services with other systems. Each system works separately, lacking any type of interoperability.

25.1.1.2 Level 1 or Technical Interoperability

This level presents basic communication protocols that allow systems to exchange data. Technical interoperability sets the foundation for systems to connect and communicate fundamentally, similar to plug-and-play compatibility by utilizing agreed-upon signal and protocol standards.

25.1.1.3 Level 2 or Syntactic Interoperability

Syntactic interoperability builds on technical interoperability by standardizing the format and structure of shared data. Systems at this level can figure out the data based on predefined syntax rules, like those found in protocols such as the high-level architecture, ensuring how data is structured and transferred between the connected systems.

25.1.1.4 Level 3 or Computational Interoperability

The use of semantics exhibits the most advanced stage of interoperability. At this level, the system has the ability to interpret and understand the meaning of that data consistently and accurately, ensuring correct and accurate interpretation of shared data by the receiving system, even if it uses a different structure, terminology, or natural language to produce meaningful and actionable results based on the exchanged information, aligning with the intended use identified by end users.

25.1.1.5 Pragmatic Interoperability

At this level, systems go beyond exchanging data using agreed-upon syntax and semantics but also possess an understanding of how the data will be used and applied within their respective contexts. This displays

DOI: 10.1201/9781003533351-25

a deeper level of coordination and mutual understanding between interoperating systems, ensuring meaningful and useful shared data to achieve specific operational goals.

25.1.1.6 Dynamic Interoperability

Dynamic interoperability means that the systems can adapt and respond to changes in presumptions, restrictions, and operational frames over time. Systems that have achieved dynamic interoperability can effectively distinguish and employ ongoing conditions and requirements, allowing flexible and adaptive behavior as situations change.

25.1.1.7 Conceptual Interoperability

Conceptual interoperability displayed the highest level of interoperability, where systems share a common understanding of the fundamental concepts and models that represent their operation domain. This level ensures the assumptions and constraints alignment to interpret and manipulate data, facilitating harmonization across different systems and environments.

In essence, while technical and syntactic interoperability focus on the mechanics of data exchange and format compatibility, semantic interoperability goes beyond by ensuring that the exchanged information retains its intended meaning and utility across disparate systems and contexts. Achieving semantic interoperability is essential for enhancing collaboration and decision-making in complex environments such as healthcare, where accurate interpretation of data is critical for patient care and operational efficiency [3]. Pragmatic, dynamic, and conceptual interoperability represent advanced stages in the evolution of system integration and collaboration.

Achieving increased levels of integration in healthcare is crucial for overcoming the fragmentation of patient information across diverse settings such as hospitals, laboratories, pharmacies, and clinics. This fragmentation results in critical patient data being dispersed across various proprietary and heterogeneous systems, leading to patient harm, unnecessary resource utilization, and even avoidable deaths due to inadequate communication between healthcare practitioners and complicating efforts to provide a comprehensive view of an individual's medical history. As Middleton in 2004 highlights, this issue leads to alarming rates of missing patient data, which impedes timely and informed decision-making. Additionally, Lau and Shakib in 2005 point out that the prevalence of non-standard, unstructured text formats further exacerbates the difficulty of information exchange among healthcare providers. Addressing these challenges requires advancements in standardization, data integration, and the structuring of unstructured data, supported by interoperability frameworks and regulatory policies to improve the standard and effectiveness of patient treatment and enable clinicians to access longitudinal patient records stored across different systems immediately and regardless of their geographical locations. Interoperability can also cnhance workflows, minimize errors, and boost the competency of healthcare delivery, leading to patient care optimization. This interoperability, in addition to aiding clinical decision-making by providing comprehensive patient information, also enables patients to manage their health more effectively by having access to their complete medical records from various providers.

Kennedy (2023) investigated the relationships between patient involvement, e-health systems in the United States, and clinical integration and compatibility. Clinical integration refers to the collaboration of patient care activities across conditions, services, settings, and providers in a seamlessly and patient-centered manner. He investigated how these factors interplay in the socioeconomic context of healthcare delivery in the United States. Also, he explored how improved clinical integration through enhanced interoperability can lead to greater patient engagement, potentially influencing healthcare outcomes and patient satisfaction. Understanding these dynamics is crucial for optimizing electronic health system capacities to better address the demands of both patients and healthcare professionals alike, ultimately aiming to enhance overall healthcare delivery effectiveness and efficiency [4].

25.2 THE ROLE OF ARTIFICIAL INTELLIGENCE (AI) IN DRIVING INTEROPERABILITY

Since the late 1990s, interoperability has been an important topic in the public sector, especially with the rise of e-government. Initially, the focus was on overcoming technological issues that would allow multiple systems to connect and share information seamlessly as online public services expanded. However, as e-government developed, the idea of interoperability grew to encompass data, procedures, and organizational structures. This larger perspective is critical for successfully adopting digital governance since it helps to break down silos and clarify diverse parts. In 2004, the European Commission published the first European Interoperability Framework to upgrade public service interoperability throughout Europe. This framework identifies four pillars of integrated public service governance: legal, organizational, semantic, and technical interoperability. As digital government continues to advance, the concept of interoperability is expected to become more complex. New technologies, including AI and blockchain, could revolutionize data exchange and processing, presenting new challenges and opportunities for interoperability [5].

25.2.1 Data Mapping

The mapping of information documents refers to the linkages and matches between the various fields in different databases. It helps owners thoroughly observe and comprehend data storage and connectivity between different data stores.

Data mapping is the initial step for conducting numerous processes related to the data and applying them to the construction of databases.

25.2.1.1 Data Integration

This entails combining and standardizing diverse data in one central hub into a cohesive flow. For example, to avoid data duplication and provide a smooth flow of data, medical firms combine patient records from several departments to ensure reliable and consistent information across the board.

25.2.1.2 Data Migration

This procedure applies to transferring data from one system to another, including an alternate storage type, layout, or IT system with unique structural characteristics.

Data Transformation: This presents the procedure of transforming unorganized information from one format to another, like converting from an extended markdown language (XML) file to a separated by commas (CSV) file.

25.2.1.3 The Value of Data Mapping in Healthcare

Data mapping is vitally crucial in healthcare to ensure the success of integration projects. Its relevance originates from the need for seamless networking and interoperability among systems to support quick decision-making through correct data sharing. Poor-quality data can cost healthcare organizations approximately $12.9 million annually, as noted by Gartner. Data mapping offers several benefits in healthcare, which could be mainly attributed to:

Improved Patient Care: By combining data from multiple sources, clinicians obtain an extensive comprehension of a patient's condition, leading to faster referrals, reduced delays, and lower administrative burdens, which in turn enhance patient satisfaction and reduce costs.

Enhanced Interoperability: Data mapping enables healthcare professionals to access a complete patient health record across different systems, leading to better-informed decisions and improved care coordination. Moreover, mapping automates data transmission and streamlines billing and insurance procedures, resulting in cost savings. Also, it ensures accurate, safe, and confidential management of health information. On the other hand, data mapping helps researchers access integrated resources for large-scale investigations, identify illness trends, and create successful solutions.

Effective data mapping enables thorough analysis, optimizes resource allocation, and improves operational efficiency [6].

25.2.1.4 The Complexities of Data Mapping in Healthcare

Healthcare data mapping is a challenging and risky task due to data loss, missed values, and complicated details. These integration initiatives have significant implications that directly affect the standard of patient care. Because of this, it's essential to recognize these risks and take precautions when conducting operations.

25.2.1.5 Data Mapping Accuracy

Accurately mapping historic information in older medical facilities presents considerable issues for medical professionals. A small mistake in this process can have a big impact, including delays in patient treatment or misdiagnosis, as well as disturbances in attempts to integrate healthcare. When creating data maps, businesses frequently make the mistake of deleting important information due to the complexity of healthcare data, which makes the maps less useful and difficult to maintain the data mapping process efficiently due to the massive volume of input. To guarantee successful healthcare integration projects, workers must be more familiar with various technologies of healthcare data mapping and comprehend the differing data mapping needs.

25.2.1.6 Data Discrepancies

Managing data disparities between source and destination systems is a major difficulty in the field of healthcare data mapping. When several data formats or terminologies are utilized, this complexity appears, which makes mapping data fields a difficult operation. Healthcare companies must use data transformation strategies like normalization and cleaning to address these disparities. Data cleaning refers to identifying and correcting mistakes in the data to guarantee its integrity and dependability. By applying these transformation approaches and putting them into practice, healthcare businesses may successfully align data pieces and achieve accurate data integration.

25.2.1.7 Transforming the Mapping Data into Useful Knowledge

In healthcare, especially in healthcare integration projects, the meticulous process of data mapping is crucial to providing actionable insights. Advances achieved by several electronics engineers have led to the automation of this process through the employment of very effective data mapping technologies. As a result, patient care quality has been raised, and medical procedures have been more simplified, ultimately contributing to improved healthcare results.

25.2.1.8 Future of Data Mapping in Healthcare

There is a lot of promise for the future of healthcare systems, especially with regard to the possibility of customized, individualized care. The key to realizing this promise in the healthcare industry is applying large-scale data mapping correctly. This will facilitate improved collaboration and a smoother exchange of information, in addition to increasing data connectivity and interoperability. Furthermore, data mapping in healthcare could soon undergo major change due to machine learning and AI. These technological advancements have facilitated disease diagnosis and supported the maintenance of the necessary standard for healthcare integration projects, in addition to validating the correctness of data mapping in healthcare operations.

Telemedicine is another significant development in the field that has revolutionized the way healthcare is delivered. By transforming data formats, AI algorithms can speed up data mapping and facilitate simple communication between telemedicine systems and electronic health records (EHRs). Businesses in the healthcare industry can guarantee efficient and successful data mapping by adopting these goals. This indicates relationships and patterns in the data that aren't always obvious. As a result, medical professionals and patients can quickly obtain vital information and make better decisions. This quickly raises the standard for healthcare quality by converting data into knowledge that can be put into practice.

25.2.2 Patients Records

This section examines how AI technologies can transform the gathering, processing, and use of patient data. It highlights the ways in which these technologies impact illness monitoring, customized therapy, and prognostic analytics.

25.2.2.1 Confidentiality of Medical Records

Algorithms with AI are employed in medical records safety to identify vulnerabilities, spot abnormalities in user behavior, and protect against online threats.

25.2.2.2 Data Gathering and Integration

Automating the aggregation of data from wearable technology, drug surveillance databases, digital health records (EHRs), health apps, and health data technologies makes data collection and assimilation easier.

25.2.2.3 Natural Language Processing

Disease diagnosis, medication discovery, pharmaceutical surveillance, and the extraction of health information from unstructured data are all enhanced by natural language processing (NLP). Accurate diagnosis and treatment are made possible by NLP of large amounts of medical literature, including research papers and clinical notes. Additionally, it promotes research by reviewing the state of the field, identifying new trends, and assessing the effectiveness of treatments using meta-analyses and literature reviews.

25.2.3 Clinical Decision Support Systems (CDSS)

By utilizing complex algorithms such as deep learning (DL), NLP, and graph neural networks (GNNs) to evaluate massive datasets, AI models greatly improve CDSS.

25.2.4 Predictive Analytics

Early disease detection and therapy optimization rely heavily on predictive analytics. For example, convolutional neural networks (CNNs) were utilized to separate lung nodules from CT scans, which aids in early lung cancer identification and treatment. Furthermore, AI-driven remote patient monitoring systems use wearable devices and remote medical monitoring technologies, such as blood pressure (BP) cuffs and ECG electrodes, to gather biological information.

25.2.5 Researching and Creating New Medications

AI accelerates drug discovery by enhancing various stages of the pipeline. Generative models can design new molecules, reinforcement learning optimizes molecular properties, and GNNs predict drug-disease correlations and drug repurposing.

25.2.6 Clinical Trials Optimization

Using AI to match patients with suitable trials according to their medical background and genetic information can speed up the progress of developing new treatments.

25.2.7 Monitoring of Illnesses and Forecasting of Outbreaks

AI supports disease surveillance by tracking patterns and forecasting outbreaks, allowing for timely intervention and public health measures. AI algorithms study large sets of data, such as network connections, individual actions, and geographical information, to detect early signs of outbreaks and evaluate their likely consequences.

25.2.8 Administrative Applications of AI in Healthcare

Administrative tasks are one of the important tasks in the healthcare industry to ensure quality patient care. However, the burden of documenting every record manually, especially with the growing number of patients, can lead to lost efficiency and potential dips in patient care quality. For instance, the average nurse in the United States spends about 25% of their work hours on regulatory and administrative tasks. AI offers a range of administrative applications in healthcare, although its impact in this area may not be as transformative as in patient care. Nonetheless, AI can greatly enhance efficiency, reduce costs, and improve overall care quality.

Here are some practical examples:

25.2.8.1 Appointment Scheduling and Management

AI-driven chatbots or virtual assistants are as effective as humans in helping patients book an appointment with the doctor. Besides aligning doctors' schedules with the patients' availability, these digital agents can quickly respond to cancellations and send reminders to patients regarding their appointments.

25.2.8.2 Patient Registration and Verification

Registering and verifying every patient is undoubtedly time-consuming, which increases the chances of errors. With AI assistants, the process can be fully automated while reducing errors and administrative overhead.

25.2.8.3 Billing and Claims Processing

Any manual error in this process can cost too much to a healthcare institution. AI can help in automating billing processes, making sure medical codes are accurate, and flagging any suspicious stuff. This means patients can get faster reimbursements and fewer headaches for administrative staff in dealing with denied claims.

25.2.8.4 Health Record Management

Keeping all those EHRs organized and secure is a challenge in the healthcare industry, within the strict regulations. While preserving patient protection and privacy, AI-powered solutions are helpful for managing health records.

25.2.8.5 Document Digitization and Extraction

In the healthcare sector, not all papers are available online; several handwritten notes and documents can require scanning and uploading to EHR systems. Handwritten papers may be quickly converted into structured data by AI-based solutions.

25.2.8.6 Chatbots for Patient Queries

In addition to answering standard patient questions, AI chatbots may offer directions, office hours, and answers to common health-related queries. AI frees up administrative personnel time to focus on more important activities like providing outstanding patient care. Robotic process automation (RPA) is one of the AI technologies most relevant to this application. Medical record handling, clinical documentation management, revenue cycle supervision, and claim processing are just a few of the administrative duties in the healthcare industry that RPA can expedite. Furthermore, a few medical institu tions have looked at the use of chatbots for telemedicine, mental health assistance, and patient contacts. These NLP-powered chatbots can help with simple tasks like making appointments and refilling prescriptions [7].

25.3 CHALLENGES OF AI-BASED INTEROPERABILITY

25.3.1 Data Privacy and Security

In the age of vast open online archives, data privacy has become increasingly important. In the medical industry in particular, patient-traceable data can be used to great purpose, opening up a myriad of alternatives. However, by definition, AI systems need a lot of data [8]. Therefore, a crucial requirement for advancing any AI research is the secure storage of patient data.

25.3.2 Ethical Considerations and Open-Source Database in AI Research

There are moral and legal issues with big data utilization in the healthcare sector, especially when it comes to privacy and autonomy. Infrastructure, security, and data heterogeneity are major technological problems. In the European Union, there is broad approval for the use of medical information for purposes other than therapy. Governments must protect private medical information in terms of anonymity, safety, and confidentiality while ensuring that technological advances benefit from the use of open data [9].

25.3.3 Big Data in AI Research

Large-scale data is the term for enormously varied data sets that can be swiftly accessed and examined, including a broad spectrum of data types and users [10]. The recent COVID-19 pandemic served as another evidence of the usefulness of these publicly accessible datasets, as it facilitated the rapid development of algorithms with a variety of applications to support medical professionals and COVID-19 patients. Therefore, hospitals need to be able to manage enormous amounts of data, allow for their security, and permit their safe use in order to establish and improve such cutting-edge diagnostic tools. This type of data must first be de-identified before being transferred to data from a third-party aggregator, although new techniques for data linkage may eventually be created that make the origins available.

25.3.4 The Risk of Re-Identification and Challenges in De-Identification and Privacy Violations in AI and Big Data

Numerous studies showing how more recent algorithms may identify individuals from both public and private data repositories have been illustrated and proven [11, 12]. In research conducted in 2018 analyzing data sets from the National Health and Nutrition Examination Survey, it was noted that although identifiers of protected health information were suggested to be removed, in a physical activity cohort study, an algorithm was still able to identify 85.6% of adults and 69.8% of children [13].

The issue of patient identification may be especially significant in domains like dermatology, where it is difficult to de-identify patient photos, even though the patient's metadata can be delinked. This is especially true if the lesion is visible on the face or another easily recognized area of the body. Such a privacy violation may have consequentialist, deontological, or both effects. Negative consequences that are quantifiable and real are of interest to consequentialists. A few typical examples would be having to deal with discrimination at work if one's medical history is disclosed or having to pay higher health insurance rates because more information became available as a result of a privacy breach [12].

This might turn out to be a more significant issue in AI-based applications that make a forecast of a future health event based on specific deductions made from an individual's observed behavioral and daily habits, which could lead to a lot of implications for clinical, social, and occupational settings [14, 15].

25.3.5 Deontological Impact of Privacy Breaches in AI

Deontological impacts can occur even in cases where there is no real misuse of personal information or no notification of the breach to the individual. These effects are subjective, unquantifiable, and evident. These worries are a result of the sense of powerlessness that follows a privacy invasion of this kind. Though it may not be quantifiable, the mental agony that such knowledge may induce is evident [12].

25.3.6 Data Integrity and Accuracy

AI has shown to have many potentials and limitless possibilities since its inception and widespread use in diverse contexts. Di Vaio claims that AI greatly improves data integrity by decreasing human error and raising processing efficiency [16]. Parikh et al. asserted that sophisticated algorithms and machine learning models—both of which are AI functions—can handle massive data sets more precisely than people can lend credence to this point of view and enhance the caliber of data analysis [17].

Data integrity remains a major challenge in the modern healthcare industry. Ensuring medical data accuracy and security is important because inaccurate data can pose critical health risks for patients and medical professionals. Furthermore, it can lead to scams, data theft, misconduct, and inadequate treatment.

On the other hand, emphasized that the efficacy of AI systems relies on the quality of the given data and the programming they receive [16]. As a result, there is a risk that AI will exacerbate faulty or biased input data. This viewpoint emphasizes the ethical duty of data curation and the necessity of AI algorithms' transparency to guarantee data integrity. To maintain data integrity, this viewpoint emphasizes the ethical duty of data curation and the necessity of algorithmic transparency in AI.

25.3.7 Data Bias

25.3.7.1 Understanding Data Bias

Data bias in AI, from a medical perspective, refers to the systematic deviation or distortion in the collection, selection, interpretation, or presentation of healthcare-related data that leads to inaccurate or unfair conclusions when used in AI algorithms and models. This bias can originate from various sources: sampling, measurement, algorithmic, reporting, and confounding bias [17].

According to sampling bias, it takes place when trained AI models use data that doesn't accurately represent the entire patient population, leading to skewed results for underrepresented groups [18]. Moreover, measurement bias results from inconsistencies or errors in how medical data is collected, recorded, or measured across different healthcare settings or demographic groups. On the other hand, algorithmic bias takes place when AI models inadvertently perpetuate or amplify existing prejudices or disparities in healthcare due to biased training data or flawed model design [19]. Finally, reporting and confounding bias occurs when the tendency to selectively publish or report certain types of medical outcomes or studies leads to an incomplete representation of the true clinical landscape and when AI models fail to account for important confounding factors in medical data, potentially leading to spurious correlations or incorrect causal inferences, respectively [20].

Data bias in AI can have significant implications in healthcare, potentially leading to misdiagnoses, inappropriate treatment recommendations, or the exacerbation of health disparities among different patient populations. Recognizing and mitigating these biases is important for developing fair, accurate, and reliable AI systems in medicine that can improve patient outcomes across diverse populations [21].

25.3.7.2 Challenges of Bias in Medical AI Applications

AI has shown promise in some domains, such as the diagnosis and treatment of some diseases by using AI. There have been multiple examples to date of algorithms that discriminate against marginalized populations. As one of many instances, CNNs that classify skin lesions accurately are frequently trained on datasets with an estimated 5%–10% Black patient proportion, employing images of skin lesion samples from White patients [22]. Consequently, the network's diagnostic accuracy is around half of what its developers initially stated when evaluated using Black patients' photos [22]. Automated algorithms may thus have a lower diagnosis accuracy rate for Black patients whose lesions may differ from those of White patients. This omission should not be ignored, as Black patients have the highest 5-year survival rate of only 70% as opposed to 94% for patients who are White, making them the most likely to die from melanoma. Skin cancer in Black people may be diagnosed at a later stage, making treatment more difficult as a result of misdiagnoses and socioeconomic constraints preventing access to healthcare [23].

25.3.8 Financial Challenges

It is important to recognize that integrating AI into healthcare will provide cost issues. One major obstacle is the high initial investment expenses, which include the cost of purchasing AI systems, employee training, computer equipment, and data management tools. Costs associated with data management are especially important because they include a lot of resources to be used for gathering, storing, sanitizing, and safeguarding private patient information, as well as developing algorithms that are reliable [24].

The financial burden is further increased by the need to integrate AI into current systems, train medical personnel, and maintain regulatory compliance. To overcome these financial obstacles, strategic investment, teamwork, and meticulous preparation are needed. Making AI integration in healthcare more financially realistic can be achieved through government backing, public-private collaborations, and creative finance structures, ensuring that the advantages of AI are available to all healthcare facilities [24].

25.3.9 Resistance to Adoption

Adoption resistance represents one of the major challenges faced by healthcare that uses AI. Because of their unfamiliarity with the technology or worries about how AI-driven technologies may affect their job security, healthcare professionals may be reluctant to adopt AI-driven technologies. Training, education, and effective change management techniques are necessary to overcome adoption resistance. To promote acceptance, healthcare companies should involve frontline personnel in the implementation process, address their concerns, and highlight the advantages of AI integration. Overcoming opposition to adoption can be facilitated by cultivating a culture that values innovation and ongoing development. Continuing education and assistance can also make healthcare workers feel more at ease and certain when utilizing AI-driven solutions [24].

25.3.10 Regulatory Compliance

Regulatory compliance is another challenge in the healthcare system based on AI. Healthcare institutions should follow strict laws and regularities, such as Health Insurance Portability and Accountability Act (HIPAA), to ensure patient privacy. It can be difficult to integrate AI while adhering to these rules.

It is imperative for organizations to remain current with the most recent regulatory standards and to put measures in place to guarantee compliance. Putting in place effective data security safeguards is necessary for this, making sure AI-driven decision-making processes are transparent and holding AI-driven decision-makers accountable for their actions. Moreover, frequent training initiatives and audits can guarantee that medical personnel understand regulatory obligations [24].

25.3.11 Interoperability Issues

Since integrating AI frequently necessitates exchanging data across several platforms and systems, interoperability concerns pose serious barriers to the effortless integration of AI into healthcare systems. Thus, there is a tremendous deal of difficulty in guaranteeing the safe transport of this data while preserving its confidentiality and integrity. Healthcare companies need to make investments in systems that can successfully communicate with one another to address interoperability challenges. Standardized formats and protocols can assist in enabling data flow and communication amongst diverse systems and technologies [24].

25.4 SOLUTION PERTAINING TO HEALTH CARE INTEROPERABILITY WITH AI

25.4.1 Privacy and Security Protection

Privacy and security are vital in using AI in healthcare. Due to the high sensitivity of health data, robust protection mechanisms are necessary to prohibit unauthorized access and data violations to ensure patient confidentiality and information integrity. Current challenges include the risk of data breaches, the need for secure data-sharing mechanisms, and compliance with stringent regulatory requirements.

25.4.1.1 Technical Solutions

Several technological advancements have emerged to address these challenges. Data security is increased by encryption techniques like homomorphic encryption, which enable calculations on encrypted data without requiring its first decryption. Additionally, blockchain technology offers a decentralized approach to secure data transactions, ensuring that only authorized parties can access the data and that all transactions are transparently recorded.

25.4.1.2 Policy and Regulatory Measures

Compliance with rules such as the General Data Protection Regulation and the US HIPAA in Europe is essential for any AI system handling healthcare data. These regulations provide advice for data security precautions and patient privacy rights. Adopting best practices for compliance, such as regular audits and staff training on data protection, is crucial for healthcare organizations [25, 26].

25.4.2 Data Quality Standards

25.4.2.1 Importance of Data Quality

Data quality is critical for the effective performance of AI in healthcare. High-quality data ensures that AI models can generate accurate and reliable predictions and insights. However, healthcare data often suffers from issues such as incompleteness, inaccuracies, and inconsistencies, which can adversely affect AI outcomes.

25.4.2.2 Standardization Effort

The objective of standardization efforts is to enhance the quality of healthcare data. Standards such as Health Level Seven (HL7) and Fast Healthcare Interoperability Resources (FHIR) provide frameworks for consistent data exchange. Organizations like the International Organization for Standardization (ISO) and the Institute of Electrical and Electronics Engineers are also working to establish comprehensive data quality standards.

Technological Solutions: Technological solutions to improve data quality include machine learning algorithms that can clean and standardize data. These algorithms can identify and correct errors, fill in missing values, and harmonize data from different sources. Implementing real-time data validation systems can also help maintain high data quality by continuously monitoring and verifying incoming data [27, 28].

25.4.3 Strategies for Mitigating Bias

Mitigating bias requires a multifaceted approach. Ensuring diverse and representative data sets is crucial to capturing the full spectrum of patient populations. Regular auditing and updating of AI models can aid in identifying and correcting biases that may occur over time. Additionally, incorporating fairness metrics into model evaluation can provide insights into potential biases and guide corrective actions [29, 30].

25.4.4 Explainability and Transparency

AI-based explainability is the capacity to comprehend and evaluate the procedures used in making decisions about AI models. In healthcare, explainability is essential for building trust and accountability, as patients as well as physicians need to evaluate how and why an AI system arrived at a specific choice.

There are various methods to enhance explainability in AI. Interpretable models, like decision trees and linear models, offer straightforward explanations of their predictions. For more complex models, techniques like SHapley Additive exPlanations (SHAP) and local interpretable model-agnostic explanations (LIME) might shed light on the variables driving particular forecasts. The principle behind this statement lies in the concept of model interpretability in machine learning. Interpretable models, such as linear and decision trees, are valued for their ability to provide clear and understandable explanations for their predictions. For instance, in a decision tree, you can easily trace the decision path from the root to the leaf to understand why a particular prediction was made. Similarly, linear models allow you to see the involvement of each feature in the final prediction, making it easy to interpret the results.

However, in the case of models with greater complexity, such as deep neural networks or ensemble methods, the decision-making process is not as transparent. This is where interpretability techniques like LIME and SHAP come into play. LIME works by approximating the complex model locally around the prediction, helping to explain why a specific decision was made. SHAP, on the other hand, is based on collaborative game theory and assigns a contribution value to each feature, reflecting its importance in the final prediction. These techniques help explain the inner workings of complex models, making them more understandable and trustworthy [31, 32].

25.4.5 Regulatory and Ethical Consideration

Regulatory guidelines emphasize the importance of transparency in AI systems. For example, the AI Act of the European Union includes provisions for ensuring the explainability of high-risk AI applications. Ethical guidelines also stress the need for transparency to ensure that AI systems are used responsibly and ethically, sensibly, and morally in healthcare.

25.4.5.1 *Ethical Challenges in AI*

AI in healthcare presents various ethical challenges, such as ensuring informed consent, respecting patient autonomy, and addressing potential harms. Balancing the innovative potential of AI with ethical considerations is crucial for responsible AI deployment [33, 34].

25.4.5.2 *Regulatory Frameworks*

Regulatory frameworks provide guidelines for addressing ethical challenges in AI. The World Health Organization (WHO) and the Food and Drug Administration have established regulations to ensure the ethical use of AI in healthcare. These frameworks emphasize the importance of transparency, accountability, and patient-centered approaches.

25.4.5.3 *Best Practice*

Adopting best practices for ethical AI involves several key steps. Interdisciplinary collaboration among healthcare professionals, AI developers, ethicists, and policymakers can help ensure comprehensive ethical considerations. Continuous monitoring and evaluation of AI systems, along with mechanisms for patient feedback, are also essential for maintaining ethical standards [35].

25.5 FUTURE DIRECTIONS AND INNOVATIONS IN INTEROPERABILITY

The evolution of humankind has demonstrated time and again that the primary impetus for scientific and technological advancement has always come from healthcare applications. Humanity has dreamed of remote disease diagnosis and treatment since the beginning of time.

This influenced the emergence of intelligent domains and gave rise to the idea of the Internet of Things (IoT). The fields of medicine and healthcare are emerging as some of the most important ones [36]. The IoT is a network that uses the Internet to link different ubiquitous objects, including RFID tags, sensors, actuators, and cell phones. It becomes indispensable to daily life by bridging the gap between the real and virtual worlds [37]. The phrase "smart healthcare" describes platforms for health systems that link individuals, resources, and organizations through the use of devices such as wearable electronics, mobile Internet, and the IoT. This makes it easier for patients to submit health records [38]. In the field of smart healthcare, a number of cutting-edge technologies and solutions—including virtual reality, autonomous devices, robots, intelligent computing solutions, and smart sensors—have been studied, created, and implemented. Here are some advanced core technologies: [36].

25.5.1 Smart Sensors

A smart sensor is a device that allows users to record changes in the physical and chemical properties of the area it is fixed on precisely and automatically. Intelligent sensors achieve highly accurate data collection [36].

Sensors can be categorized into wearable, implantable, and other types. Wearables can be used in the smart healthcare sector to monitor and record patients' vital signs and activities in order to notify or send data to healthcare systems remotely. The following are examples of a patient's vitals that require wearable sensor monitoring: [39].

Pulse: The human body's pulse is tracked by a pulse sensor, which can be used to track life-threatening situations like pulmonary embolism and cardiac arrest. Positions for the pulse signal installation include the wrist, earlobe, chest, fingertip, etc.

Respiratory Rate: When monitoring a patient with an airway disease, such as asthma, lung cancer, respiratory failure, tuberculosis, etc., a respiratory rate sensor counts the breaths the patient takes in a minute.

Body Temperature: A patient's temperature is taken using a body temperature sensor. The proper temperature range can be read by these sensors, allowing you to keep an eye on the body's temperature. The location of it in the human body affects the body temperature sensor's accuracy.

Blood Pressure: High BP is one of the risk factors for cardiovascular diseases. A BP sensor is used to determine and monitor BP in patients, which will help healthcare providers and patients themselves easily assess their BP. The BP sensors' accuracy depends on their position in the patient's body; therefore, many different approaches have been suggested, such as using two photoplethysmography optical sensors placed at different points on the patient's arm to provide more accurate BP estimation.

Oxygen: The blood oxygen volume is measured by an oxygen sensor. This is a crucial parameter that helps physicians diagnose the body's oxygen supply accurately.

25.5.2 Metaverse

Healthcare is about to undergo a revolution thanks to the Metaverse, which is expected to be the Web 3.0 application interface. Within this framework, Li et al. have outlined their vision for the Metaverse's next-generation, interoperable healthcare ecosystem [40]. Avatar-based meetings that enable consultations remotely, immersive simulations that support medical education, and social interactions that promote patient engagement are just a few of the many healthcare-related opportunities presented by the Metaverse.

There are a few important issues that need to be resolved before we can fully utilize the metaverse in healthcare. These include the need for strong regulatory oversight to guarantee data security and moral use cases, the need for ongoing technological innovation to improve and broaden Metaverse capabilities, and the establishment of strong governance frameworks to steer this changing digital environment [40].

Li et al.'s proposed metaverse concept and framework hold the potential to revolutionize conventional healthcare systems and expedite the digital transformation of the industry. Although the metaverse is still in its infancy, it has enormous potential; it's similar to AI technology in the early 2010s. An interoperable healthcare ecosystem in the metaverse cannot be developed in the real world without more research into technology, security, and applications [40].

25.5.3 Cyber-Physical System (CPS)

CPS is a system of tiny sensors that are in close proximity to a person's body and work together to support each patient on an individual basis. Ali Hassan et al. [41] suggest an intelligent decision-making method based on fuzzy logic for determining which of the performance indicators in CPS is the most important parameter. Furthermore, a framework for interoperability is suggested in order to choose the relevant and vital metrics for improving CPS performance while utilizing a cost-effective and maximum-profit approach.

25.5.4 Blockchain

A sophisticated database management system called blockchain technology makes it possible to share transparently across a dispersed network environment. Blocks of data are linked together in a consistent chronological order for storage. As a result, without the approval of all other users, the user cannot remove or alter the chain. Consequently, blockchain technology generates an unchangeable ledger to record data records.

Blockchain technology is used in the smart Internet of Healthcare Things (IoHT) space to manage patient medical records securely and privately. The academic community has shown a great deal of interest in incorporating blockchain technology into healthcare applications in recent years, and some encouraging outcomes have been observed [42].

25.5.5 Applications

25.5.5.1 Real-Time Patient Monitoring

Keeping an eye on vital signs like blood oxygen levels, heart rate, and temperature is essential for providing healthcare in real time. The IoT is used to connect sensors to the human body, which measure a variety of indicators. These data is then analyzed to suggest medications for patients in need of emergency care. Real-time patient monitoring can be achieved through the use of blockchain and IoT. Health insurance companies maintain accurate records of patient claims, which are entered into a ledger that cannot be altered once it has been added [43]. Shain et al. created a medical electronic system that tracks temperature, pulse rate, foot pressure, and ECG. In addition to using RFID to uniquely recognize each patient, the Arduino Uno mainboard serves as the microcontroller, and ThingSpeak acts as middleware for handling medical signals. The system uses the patient's GPS location to provide urgent care [43].

The Systematized Nomenclature of Medicine Clinical Terms or the International Classification of

Diseases by the WHO will be extremely helpful in creating the interoperability terminology interface that the digital health strategy suggests. Blockchain is an emerging technology that can be used to mitigate data blocking as defined by the US HIPAA. Blockchain allows for shared ownership and administration of data. Limited interoperability has been caused by data blocking, and low- and middle-income nations frequently use Western laws like the HIPAA as a benchmark for best practices. The surveys revealed a deficiency in the feedback loop, particularly in healthcare facilities [44].

Within the SIRIC (Integrated Cancer Research Site) research network, the OSIRIS (Interoperability and data sharing of clinical and biological data in oncology) set serves as a helpful standardization tool for clinical cancer research data. Given that it is predicated on established terminologies, it could also hold value in the context of a broader global community spanning various networks. Furthermore, data model harmonization and data integration with other repositories are made much easier by the OSIRIS set. Healthcare data interoperability is a significant global challenge, and several standards, like FHIR, have been developed [45].

25.5.5.2 Chronic Diseases Detection and Prevention

Many sufferers of severe illnesses like cancer, diabetes, heart disease, etc., experience depression as a result of their conditions. In order to address this issue, Sundhara Kumar et al. created an IoHT system to keep an eye on individuals with autism [46]. Through neural sensors, it measures and gathers EEG waveform signals. If any abnormal results are found, caregivers receive alerts [47]. Various designs and frameworks supporting IoT-based healthcare solutions for cancer patients have been proposed by Onasanya et al.

The main attention is on cloud services, which analyze data remotely using big data technologies.

An IoHT system was created by Sood et al. to track the chikungunya virus pandemic. Medical data is gathered by the system through medical sensors, and it then uploads this data to a private cloud. From there, fuzzy logic systems are combined with fog computing techniques to identify potential infections in patients and promptly notify hospitals and caregivers. For security and privacy purposes, private cloud storage is used to store sensitive medical records pertaining to individual patients [48].

25.5.5.3 Home Healthcare and Healthcare for the Elderly

For continuous monitoring of elderly patients who have slowly moved and take longer to get to the hospital for routine or emergency medical care, IoT and virtual reality technologies can be installed in their homes. A health monitoring system for life support was proposed by Abdelgawad et al. In fact, the authors collected medical signals using a variety of sensor types, and then they moved the data to the cloud infrastructure for processing and big data analysis. Also, they created a prototype to demonstrate the viability of the suggested remedy. A Bluetooth-based communication module, a Raspberry Pi-based microprocessor, and a Wi-Fi-based communication module make up this system's six different types of medical sensors. Cloud computing infrastructure stores all of the medical signals [49].

Yang et al. suggested a system of personalized health care for wheelchair-using individuals living alone. The system monitors BP, heart rate, ECGs, and environmental indicators in real time and links IoT and Wireless Body Area Network technologies to offer wheelchair patients effective medical solutions [50].

25.6 SUMMARY

AI being integrated into the smart healthcare system is going to dramatically change many facets of healthcare delivery, particularly with respect to increasing interoperability and improving patient care. Interoperability within healthcare delivery allows the smooth transfer of patient information over different systems for decision-making and care continuity. AI technologies enable interoperability not only on technical and syntactic levels but also on semantic, pragmatic, dynamic, and conceptual levels through the challenge of integrating data. This is quite vital in avoiding scattering of information related to patients and ensuring full, precise, and informative data within any medical context. AI has a variety of functions that promote interoperability and improve patient care. AI is increasing the effectiveness and precision of healthcare procedures by automating data integration and collection, as well as by utilizing NLP to enhance clinical decision support and predictive analytics. The use of AI for administrative duties, such as appointment scheduling, patient registration, billing, and health record management, highlights the technology's revolutionary potential. These innovations contribute to more effective healthcare operations and resource utilization in addition to improving clinical outcomes and patient experiences.

There is huge potential and difficult challenges in the integration of AI into health care. Development in AI technology holds, at the same time, the possibility of improved patient outcomes, increased diagnostic capacity, and streamlined healthcare delivery. Such advantages are balanced against the significant challenges posed by bias, integrity of data, and privacy and security concerns. Of course, strong protection measures are needed in view of privacy concerns, and re-identification and security breaches, in particular, should be taken into consideration. Accuracy and high-quality data cannot be foregone since AI systems depend on these for consistent functioning. Moreover, biases within AI models must be ironed out to prevent health disparities and ensure equal care. Difficulties in deploying AI into production arise due to financial constraints and resistance to adoption, while regulatory compliance guarantees the responsible use of the technologies.

Some examples of such new technologies that can transmute healthcare include blockchain, cyber-physical systems, and the metaverse. In this book, it is further explained how the metaverse can change healthcare by allowing an immersive experience in medical education, avatar-based consulting, and richer patient engagement. But to do so, there will have to be improvements in governance, technological innovation, and regulatory frameworks. Similarly, CPS provides huge enhancements to intelligent decision-making and real-time patient monitoring—powered by IoT devices and AI—while simultaneously ensuring medical records are safely and openly administered using blockchain. These technologies can be used to develop and enhance home healthcare, management of chronic illnesses, and general patient monitoring.

REFERENCES

1. Big Data: A Survey | Mobile Networks and Applications: https://link.springer.com/article/10.1007/s11036-013-0489-0. Accessed: 2024-10-09.
2. Pollock, J.T. and Hodgson, R. 2004. *Adaptive Information: Improving Business Through Semantic Interoperability, Grid Computing, and Enterprise Integration*. John Wiley & Sons.
3. (PDF) Interoperability in Healthcare: Benefits, Challenges and Resolutions: https://www.researchgate.net/publication/282322922_Interoperability_in_Healthcare_Benefits_Challenges_and_Resolutions. Accessed: 2024-10-09.
4. Shaping the Future of Healthcare | Ken Kennedy Institute | Rice University: https://kenkennedy.rice.edu/news/current-news/ai-health-conference-2023-insights. Accessed: 2024-10-10.
5. The power of data mapping in healthcare: https://innowise.com/blog/data-mapping-in-healthcare/. Accessed: 2024-10-10.
6. Data Mapping in Healthcare: Challenges & Vision: https://kms-healthcare.com/blog/data-mapping-in-healthcare/. Accessed: 2024-10-10.
7. Zuocheng Wen and Hua Huang 2023. The potential for artificial intelligence in healthcare. *Journal of Commercial Biotechnology*. 27, 4. DOI:https://doi.org/10.5912/jcb1327.
8. Hosny, A., Parmar, C., Quackenbush, J., Schwartz, L.H. and Aerts, H.J.W.L. 2018. Artificial intelligence in radiology. *Nature Reviews. Cancer*. 18, 8, 500–510. DOI:https://doi.org/10.1038/s41568-018-0016-5.
9. Considerations for ethics review of big data health research: A scoping review - PubMed: https://pubmed.ncbi.nlm.nih.gov/30308031/. Accessed: 2024-10-10.
10. A review of attitudes towards the reuse of health data among people in the European Union: The primacy of purpose and the common good - PubMed: https://pubmed.ncbi.nlm.nih.gov/30961905/. Accessed: 2024-08-15.
11. Murdoch, B. 2021. Privacy and artificial intelligence: challenges for protecting health information in a new era. *BMC medical ethics*. 22, 1, 122. DOI:https://doi.org/10.1186/s12910-021-00687-3.
12. Privacy in the age of medical big data - PubMed: https://pubmed.ncbi.nlm.nih.gov/30617331/. Accessed: 2024-10-10.
13. Na, L., Yang, C., Lo, C.-C., Zhao, F., Fukuoka, Y. and Aswani, A. 2018. Feasibility of reidentifying individuals in large national physical activity data sets from which protected health information has been removed with use of machine learning. *JAMA Network Open*. 1, 8, e186040. DOI:https://doi.org/10.1001/jamanetworkopen.2018.6040.
14. Development and validation of a machine learning algorithm and hybrid system to predict the need for life-saving interventions in trauma patients - *PubMed:* https://pubmed.ncbi.nlm.nih.gov/24263362/. Accessed: 2024-10-10.
15. The Legal and Ethical Concerns That Arise From Using Complex Predictive Analytics In Health Care | Health Affairs: https://www.healthaffairs.org/doi/10.1377/hlthaff.2014.0048. Accessed: 2024-10-10.
16. Interplay of Islam and Economic Growth: Unveiling the Long-run Dynamics in Muslim and Non-Muslim Countries by Faruq Quadri, Oluwaseun Oladeji

Olaniyi, Olugbenga O. Olaoye :: SSRN: https://papers.ssrn.com/sol3/papers.cfm?abstract_id=4683045. Accessed: 2024-08-08.

17. Parikh, R.B., Teeple, S. and Navathe, A.S. 2019. Addressing bias in artificial intelligence in health care. *JAMA*. 322, 24 (Dec. 2019), 2377–2378. DOI:https://doi.org/10.1001/jama.2019.18058.
18. Overcoming AI Bias: Understanding, Identifying and Mitigating Algorithmic Bias in Healthcare - Accuray: https://www.accuray.com/blog/overcoming-ai-bias-understanding-identifying-and-mitigating-algorithmic-bias-in-healthcare/. Accessed: 2024-10-10.
19. Biases in AI Systems: A Survey for Practitioners: Queue: Vol 19, No 2: https://dl.acm.org/doi/10.1145/3466132.3466134. Accessed: 2024-10-10.
20. Abràmoff, M.D., Tarver, M.E., Loyo-Berrios, N., Trujillo, S., Char, D., Obermeyer, Z., Eydelman, M.B., Foundational Principles of Ophthalmic Imaging and Algorithmic Interpretation Working Group of the Collaborative Community for Ophthalmic Imaging Foundation, Washington, D.C. and Maisel, W.H. 2023. Considerations for addressing bias in artificial intelligence for health equity. *NPJ Digital Medicine*. 6, 1, 170. DOI:https://doi.org/10.1038/s41746-023-00913-9.
21. The Potential for Artificial Intelligence in Healthcare - PubMed: https://pubmed.ncbi.nlm.nih.gov/31363513/. Accessed: 2024-10-10.
22. Kamulegeya, L. H., Okello, M., Bwanika, J. M., Musinguzi, D., Lubega, W., Rusoke, D., Nassiwa, F., & Börve, A. (2019). Using artificial intelligence on dermatology conditions in Uganda: A case for diversity in training data sets for machine learning. *bioRxiv*, 826057. https://doi.org/10.1101/826057.
23. Skin Cancer Concerns in People of Color: Risk Factors and Prevention - PubMed: https://pubmed.ncbi.nlm.nih.gov/28125871/. Accessed: 2024-10-10.
24. Challenges of AI Integration in Healthcare: https://www.ominext.com/en/blog/challenges-of-ai-integration-in-healthcare. Accessed: 2024-10-10.
25. Challenges of AI and Data Privacy—And How to Solve Them: https://www.isaca.org/resources/news-and-trends/newsletters/atisaca/2021/volume-32/challenges-of-ai-and-data-privacy-and-how-to-solve-them. Accessed: 2024-10-10.
26. Privacy and Artificial Intelligence: Challenges for Protecting Health Information in a New Era - PMC: https://www.ncbi.nlm.nih.gov/pmc/articles/PMC8442400/. Accessed: 2024-10-10.
27. (PDF) Artificial Intelligence in Cyber Security: https://www.researchgate.net/publication/377235308_Artificial_Intelligence_in_Cyber_Security. Accessed: 2024-10-10.
28. Saeed, H., Malik, H., Bashir, U., Ahmad, A., Riaz, S., Ilyas, M., Bukhari, W.A. and Khan, M.I.A. 2022. Blockchain technology in healthcare: a systematic review. *PLoS ONE*. 17, 4, e0266462. DOI:https://doi.org/10.1371/journal.pone.0266462.
29. 5 Best Practices to Protect Your Data Privacy When Integrating AI: https://techinformed.com/five-best-practices-to-protect-your-data-privacy-when-implementing-gen-ai/. Accessed: 2024-10-10.
30. (PDF) HIPAA and GDPR Compliance in IoT Healthcare Systems: https://www.researchgate.net/publication/379129933_HIPAA_and_GDPR_Compliance_in_IoT_Healthcare_Systems. Accessed: 2024-10-10.
31. (PDF) HL7 Fast Healthcare Interoperability Resources (HL7 FHIR) in digital healthcare ecosystems for chronic disease management: Scoping review: https://www.researchgate.net/publication/381212965_HL7_Fast_healthcare_interoperability_resources_HL7_FHIR_in_digital_healthcare_ecosystems_for_chronic_disease_Management_Scoping_review. Accessed: 2024-10-10.
32. Dilmegani C. Data Quality in AI: Challenges, Importance & Best Practices. AIMultiple Research. November 13 2025. Available from: https://research.aimultiple.com/data-quality-ai/.
33. AI Quality Standards in Health Care: Rapid Umbrella Review - PMC: https://www.ncbi.nlm.nih.gov/pmc/articles/PMC11153979/. Accessed: 2024-10-10.
34. Balagopalan, A., Baldini, I., Celi, L.A., Gichoya, J., McCoy, L.G., Naumann, T., Shalit, U., van der Schaar, M. and Wagstaff, K.L. 2024. Machine learning for healthcare that matters: reorienting from technical novelty to equitable impact. *PLOS Digital Health*. 3, 4, e0000474. DOI:https://doi.org/10.1371/journal.pdig.0000474.
35. Hassan, F., Shaheen, M. and Sahal, R. 2020. Real-time healthcare monitoring system using online machine learning and Spark Streaming. *International Journal of Advanced Computer Science and Applications*. 11. DOI:https://doi.org/10.14569/IJACSA.2020.0110977.
36. Intelligent Healthcare: Integration of Emerging Technologies and Internet of Things for Humanity: https://www.mdpi.com/1424-8220/23/9/4200. Accessed: 2024-10-10.
37. A Quality-of-Service-Aware Service Composition Method in the Internet of Things Using a Multi-Objective Fuzzy-Based Hybrid Algorithm: https://www.mdpi.com/1424-8220/23/16/7233. Accessed: 2024-10-10.
38. The Applications of Internet of Things in Smart Healthcare Sectors: A Bibliometric And Deep Study - PubMed: https://pubmed.ncbi.nlm.nih.gov/38356528/. Accessed: 2024-10-10.
39. (PDF) Internet of Things for Smart Healthcare: Technologies, Challenges, and Opportunities: https://www.researchgate.net/publication/321383073_Internet_of_Things_for_Smart_Healthcare_Technologies_Challenges_and_Opportunities. Accessed: 2024-10-10.
40. The Next Generation of Healthcare Ecosystem in the metaverse - PubMed: https://pubmed.ncbi.nlm.nih.gov/38048990/. Accessed: 2024-10-10.

41. AI-Enabled Framework for Fog Computing Driven E-Healthcare Applications: https://www.mdpi.com/1424-8220/21/23/8039. Accessed: 2024-10-10.
42. Nguyen DC, Nguyen V-D, Ding M, Chatzinotas S, Pathirana PN, Seneviratne A, Dobre O, Zomaya AY. Intelligent Blockchain-based Edge Computing via Deep Reinforcement Learning: Solutions and Challenges. *arXiv* [Preprint]. 2022 Jun 17. Available from: https://doi.org/10.48550/arxiv.2206.09009.
43. A health Monitoring System for Vital Signs Using IoT | Semantic Scholar: https://www.semanticscholar.org/paper/A-health-monitoring-system-for-vital-signs-using-Swaroop-Chandu/8b373e0babb7c3e2a3f938dc2d257c6a44ef6d52. Accessed: 2024-10–10.
44. Chukwu, E., Garg, L., Foday, E., Konomanyi, A., Wright, R. and Smart, F. 2022. Digital health solutions and state of interoperability: landscape analysis of Sierra Leone. *JMIR Formative Research*. 6, 6, e29930. DOI:https://doi.org/10.2196/29930.
45. OSIRIS: A Minimum Data Set for Data Sharing and Interoperability in Oncology - PubMed: https://pubmed.ncbi.nlm.nih.gov/33720747/. Accessed: 2024-10-10.
46. Abdulmalek, S., Nasir, A., Jabbar, W.A., Almuhaya, M.A.M., Bairagi, A.K., Khan, M.A.-M. and Kee, S.-H. 2022. IoT-based healthcare-monitoring system towards improving quality of life: a review. *Healthcare (Basel, Switzerland)*. 10, 10, 1993. DOI:https://doi.org/10.3390/healthcare10101993.
47. Smart Integrated IoT Healthcare System for Cancer Care | Semantic Scholar: https://www.semanticscholar.org/paper/Smart-integrated-IoT-healthcare-system-for-cancer-Onasanya-Elshakankiri/78288d1a273b46e220e1c38cffb8525691ab3911. Accessed: 2024-10-10.
48. Wearable IoT Sensor Based Healthcare System for Identifying and Controlling Chikungunya Virus - PubMed: https://pubmed.ncbi.nlm.nih.gov/32287550/. Accessed: 2024-10-10.
49. (PDF) IoT-Based Health Monitoring System for Active and Assisted Living: https://www.researchgate.net/publication/318426723_IoT-Based_Health_Monitoring_System_for_Active_and_Assisted_Living. Accessed: 2024-10-10.
50. (PDF) A Home Mobile Healthcare System for Wheelchair Users: https://www.researchgate.net/publication/269297744_A_home_mobile_healthcare_system_for_wheelchair_users. Accessed: 2024-10-10.

Index

Note: Page numbers in **bold** and *italics* refer to tables and figures, respectively.

D

E

I

J

K

L

M

N

Q

R

S